ELEMENTS OF DISCRETE MATHEMATICS

McGRAW-HILL COMPUTER SCIENCE SERIES

RICHARD W. HAMMING, *Bell Telephone Laboratories*
EDWARD A. FEIGENBAUM, *Stanford University*

ELEMENTS OF DISCRETE MATHEMATICS

C. L. Liu
Department of Computer Science
University of Illinois at Urbana-Champaign

McGRAW-HILL BOOK COMPANY

New York   St. Louis   San Francisco   Auckland   Bogotá   Düsseldorf
Johannesburg   London   Madrid   Mexico   Montreal   New Delhi   Panama
Paris   São Paulo   Singapore   Sydney   Tokyo   Toronto

ELEMENTS OF DISCRETE MATHEMATICS

Copyright © 1977 by McGraw-Hill, Inc. All rights reserved. Printed in the United States of America. No part of this publication may be reproduced, stored in a retrieval system, or transmitted, in any form or by any means, electronic, mechanical, photocopying, recording, or otherwise, without the prior written permission of the publisher.

4567890 DODO 83210

This book was set in Times. The editors were Peter D. Nalle and Madelaine Eichberg; the production supervisor was Robert C. Pedersen. The drawings were done by Oxford Illustrators Limited. R. R. Donnelley & Sons Company was printer and binder.

Library of Congress Cataloging in Publication Data

Liu, Chung Laung.
    Elements of discrete mathematics.

    (McGraw-Hill computer science series)
    Includes bibliographies and index.
    1. Combinatorial analysis. 2. Algebra, Abstract.
I. Title.
QA164.L57   511   76-50063
ISBN 0-07-038131-3

CONTENTS

* All sections marked with an asterisk can be omitted without disrupting the continuity.

PREFACE

This book presents a selection of topics from set theory, combinatorics, graph theory, and algebra which I consider basic and useful to students in Applied Mathematics, Computer Science, and Engineering. It is intended to be a textbook for a course in Discrete Mathematics at the sophomore-junior level, although it can also be used in a freshman level course since the presentation does not assume any background beyond high school mathematics. The material in this book can be covered in a one-semester course at a rather brisk pace. On the other hand, it is quite possible to omit some of the topics for a slower course. A section marked with an asterisk can be omitted without disrupting the continuity.

This book is an outgrowth of a set of lecture notes I wrote for a course I taught in the Department of Computer Science at the University of Illinois at Urbana-Champaign. I hope it is not only a record of what I covered in the course but also a reflection of how I lectured the material in the classroom. I have tried to be rigorous and precise in presenting the mathematical concepts and, in the meantime, to avoid complicated formalisms and notations. As a rule I would not state a definition or a fact if I could not illustrate the utilization of the definition or the fact in some meaningful way later on. Consequently, it is quite possible that I have omitted some "important" definitions or facts in this book. I trust, however, that the students will be quite capable of looking them up somewhere else when such definitions and facts are needed. I have attempted to teach my students some useful mathematics in an interesting and exciting way. I hope that I have demonstrated to them how mathematics can be applied to solve nontrivial real life problems. Furthermore, I hope that my students not only learned in the course some powerful mathematical tools but also developed their ability to perceive, to formulate, and to solve mathematical problems. I have tried to bring in an algorithmic point of view in the treatment of several topics, although I decided not to include explicit computer programs, mainly because of time consideration. I hope that some of these personal views and tastes can be shared to some extent by the instructor using this book.

I would like to thank James N. Synder, my department head, for his encouragement and support; Murray Edelberg, Jane W. S. Liu, and Andrew H. Sherman for their careful review of the manuscript; Donald K. Friesen for his contribution to the preparation of the Instructor's Manual; and Edward M. Reingold and F. Frances Yao for their many helpful suggestions. Several years ago, I had an opportunity to serve on a panel on the impact of computing on mathematics sponsored by the Committee on the Undergraduate Program in Mathematics of the Mathematical Association of America. I was greatly benefited by the panel's discussion on the teaching of discrete mathematics, and I am much indebted to the members of the panel. I also thank Glenna Gochenour, Connie Nosbisch, Judy Watkins, and June Wingler for their typing and editorial assistance. Finally, thanks to Kathleen D. Liu for her assistance in the preparation of the index.

There is a certain amount of overlap between this book and the book *Introduction to Combinatorial Mathematics* I wrote a few years ago. In a number of instances, I follow quite closely the presentation in *Introduction to Combinatorial Mathematics*.

C. L. Liu

SETS AND PROPOSITIONS

1.1 INTRODUCTION

A major theme of this book is to study discrete objects and relationships among them. The term *discrete objects* is a rather general one. It includes a large variety of items such as people, books, computers, transistors, computer programs, and so on. In our daily lives as well as our technical work we frequently deal with these items, making statements such as, "The people in this room are Computer Science majors in their second year of study," "All the books I bought are detective stories written by A. B. Charles," and "We want to select and buy a computer among those that are suitable for both scientific and business applications at a price not exceeding $200,000." We would like to abstract some of the basic concepts dealing with the many different kinds of discrete objects and establish certain common terminology for dealing with them.

A hint of the possibility of such an abstraction is quite evident when we observe that these three statements all have "something" in common. To be specific, in the first statement we are referring to people who possess the two attributes of being a Computer Science major and of being a sophomore; in the second statement we are referring to books that possess the two attributes of being a detective story and of being written by A. B. Charles; and in the third statement we are referring to computers that possess the three attributes of being suitable for scientific applications, of being suitable for business applications, and of being priced at no more than $200,000. To put it in another way, consider the group of all the Computer Science majors and the group of all the sophomores in the university. In our first statement we are then referring to those students who belong to both of these groups. Also, consider the collection of all detective stories and the collection of all books written by A. B. Charles. In our second statement we are then referring to those books which belong to both of these collections. Finally, in our third statement we are referring to all computers which belong to the three categories of computers that are suitable for business applications, that are suitable for scientific applications, and that are priced at no more than $200,000.

Our example illustrates the many occasions on which we deal with several classes of objects and wish to refer to those objects that belong to all the classes. Similarly, one would immediately perceive occasions on which we refer to objects that belong to one of several classes of objects, such as in the statement, "I want to interview all the students who speak either German or French," where we refer to those who belong either to the group of German-speaking students or to the group of French-speaking students.

We begin with the introduction of some basic terminology and concepts in elementary set theory.

A *set* is a collection of *distinct* objects. Thus, the group of all sophomores in the university is a set. So is the group of all Computer Science majors in the university, and so is the group of all second-year Computer Science majors. We use the notation $\{a, b, c\}$ to denote the set which is the collection of the objects a, b, and c. The objects in a set are also called the *elements* or the *members* of the set. We usually also give names to sets. For example, we write $S = \{a, b, c\}$ to mean that the set named S is the collection of the objects a, b, and c. Consequently, we can refer to the set S as well as to the set $\{a, b, c\}$. As another example, we may have

Second-year-Computer-Science-majors
$$= \{\text{Smith, Jones, Wong, Yamamoto, Vögeli}\}$$

(The name of the set {Smith, Jones, Wong, Yamamoto, Vögeli} is Second-year-Computer-Science-majors, which is rather long. The reader probably would want to suggest alternative names such as S or CS. However, there is nothing wrong conceptually with having a "long" name.) We use the notation $a \in S$ to mean that a is an element in the set S. In that case, we also say that S *contains* the element a. We use the notation $d \notin S$ to mean that d is not an element in the set S. In that case, we also say that S does not contain the element d. Thus, in the example above, Jones $\in$ Second-year-Computer-Science-majors, while Kinkaid $\notin$ Second-year-Computer-Science-majors.

Note that a set contains only distinct elements. Thus, $\{a, a, b, c\}$ is a redundant representation of the set $\{a, b, c\}$. Similarly, {The-Midnight-Visitor, The-Midnight-Visitor, The-Missing-Witness, 114-Main-Street} is a redundant representation of the detective stories written by A. B. Charles. One might ask the question: What should we do if our collection of detective stories by A. B. Charles in the library indeed contains two copies of the book The-Midnight-Visitor? In that case, the set {The-Midnight-Visitor, The-Missing-Witness, 114-Main-Street} is a set of distinct titles of detective stories by A. B. Charles in our library, while the set {The-Midnight-Visitor-1, The-Midnight-Visitor-2, The-Missing-Witness, 114-Main-Street} is the set of detective stories by A. B. Charles in our library where The-Midnight-Visitor-1 is copy 1 of the book The-Midnight-Visitor, and The-Midnight-Visitor-2 is copy 2 of the book. Note that The-Midnight-Visitor-1 and The-Midnight-Visitor-2 are two distinct elements in the latter set.

Note also that the elements in a set are not ordered in any fashion. Thus, $\{a, b, c\}$ and $\{b, a, c\}$ represent the same collection of elements. Later on, we shall introduce the notion of *ordered sets*.

As was introduced above, one way to describe the membership of a set is to list exhaustively all the elements in that set. In many cases, when the elements in a set share some common properties, we can describe the membership of the set by stating the properties that uniquely characterize the elements in the set. For example, let $S = \{2, 4, 6, 8, 10\}$. We can also specify the elements of S by saying that S is the set of all even positive integers that are not larger than 10. Indeed, we can use the notation

$$S = \{x \mid x \text{ is an even positive integer not larger than } 10\}$$

for the set {2, 4, 6, 8, 10}. In general, we use the notation

$$\{x \mid x \text{ possesses certain properties}\}$$

for a set of objects that share some common properties. Thus,

$$S = \{\text{Smith, Jones, Wong, Yamamoto, Vögeli}\}$$

and

$$S = \{x \mid x \text{ is a second year Computer Science major}\}$$

are two different ways to describe the same set of elements.

It should be pointed out that our definition of a set does not preclude the possibility of having a set containing *no* elements. The set that contains no element is known as the *empty set*, and is denoted by $\emptyset$. (Following our established notation of using a pair of braces to enclose all the elements in a set, we could also denote the empty set by { }.) For example, let S denote the set of all detective stories by A. B. Charles that were published in 1924. Clearly, S is the empty set if A. B. Charles was born in 1925.

Let us note that we did not place any restriction on the elements in a set. Thus, $S = \{\text{Smith, The-Midnight-Visitor, CDC-6600}\}$ is a well-defined set. That the elements, Smith (a person), The-Midnight-Visitor (the title of a book), and CDC-6600 (a computer) do not seem to share anything in common does not prohibit them from being elements of the same set. Indeed, we should point out that it is perfectly all right to have sets as members of a set. Thus, for example, the set $\{\{a, b, c\}, d\}$ contains the two elements $\{a, b, c\}$ and d, and the set $\{\{a, b, c\}, a, b, c\}$ contains the four elements $\{a, b, c\}$, a, b, and c. The set of all committees in the U.S. Senate could be represented by $\{\{a, b, c\}, \{a, d, e, f\}, \{b, e, g\}\}$ where each element of the set is a committee which in turn is a set with the senators in the committee as elements. Similarly, $\{a, \{a\}, \{\{a\}\}\}$ is a set with three *distinct* elements $a, \{a\}, \{\{a\}\}$. Also, the set $\{\emptyset\}$ contains one element—the empty set—and the set $\{\emptyset, \{\emptyset\}\}$ contains two elements—the empty set and a set that contains the empty set as its only element.

Given two sets P and Q, we say that P is a *subset* of Q if every element of P is also an element in Q. We shall use the notation $P \subseteq Q$ to denote that P is a subset of Q. For example, the set $\{a, b\}$ is a subset of the set $\{y, x, b, c, a\}$, but it is not a subset of the set $\{a, c, d, e\}$. The set of all second-year Computer Science majors is a subset of the set of all sophomores. It is also a subset of the set of all Computer Science majors. On the other hand, the set of all Computer Science majors is not a subset of the set of all sophomores nor is the set of all sophomores a subset of the set of all Computer Science majors. Let $A = \{a, b, c\}$ and $B = \{\{a, b, c\}, a, b, c\}$. We note that it is indeed possible to have both $A \in B$ and $A \subseteq B$. As further examples, we ask the reader to check the following statements:

For any set P, P is a subset of P.
The empty set is a subset of any set.
The set $\{\emptyset\}$ is not a subset of the set $\{\{\emptyset\}\}$.

Two sets P and Q are said to be *equal* if they contain the same collection of elements. For example, the two sets

$$P = \{x \mid x \text{ is an even positive integer not larger than } 10\}$$

$$Q = \{x \mid x = y + z \text{ where } y \in \{1, 3, 5\}, z \in \{1, 3, 5\}\}$$

are equal. In a seemingly roundabout way, we can also say that two sets P and Q are equal if P is a subset of Q, and Q is a subset of P. We shall see later that on some occasions, this is a convenient way to define the equality of two sets.

Let P be a subset of Q. We say that P is a *proper* subset of Q if P is not equal to Q, that is, there is at least one element in Q that is not in P. For example, the set $\{a, b\}$ is a proper subset of the set $\{y, x, b, c, a\}$. We use the notation $P \subset Q$ to denote that P is a proper subset of Q.

1.2 COMBINATIONS OF SETS

We show now how sets can be *combined* in various ways to yield new sets. For example, let P be the set of students taking the course Theory of Computation and Q be the set of students taking the course Music Appreciation. If a certain announcement was made in both the Theory of Computation and the Music Appreciation classes, what is the set of students who know about the news announced? Clearly, it is the set of students who are taking either Theory of Computation or Music Appreciation, or both. If both these courses have their final examinations scheduled in the same hours, what is the set of students who will have conflicting final examinations? Clearly, it is the set of students who are taking both Theory of Computation and Music Appreciation. To formalize these notions, we define the union and the intersection of sets. The *union* of two sets P and Q, denoted $P \cup Q$, is the set whose elements are exactly the elements in either P or Q (or both).† For example,

$$\{a, b\} \cup \{c, d\} = \{a, b, c, d\}$$

$$\{a, b\} \cup \{a, c\} = \{a, b, c\}$$

$$\{a, b\} \cup \emptyset = \{a, b\}$$

$$\{a, b\} \cup \{\{a, b\}\} = \{a, b, \{a, b\}\}$$

The *intersection* of two sets P and Q, denoted $P \cap Q$, is the set whose elements are exactly those elements that are in both P and Q. For example,

$$\{a, b\} \cap \{a, c\} = \{a\}$$

$$\{a, b\} \cap \{c, d\} = \emptyset‡$$

$$\{a, b\} \cap \emptyset = \emptyset$$

If the elements of P are characterized by a common property and the elements in Q are characterized by another common property, then the union of P and Q is the set of elements possessing at least one of these properties, and the intersection of P and Q is the set of elements possessing both of these properties. According to the definitions, $P \cup Q$ and $Q \cup P$ denote the same set, as do $P \cap Q$ and $Q \cap P$.

In general, the union of k sets $P_1, P_2, \ldots, P_k$, denoted $P_1 \cup P_2 \cup \cdots \cup P_k$, is the set containing exactly the elements in P_1, the elements in $P_2, \ldots$, and the elements in P_k. Similarly, the intersection of k sets $P_1, P_2, \ldots, P_k$, denoted $P_1 \cap P_2 \cap \cdots \cap P_k$, is the set containing exactly the elements that are in P_1 and in $P_2 \ldots$ and in P_k. For example, the set of all undergraduate students in a university is the union of the sets of freshmen, sophomores, juniors, and seniors, and the set of graduating seniors is the intersection of the set of seniors, the set of students who have accumulated 144 or more credit hours, and the set of students who have a C or better grade-point average.

Let P denote the set of students taking Theory of Computation, Q denote the set of students taking Music Appreciation, and R denote the set of students having type AB blood. Suppose an emergency announcement was made in the classes of Theory of Computation and Music Appreciation calling for type AB blood donors. We want to determine the members of the set of potential donors who heard about the emergency call. Since $S = P \cup Q$ is the set of students who heard about the emergency call, $R \cap S$ is the set of potential donors who heard about the emergency call. Instead of using a new name S for the set $P \cup Q$, we can simply write $R \cap (P \cup Q)$, where the parentheses are used as delimiters to

---

† We do not wish to introduce the notion of algebraic operations until Chap. 8. Thus, at this moment, $P \cup Q$ is simply a name we have chosen for a set.

‡ Two sets are said to be *disjoint* if their intersection is the empty set.

avoid confusion. Note that the set of potential donors who heard about the emergency call is also the set of students with type AB blood in the Theory of Computation class together with the set of students with type AB blood in the Music Appreciation class. That is, the set $(R \cap P) \cup (R \cap Q)$. This example suggests very strongly that for any sets P, Q, R, the two sets $R \cap (P \cup Q)$ and $(R \cap P) \cup (R \cap Q)$ are equal. Indeed, this is the case, as we now show.

We show first that $R \cap (P \cup Q)$ is a subset of $(R \cap P) \cup (R \cap Q)$ by showing that every element in $R \cap (P \cup Q)$ is also in $(R \cap P) \cup (R \cap Q)$. Let x be an element in $R \cap (P \cup Q)$. The element x must be in R and must be either in P or Q. If x is in P, x is in $R \cap P$. If x is in Q, x is in $R \cap Q$. Consequently, x is in $(R \cap P) \cup (R \cap Q)$, and we conclude that $R \cap (P \cup Q)$ is a subset of $(R \cap P) \cup (R \cap Q)$. Second, we show that $(R \cap P) \cup (R \cap Q)$ is a subset of $R \cap (P \cup Q)$. Let x be an element in $(R \cap P) \cup (R \cap Q)$. Thus, x must either be in $R \cap P$ or be in $R \cap Q$. That is, x must either be in both R and P or be in both R and Q. In other words, x must be in R and must be either in P or in Q. Consequently, x is in $R \cap (P \cup Q)$, and we can conclude that $(R \cap P) \cup (P \cap Q)$ is a subset of $R \cap (P \cup Q)$. It follows that the two sets $R \cap (P \cup Q)$ and $(R \cap P) \cup (R \cap Q)$ are equal.

In a similar manner we can show that for any sets P, Q, R, the two sets $R \cup (P \cap Q)$ and $(R \cup P) \cap (R \cup Q)$ are equal. Furthermore, we have

$$R \cap (P_1 \cup P_2 \cup \cdots \cup P_k) = (R \cap P_1) \cup (R \cap P_2) \cup \cdots \cup (R \cap P_k)$$

$$R \cup (P_1 \cap P_2 \cap \cdots \cap P_k) = (R \cup P_1) \cap (R \cup P_2) \cap \cdots \cap (R \cup P_k)$$

We leave the details to the reader.†

The difference of two sets P and Q, denoted $P - Q$, is the set containing exactly those elements in P that are not in Q. For example,

$$\{a, b, c\} - \{a\} = \{b, c\}$$

$$\{a, b, c\} - \{a, d\} = \{b, c\}$$

$$\{a, b, c\} - \{d, e\} = \{a, b, c\}$$

If P is the set of people who have tickets to a ball game and Q is the set of people who are ill on the day of the game, then $P - Q$ is the set of people who will go to the game. Note that Q might contain some or none of the elements of the set P. However, these elements will not appear in $P - Q$ in any case, just as in the example, those people who are ill but do not have tickets to the ball game will not go to the game anyway. Indeed, if the elements in Q are characterized by some common property, then $P - Q$ is the set of elements in P that do not possess this property. If Q is a subset of P, the set $P - Q$ is also called the *complement of Q with respect to P*. For example, let P be the set of all students

---

† Again, we do not wish to introduce the notions of algebraic operations, associativity, and distributivity until Chap. 8. Note, however, these notions are not needed here because $P \cap Q$, $P \cup Q$, $P_1 \cup P_2 \cup \cdots \cup P_k$ are simply names for sets obtained according to our definitions.

in the course Theory of Computation and Q be the set of those students who have passed the course. Then $P - Q$ is the set of students who failed the course. On many occasions, when the set P is clear from the context, we shall abbreviate the *complement of Q with respect to P* as *the complement of Q*, which will be denoted $\bar{Q}$. For example, let P be the set of all students in the course Theory of Computation. Let Q be the set of Computer Science majors in the course, and R be the set of sophomores in the course. Then the complement of Q refers to the set of students in the course who are not Computer Science majors, and the complement of R refers to the set of those students who are not sophomores, if it is understood that in our discussion we always restrict ourselves to students in the course Theory of Computation. Indeed, when our discussion is always restricted to the subsets of a set P, P is referred to as the *universe*.

The *symmetric difference* of two sets P and Q, denoted $P \oplus Q$, is the set containing exactly all the elements that are in P or in Q but not in both. In other words, $P \oplus Q$ is the set $(P \cup Q) - (P \cap Q)$. For example,

$$\{a, b\} \oplus \{a, c\} = \{b, c\}$$

$$\{a, b\} \oplus \emptyset = \{a, b\}$$

$$\{a, b\} \oplus \{a, b\} = \emptyset$$

If we let P denote the set of cars that have defective steering mechanisms and Q denote the set of cars that have defective transmission systems, then $P \oplus Q$ is the set of cars that have one but not both of these defects. Suppose that a student will get an A in a course if she did well in both quizzes, will get a B if she did well in one of the two quizzes, and will get a C if she did poorly in both quizzes. Let P be the set of students who did well in the first quiz and Q be the set of students who did well in the second quiz. Then $P \cap Q$ is the set of students who will get A's, $P \oplus Q$ is the set of students who will get B's, and $S - (P \cup Q)$ is the set of students who will get C's, where S is the set of all students in the course. We define $P_1 \oplus P_2 \oplus \cdots \oplus P_k$ to be the set of elements that are in an odd number of the sets $P_1, P_2, \ldots, P_k$.

The *power set* of a set A, denoted $\mathscr{P}(A)$, is the set that contains exactly all the subsets of A. Thus $\mathscr{P}(\{a, b\}) = \{\emptyset, \{a\}, \{b\}, \{a, b\}\}$, and $\mathscr{P}(\emptyset) = \{\emptyset\}$. For example, let $A = \{\text{novel, published-in-1975, paperback}\}$ be the three attributes concerning the books in the library in which we are interested. Then $\mathscr{P}(A)$ is the set of all possible combinations of these attributes the books might possess, ranging from books that have none of these attributes [the empty set in $\mathscr{P}(A)$] to books that have all three of these attributes [the set A in $\mathscr{P}(A)$].

Sets obtained from combinations of given sets can be represented pictorially. If we let P and Q be the sets represented by the cross-hatched areas in Fig. 1.1a, then the cross-hatched areas in Fig. 1.1b represent the sets $P \cup Q$, $P \cap Q$, $P - Q$, and $P \oplus Q$, respectively. These diagrams are known as *Venn diagrams*.

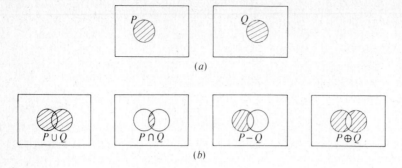

Figure 1.1

1.3 FINITE AND INFINITE SETS

Intuitively, it is quite clear that by the size of a set we mean the number of distinct elements in the set. Thus, there is little doubt when we say the size of the set $\{a, b, c\}$ is 3, the size of the set $\{a, \emptyset, d\}$ is also 3, the size of the set $\{\{a, b\}\}$ is 1, and the size of the set $\emptyset$ is 0. Indeed, we could stop our discussion on the size of sets at this point if we were only interested in the size of "finite" sets. However, a much more intriguing topic is the size of "infinite" sets. At this point, a perceptive reader will probably ask the question, "What is an infinite set in the first place?" An evasive answer such as, "An infinite set is not a finite set," is no answer at all, because if we start to think about it, we should also ask the question, "What is a finite set anyway?"

Let us begin by declaring that we have not yet committed ourselves to the precise definitions of finite sets and infinite sets. As the basis of our discussion, we want to construct an example of an infinite set. For a given set A, we define the *successor* of A, denoted A^+, to be the set $A \cup \{A\}$. Note that $\{A\}$ is a set that contains A as the only element. In other words, A^+ is a set that consists of all the elements of A together with an additional element which is the set A. For example, if $A = \{a, b\}$, then $A^+ = \{a, b\} \cup \{\{a, b\}\} = \{a, b, \{a, b\}\}$; and if $A = \{\{a\}, b\}$, then $A^+ = \{\{a\}, b, \{\{a\}, b\}\}$. Let us now construct a sequence of sets starting with the empty set $\emptyset$. The successor of the empty set is $\{\emptyset\}$, whose successor is $\{\emptyset, \{\emptyset\}\}$, and whose successor, in turn, is $\{\emptyset, \{\emptyset\}, \{\emptyset, \{\emptyset\}\}\}$. It is clear that we can go on to construct more and more successors. Let us also assign names to these sets. Let

$$0 = \emptyset$$

$$1 = \{\emptyset\}$$

$$2 = \{\emptyset, \{\emptyset\}\}$$

$$3 = \{\emptyset, \{\emptyset\}, \{\emptyset, \{\emptyset\}\}\}$$

. . . .

We have, clearly, $1 = 0^+$, $2 = 1^+$, $3 = 2^+$, and so on.† Let us now define a set N such that

1. N contains the set 0.
2. If the set n is an element in N so is the set n^+.
3. N contains no other sets.

Since for every set in N its successor is also in N, the reader probably would agree that N is indeed an "infinite set." However, let us proceed in a more precise way.

  We shall talk about the sizes of sets in a comparative manner. To this end, let us introduce a definition: Given two sets P and Q, we say that there is a *one-to-one correspondence* between the elements in P and the elements in Q if it is possible to pair off the elements in P and Q such that every element in P is paired off with a distinct element in Q.‡ Thus, there is a one-to-one correspondence between the elements in the set $\{a, b\}$ and the elements in the set $\{c, d\}$, because we can pair a with c and b with d, or we can pair a with d and b with c. There is also a one-to-one correspondence between the elements in the set $\{a, b, c\}$ and the elements in the set $\{\emptyset, a, d\}$. On the other hand, there is no one-to-one correspondence between the elements in the sets $\{a, b, c\}$ and $\{a, d\}$. The intention of introducing the notion of one-to-one correspondence between the elements of two sets is quite obvious, because we can now compare two sets and say that they are of the same size or that they are of different sizes. The basis of our comparison is indeed the sets we constructed above, namely, 0, 1, 2, 3, …, and N. We are now ready to introduce some formal definitions. A set is said to be a *finite* set if there is a one-to-one correspondence between the elements in the set and the elements in some set n, where $n \in N$; n is said to be the *cardinality* of the set. Thus, for example, the cardinalities of the sets $\{a, b, c\}$, $\{a, \emptyset, d\}$, $\{\emptyset, \{\emptyset\}, \{\emptyset, \{\emptyset\}\}\}$ are all equal to 3. Note that it is now precise for us to say that a set is an infinite set if it is not a finite one. We can, however, be more precise about the "size" of infinite sets: A set is said to be a *countably infinite* set (or the cardinality of the set is countably infinite§) if there is a one-to-one correspondence between the elements in the set and the elements in N. We observe first of all that the set of all natural numbers $\{0, 1, 2, 3, …\}$¶ is a countably infinite set. It follows that the set of all nonnegative even integers $\{0, 2, 4, 6, 8, …\}$ is a countably infinite set because there is an obvious one-to-one correspondence between all nonnegative even integers and all natural numbers, namely, the even integer $2i$ corresponds to the natural number i for $i = 0, 1,$

---

† Using $0, 1, 2, 3, …$ as names of sets is just as good as using $A, B, C, D, ….$ As will be seen, it is intentional that we choose $0, 1, 2, 3, …$ as names.

‡ Such an intuitive definition will be made more formal later on.

§ In the literature, the cardinality of a countably infinite set is also referred to as $\aleph_0$. ($\aleph$ is the first letter in the Hebrew alphabet.)

¶ The notation is perhaps confusing. However, it is intentional, because the set N is *indeed* a precise definition of the set of natural numbers.

2, ... Similarly, the set of all nonnegative multiples of 7 $\{0, 7, 14, 21, ...\}$ is also a countably infinite set. So is the set of all positive integers $\{1, 2, 3, ...\}$. We note that a set is a countably infinite set if starting from a certain element we can sequentially list all the elements in the set one after another, because such a listing will yield a one-to-one correspondence between the elements in the set and the natural numbers. For example, the set of all integers $\{..., -2, -1, 0, 1, 2, ...\}$ is a countably infinite set, since its elements can be listed sequentially as $\{0, 1, -1, 2, -2, 3, -3, ...\}$. This example suggests that the union of two countably infinite sets is also a countably infinite set. It indeed is the case. As a matter of fact, the union of a finite number of countably infinite sets is a countably infinite set and, furthermore, so is the union of a countably infinite number of countably infinite sets (see Prob. 1.19).

Finally, we show that there are infinite sets whose cardinalities are not countably infinite. In particular, we shall show that the set of real numbers between 0 and 1 is not a countably infinite set. Our proof procedure is to assume that the set is a countably infinite set and then show the existence of a contradiction. If the cardinality of the set of real numbers between 0 and 1 is countably infinite, there is a one-to-one correspondence between these real numbers and the natural numbers. Consequently, we can exhaustively list them one after another in decimal form as in the following:†

$$0.a_{11}a_{12}a_{13}a_{14} \cdots$$

$$0.a_{21}a_{22}a_{23}a_{24} \cdots$$

$$0.a_{31}a_{32}a_{33}a_{34} \cdots$$

$$\cdots \cdots \cdots \cdots \cdots \cdots$$

$$0.a_{i1}a_{i2}a_{i3}a_{i4} \cdots$$

$$\cdots \cdots \cdots \cdots \cdots \cdots$$

where a_{ij} denotes the jth digit of the ith number in the list. Consider the number

$$0.b_1b_2b_3b_4 \cdots$$

where

$$b_i = \begin{cases} 1 & \text{if} \quad a_{ii} = 9 \\ 9 - a_{ii} & \text{if} \quad a_{ii} = 0, 1, 2, ..., 8 \end{cases}$$

for all i. Clearly, the number $0.b_1b_2b_3b_4 \cdots$ is a real number between 0 and 1 that does not have an infinite string of trailing 0's (i.e., $0.34000 \cdots$). Moreover, it is different from each of the numbers in the list above because it differs from the first number in the first digit, the second number in the second digit, ..., the ith number in the ith digit, ..., and so on. Consequently, we conclude that the list above is not an exhaustive listing of the set of all real numbers between 0 and 1, contradicting the assumption that this set is a countably infinite set.

† A number such as 0.34 can be written in two different forms, namely, 0.34000 ... or 0.339999 ... We follow an arbitrarily chosen convention of writing it in the latter form.

It is possible to continue in this direction to classify infinite sets so that notions such as some infinite sets are "more infinite" than other infinite sets can be made precise. This, however, will be beyond our scope of discussion.

1.4 MATHEMATICAL INDUCTION

Let us consider some illustrative examples:

Example 1.1 Suppose we have stamps of two different denominations, 3 cents and 5 cents. We want to show that it is possible to make up exactly any postage of 8 cents or more using stamps of these two denominations. Clearly, the approach of showing case by case how to make up postage of 8 cents, 9 cents, 10 cents, ..., using 3-cent and 5-cent stamps will not be a fruitful one, because there is an infinite number of cases to be examined.† Let us consider an alternative approach. We want to show that if it is possible to make up exactly a postage of k cents using 3-cent and 5-cent stamps, then it is also possible to make up exactly a postage of $k + 1$ cents using 3-cent and 5-cent stamps. We examine two cases: Suppose we make up a postage of k cents using at least one 5-cent stamp. Replacing a 5-cent stamp by two 3-cent stamps will yield a way to make up a postage of $k + 1$ cents. On the other hand, suppose we make up a postage of k cents using 3-cent stamps only. Since $k \geqq 8$, there must be at least three 3-cent stamps. Replacing three 3-cent stamps by two 5-cent stamps will yield a way to make up a postage of $k + 1$ cents. Since it is obvious how we can make up a postage of 8 cents, we conclude that we can make up a postage of 9 cents, which, in turn, leads us to conclude that we can make up a postage of 10 cents, which, in turn, leads us to conclude that we can make up a postage of 11 cents and so on. ☐

Example 1.2 Suppose we remove a square from a standard 8×8 chessboard as shown in Fig. 1.2*a*. Given 21 L-shaped triominoes‡ as shown in Fig. 1.2*b*, we want to know whether it is possible to *tile* the 63 remaining squares of the chessboard with the triominoes. (By tiling the remaining squares of the chessboard, we mean covering each of them exactly once without parts of the triominoes extending over the removed square or the edges of the board.) The answer to our question is affirmative, as Fig. 1.3 shows. We can actually prove a more general result as we shall proceed to do.

---

† See, however, Prob. 1.20.

‡ The word *tri*omino is derived from the word domino. Also, there are *tet*rominoes, *pent*ominoes, *hex*ominoes, ... and, in general *poly*ominoes. For many interesting results in connection with polyominoes, see Golomb[4].

(a)   (b)

Figure 1.2

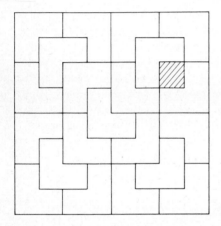

Figure 1.3

A chessboard with one of its squares removed will be referred to as a *defective* chessboard. We want to show that any defective $2^n \times 2^n$ chessboard can be tiled with L-shaped triominoes.† It is trivially obvious that a defective 2×2 chessboard can be tiled with an L-shaped triomino. Let us now assume that any defective $2^k \times 2^k$ chessboard can be tiled with L-shaped triominoes and proceed to show that any defective $2^{k+1} \times 2^{k+1}$ chessboard can also be tiled with L-shaped triominoes. Consider a defective $2^{k+1} \times 2^{k+1}$ chessboard as shown in Fig. 1.4a. Let us divide the chessboard into four quadrants, each of which is a $2^k \times 2^k$ chessboard, as shown in Fig. 1.4b. One of these $2^k \times 2^k$ chessboards is a defective one.

---

† One would immediately question whether $2^n \times 2^n - 1$ is always divisible by 3. The answer is affirmative. (See Prob. 1.23.)

Furthermore, by placing an L-shaped triomino at the center of the $2^{k+1} \times 2^{k+1}$ chessboard as shown in Fig. 1.4c, we can imagine that the other three quadrants are also defective $2^k \times 2^k$ chessboards. Since we assume that any defective $2^k \times 2^k$ chessboard can be tiled with L-shaped triominoes, we can tile each of the quadrants with L-shaped triominoes, and conclude that any defective $2^{k+1} \times 2^{k+1}$ chessboard can be tiled with L-shaped triominoes. Thus, starting with the tiling of any defective 2×2 chessboard, we have proved that we can tile any $2^n \times 2^n$ defective chessboard. □

These two examples illustrate a very powerful proof technique in mathematics known as the principle of *mathematical induction.* For a given

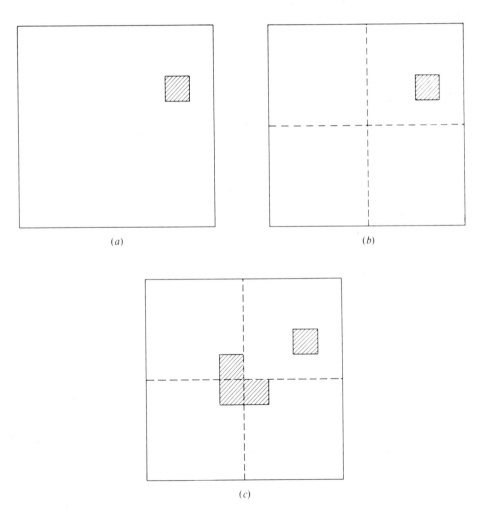

(a)

(b)

(c)

Figure 1.4

statement involving a natural number n, if we can show that:

1. The statement is true for $n = n_0$; and
2. The statement is true for $n = k + 1$, assuming that the statement is true for $n = k$, $(k \geq n_0)$,

then we can conclude that the statement is true for all natural numbers $n \geq n_0$. (1) is usually referred to as the *basis of induction*, and (2) is usually referred to as the *induction step*. For example, in the postage-stamp problem, we want to prove the statement, "It is possible to make up exactly any postage of n cents using 3-cent stamps and 5-cent stamps for $n \geq 8$." In order to prove the statement we show that:

1. *Basis of induction.* It is possible to make up exactly a postage of 8 cents.
2. *Induction step.* It is possible to make up exactly a postage of $k + 1$ cents, assuming it is possible to make up exactly a postage of k cents.

We note that the principle of mathematical induction is a direct consequence of the definition of natural numbers. Consider a set S such that

1. The natural number n_0 is in S.
2. If the natural number k is in S, then the natural number $k + 1$ is also in S.

According to the definition of the set of natural numbers, we can conclude that S contains all the natural numbers larger than or equal to k. However, this is exactly the statement of the principle of mathematical induction when we consider S to be the set of natural numbers for which a given statement is true.

We consider now more examples:

Example 1.3 Show that

$$1^2 + 2^2 + \cdots + n^2 = \frac{n(n+1)(2n+1)}{6} \qquad n \geq 1$$

by mathematical induction.

1. *Basis of induction.* For $n = 1$, we have

$$1^2 = \frac{1(1+1)(2+1)}{6}$$

2. *Induction step.* Assume that

$$1^2 + 2^2 + \cdots + k^2 = \frac{k(k+1)(2k+1)}{6}$$

We have

$$1^2 + 2^2 + \cdots + k^2 + (k+1)^2 = \frac{k(k+1)(2k+1)}{6} + (k+1)^2$$

$$= \frac{(k+1)[k(2k+1) + 6(k+1)]}{6}$$

$$= \frac{(k+1)(2k^2 + 7k + 6)}{6}$$

$$= \frac{(k+1)(k+2)(2k+3)}{6}$$

$$= \frac{(k+1)[(k+1)+1][2(k+1)+1]}{6} \qquad \square$$

Example 1.4 Show that any integer composed of 3^n identical digits is divisible by 3^n. (For example, 222 and 777 are divisible by three, 222,222,222 and 555,555,555 are divisible by nine.) We shall prove the result by induction on n.

1. *Basis of induction.* For $n = 1$, we note that any 3-digit integer with three identical digits is divisible by three.†
2. *Induction step.* Let x be an integer composed of 3^{k+1} identical digits. We note that x can be written as

$$x = y \cdot z$$

where y is an integer composed of 3^k identical digits and

$$z = 10^{2 \cdot 3^k} + 10^{3^k} + 1 = 1\underbrace{000000 \cdots 0}_{3^k - 1 \ 0s}1\underbrace{000000 \cdots 0}_{3^k - 1 \ 0s}1$$

Since we assume that y is divisible by 3^k, and z is clearly divisible by three, we conclude that x is divisible by 3^{k+1}. $\qquad \square$

We shall see more examples illustrating the application of the principle of mathematical induction later on.

1.5 PRINCIPLE OF INCLUSION AND EXCLUSION

We present in this section some results related to the cardinality of finite sets. We shall use the notation $|P|$ to denote the cardinality of the set P. Some simple results, the derivation of which is left to the reader, are:

$$|P \cup Q| \leq |P| + |Q|$$

† We remind the reader of the elementary result that an integer is divisible by three if the sum of its digits is divisible by three.

$$|P \cap Q| \leqq \min(|P|, |Q|)$$

$$|P \oplus Q| = |P| + |Q| - 2|P \cap Q|$$

$$|P - Q| \geqq |P| - |Q|$$

We show in the following a less obvious result. Let A_1 and A_2 be two sets. We want to show that

$$|A_1 \cup A_2| = |A_1| + |A_2| - |A_1 \cap A_2| \tag{1.1}$$

Note that the sets A_1 and A_2 might have some common elements. To be specific, the number of common elements between A_1 and A_2 is $|A_1 \cap A_2|$. Each of these elements is counted twice in $|A_1| + |A_2|$ (once in $|A_1|$ and once in $|A_2|$), although it should be counted as one element in $|A_1 \cup A_2|$. Therefore, the *double count* of these elements in $|A_1| + |A_2|$ should be adjusted by the subtraction of the term $|A_1 \cap A_2|$ in the right-hand side of (1.1). As an example, suppose that among a set of 12 books, 6 are novels, 7 were published in the year 1975, and 3 are novels published in 1975. Let A_1 denote the set of books that are novels, and A_2 denote the set of books published in 1975. We have

$$|A_1| = 6 \qquad |A_2| = 7 \qquad |A_1 \cap A_2| = 3$$

Consequently, according to (1.1),

$$|A_1 \cup A_2| = 6 + 7 - 3 = 10$$

That is, there are ten books which are either novels or 1975 publications, or both. Consequently, among the 12 books there are 2 nonnovels that were not published in 1975.

Extending the result in (1.1), we have, for three sets A_1, A_2, and A_3,

$$|A_1 \cup A_2 \cup A_3| = |A_1| + |A_2| + |A_3| - |A_1 \cap A_2| - |A_1 \cap A_3|$$

$$- |A_2 \cap A_3| + |A_1 \cap A_2 \cap A_3| \tag{1.2}$$

As we shall prove a more general result in the following, we shall not prove the result in (1.2) here. On the other hand, we suggest that a reader check the result in (1.2) by examining the Venn diagram in Fig. 1.5.

Let us consider some illustrative examples:

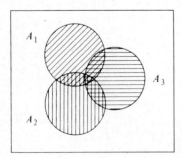

Figure 1.5

Example 1.5 Suppose we have six computers with the following specifications:

Computer	Floating point arithmetic unit	Magnetic disk memory	Graphical display terminal
I	Yes	Yes	No
II	Yes	Yes	Yes
III	No	No	No
IV	No	Yes	Yes
V	No	Yes	No
VI	No	Yes	Yes

Let A_1, A_2, and A_3 be the sets of computers with a floating-point arithmetic unit, a magnetic-disk storage, and a graphical-display terminal, respectively. We have

$$|A_1| = 2 \qquad |A_2| = 5 \qquad |A_3| = 3$$

$$|A_1 \cap A_2| = 2 \qquad |A_1 \cap A_3| = 1 \qquad |A_2 \cap A_3| = 3$$

$$|A_1 \cap A_2 \cap A_3| = 1$$

Consequently,

$$|A_1 \cup A_2 \cup A_3| = 2 + 5 + 3 - 2 - 1 - 3 + 1 = 5$$

That is, five of the six computers have one or more of the three kinds of hardware considered. ☐

Example 1.6 Thirty cars were assembled in a factory. The options available were a radio, an air conditioner, and white-wall tires. It is known that 15 of the cars have radios, 8 of them have air conditioners, and 6 of them have white-wall tires. Moreover, three of them have all three options. We want to know at *least* how many cars do not have any options at all. Let A_1, A_2, and A_3 be the sets of cars with a radio, an air conditioner, and white-wall tires, respectively. Since

$$|A_1| = 15 \qquad |A_2| = 8 \qquad |A_3| = 6$$

and

$$|A_1 \cap A_2 \cap A_3| = 3$$

according to (1.2)

$$|A_1 \cup A_2 \cup A_3| = 15 + 8 + 6 - |A_1 \cap A_2| - |A_1 \cap A_3| - |A_2 \cap A_3| + 3$$

$$= 32 - |A_1 \cap A_2| - |A_1 \cap A_3| - |A_2 \cap A_3|$$

Since

$$|A_1 \cap A_2| \geq |A_1 \cap A_2 \cap A_3|$$

$$|A_1 \cap A_3| \geq |A_1 \cap A_2 \cap A_3|$$

$$|A_2 \cap A_3| \geq |A_1 \cap A_2 \cap A_3|$$

we have

$$|A_1 \cup A_2 \cup A_3| \leq 32 - 3 - 3 - 3 = 23$$

That is, there are *at most* 23 cars that have one or more options. Consequently, there are *at least* seven cars that do not have any options. □

In the general case, for the sets $A_1, A_2, \ldots, A_r$, we have

$$|A_1 \cup A_2 \cup \cdots \cup A_r| = \sum_i |A_i| - \sum_{1 \leq i < j \leq r} |A_i \cap A_j| + \sum_{1 \leq i < j < k \leq r} |A_i \cap A_j \cap A_k|$$

$$+ \cdots + (-1)^{r-1} |A_1 \cap A_2 \cap \cdots \cap A_r| \qquad (1.3)$$

Although the result in (1.2) is not difficult to visualize, the result in (1.3) is not as obvious. We now prove (1.3) by induction on the number of sets r. Clearly, (1.1) can serve as the basis of induction. As the induction step, we assume that (1.3) is valid for any $r - 1$ sets. We note first that, viewing $(A_1 \cup A_2 \cup \cdots \cup A_{r-1})$ and A_r as two sets, according to (1.1) we have

$$|A_1 \cup A_2 \cup \cdots \cup A_r| = |A_1 \cup A_2 \cup \cdots \cup A_{r-1}| + |A_r|$$

$$- |A_r \cap (A_1 \cup A_2 \cup \cdots \cup A_{r-1})| \qquad (1.4)$$

Now

$$|A_r \cap (A_1 \cup A_2 \cup \cdots \cup A_{r-1})| = |(A_r \cap A_1) \cup (A_r \cap A_2) \cup \cdots \cup (A_r \cap A_{r-1})|$$

According to the induction hypothesis, for the $r - 1$ sets $A_r \cap A_1$, $A_r \cap A_2, \ldots, A_r \cap A_{r-1}$, we have

$$|(A_r \cap A_1) \cup (A_r \cap A_2) \cup \cdots \cup (A_r \cap A_{r-1})|$$

$$= |A_r \cap A_1| + |A_r \cap A_2| + \cdots + |A_r \cap A_{r-1}|$$

$$- |(A_r \cap A_1) \cap (A_r \cap A_2)| - |(A_r \cap A_1) \cap (A_r \cap A_3)|$$

$$- \cdots$$

$$+ |(A_r \cap A_1) \cap (A_r \cap A_2) \cap (A_r \cap A_3)| + \cdots$$

$$- \cdots$$

$$+ (-1)^{r-2} |(A_r \cap A_1) \cap (A_r \cap A_2) \cap \cdots \cap (A_r \cap A_{r-1})|$$

$$= |A_r \cap A_1| + |A_r \cap A_2| + \cdots + |A_r \cap A_{r-1}|$$

$$- |A_r \cap A_1 \cap A_2| - |A_r \cap A_1 \cap A_3| - \cdots$$

$$+ |A_r \cap A_1 \cap A_2 \cap A_3| + \cdots$$

$$- \cdots$$

$$+ (-1)^{r-2}|A_r \cap A_1 \cap A_2 \cap \cdots \cap A_{r-1}| \qquad (1.5)$$

Also, according to the induction hypothesis, for the $r-1$ sets $A_1, A_2, \ldots, A_{r-1}$, we have

$$|A_1 \cup A_2 \cup \cdots \cup A_{r-1}| = |A_1| + |A_2| + \cdots$$

$$- |A_1 \cap A_2| - |A_1 \cap A_3| - \cdots$$

$$+ \cdots$$

$$+ (-1)^{r-2}|A_1 \cap A_2 \cap \cdots \cap A_{r-1}| \qquad (1.6)$$

Substituting (1.5) and (1.6) into (1.4), we obtain (1.3).

Example 1.7 Let us determine the number of integers between 1 and 250 that are divisible by any of the integers 2, 3, 5, and 7. Let A_1 denote the set of integers between 1 and 250 that are divisible by 2, A_2 denote the set of integers that are divisible by 3, A_3 denote the set of integers that are divisible by 5, and A_4 denote the set of integers that are divisible by 7. Since

$$|A_1| = \left\lfloor \frac{250}{2} \right\rfloor^\dagger = 125 \qquad\qquad |A_2| = \left\lfloor \frac{250}{3} \right\rfloor = 83$$

$$|A_3| = \left\lfloor \frac{250}{5} \right\rfloor = 50 \qquad\qquad |A_4| = \left\lfloor \frac{250}{7} \right\rfloor = 35$$

$$|A_1 \cap A_2| = \left\lfloor \frac{250}{2 \times 3} \right\rfloor = 41 \qquad\qquad |A_1 \cap A_3| = \left\lfloor \frac{250}{2 \times 5} \right\rfloor = 25$$

$$|A_1 \cap A_4| = \left\lfloor \frac{250}{2 \times 7} \right\rfloor = 17 \qquad\qquad |A_2 \cap A_3| = \left\lfloor \frac{250}{3 \times 5} \right\rfloor = 16$$

$$|A_2 \cap A_4| = \left\lfloor \frac{250}{3 \times 7} \right\rfloor = 11 \qquad\qquad |A_3 \cap A_4| = \left\lfloor \frac{250}{5 \times 7} \right\rfloor = 7$$

$$|A_1 \cap A_2 \cap A_3| = \left\lfloor \frac{250}{2 \times 3 \times 5} \right\rfloor = 8 \qquad\qquad |A_1 \cap A_2 \cap A_4| = \left\lfloor \frac{250}{2 \times 3 \times 7} \right\rfloor = 5$$

$$|A_1 \cap A_3 \cap A_4| = \left\lfloor \frac{250}{2 \times 5 \times 7} \right\rfloor = 3 \qquad\qquad |A_2 \cap A_3 \cap A_4| = \left\lfloor \frac{250}{3 \times 5 \times 7} \right\rfloor = 2$$

$$|A_1 \cap A_2 \cap A_3 \cap A_4| = \left\lfloor \frac{250}{2 \times 3 \times 5 \times 7} \right\rfloor = 1$$

we have

$$|A_1 \cup A_2 \cup A_3 \cup A_4| = 125 + 83 + 50 + 35 - 41 - 25 - 17 - 16 - 11 - 7$$
$$+ 8 + 5 + 3 + 2 - 1 = 193 \qquad \square$$

† We use $\lfloor x \rfloor$ to denote the largest integer that is smaller than or equal to x.

We shall present more examples on the application of the formula (1.3) which is called the *principle of inclusion and exclusion* in later chapters.

*1.6 MULTISETS

We recall that a set is a collection of distinct objects. There are many occasions, however, when we encounter collections of nondistinct objects. For example, consider the names of the students in a class. We might have two or more students who have the same name, and we might wish to talk about the collection of the names of the students. We define a *multiset* to be a collection of objects that are not necessarily distinct. Thus, $\{a, a, a, b, b, c\}$, $\{a, a, a, a\}$, $\{a, b, c\}$, $\{\ \}$ are all examples of multisets. The *multiplicity* of an element in a multiset is defined to be the number of times the element appears in the multiset. Thus the multiplicity of the element a in the multiset $\{a, a, a, c, d, d\}$ is 3. The multiplicity of the element b is 0, the multiplicity of the element c is 1, and the multiplicity of the element d is 2. Note that sets are merely special instances of multisets in which the multiplicity of an element is either 0 or 1. The cardinality of a multiset is defined to be the cardinality of the set it corresponds to, assuming that the elements in the multiset are all distinct.

Let P and Q be two multisets. The union of P and Q, denoted $P \cup Q$, is a multiset such that the multiplicity of an element in $P \cup Q$ is equal to the maximum of the multiplicities of that element in P and in Q. Thus, for $P = \{a, a, a, c, d, d\}$ and $Q = \{a, a, b, c, c\}$

$$P \cup Q = \{a, a, a, b, c, c, d, d\}$$

For example, let the multiset $P = \{$electrical engineer, electrical engineer, electrical engineer, mechanical engineer, mathematician, mathematician, draftsman$\}$ be the personnel needed in the first phase of an engineering project, and the multiset $Q = \{$electrical engineer, mechanical engineer, mechanical engineer, mathematician, computer scientist, computer scientist$\}$ be the personnel needed in the second phase of the project. The multiset $P \cup Q$ is the personnel we should hire for the project.

The intersection of P and Q, denoted $P \cap Q$, is a multiset such that the multiplicity of an element in $P \cap Q$ is equal to the minimum of the multiplicities of the element in P and in Q. Thus, for $P = \{a, a, a, c, d, d\}$ and $Q = \{a, a, b, c, c\}$

$$P \cap Q = \{a, a, c\}$$

For the example above on the personnel of an engineering project, the multiset $P \cap Q$ is the personnel that will be involved in both phases of the project.

The difference of P and Q, denoted $P - Q$, is a multiset such that the multiplicity of an element in $P - Q$ is equal to the multiplicity of the element in P minus the multiplicity of the element in Q if the difference is positive, and is equal to 0 if the difference is zero or is negative. For example, let $P = \{a, a, a, b,$

$b, c, d, d, e\}$ and $Q = \{a, a, b, b, b, c, c, d, d, f\}$. We have,

$$P - Q = \{a, e\}$$

For the foregoing example concerning the personnel of an engineering project, the multiset $P - Q$ is the personnel to be reassigned after the first phase of the project.

Note that the definitions of union, intersection, and difference of multisets are so chosen that they are consistent with that of sets. We did not define the symmetric difference of two multisets here; the interested reader might want to see Prob. 1.40.

Finally, we define the sum of two multisets P and Q, denoted $P + Q$, to be a multiset such that the multiplicity of an element in $P + Q$ is equal to the sum of the multiplicities of the element in P and in Q. Note that there is no corresponding definition of the sum of two sets. For example, let $P = \{a, a, b, c, c\}$ and $Q = \{a, b, b, d\}$. We have $P + Q = \{a, a, a, b, b, b, c, c, d\}$. As another example, let P be a multiset containing the account numbers of all the transactions in a bank on a certain day, and Q be a multiset containing the account numbers of all the transactions on the next day. P and Q are multisets because an account might have more than one transaction in a day. Thus, $P + Q$ is a combined record of the account numbers of the transactions in these two days.

1.7 PROPOSITIONS

A *proposition* is a declarative sentence that is either true or false. "It rained yesterday," "The pressure inside of the reactor chamber exceeds the safety threshold," and "We shall have chicken for dinner" are all examples of propositions. On the other hand, "What time is it," and "Please submit your report as soon as possible" are not propositions because they are not declarative sentences, and, consequently, it is not meaningful to speak of them being true or false. Note that we do not rule out the possibility that a proposition is definitely true such as "15 is divisible by 3," and the possibility that a proposition is definitely false such as "Champaign is the state capital of Illinois." A proposition that is true under all circumstances is referred to as a *tautology*, and a proposition that is false under all circumstances is referred to as a *contradiction*.

We shall frequently refer to propositions by symbolic names. For example, if we let p denote the proposition, "Every student in the class passed the final examination," we can conveniently say that p is true or p is false depending on the outcome of the final examination. The two possibilities of a proposition being true or false are also referred to as the two possible values a proposition might assume. It is customary to use T to denote the value *true* and to use F to denote the value *false*. Consequently, instead of saying that the proposition,

"Every student in the class passed the final examination" is true, we can simply say that the value of p is T.

Two propositions p and q are said to be *equivalent* if when p is true q is also true, when p is false q is also false, and conversely. For example, the two propositions, "Water froze this morning" and "Temperature was below 0°C this morning" are equivalent. Also the two propositions, "He was born in 1934" and "He will be 50 years old in 1984" are equivalent. The two propositions, "I will go to the ball game tonight" and "There is no class tomorrow" might or might not be equivalent. On the other hand, the two propositions, "x is a prime number" and "x is not divisible by 2" are not equivalent because that x is not divisible by 2 does not necessarily mean that x is a prime number.

Propositions can be combined to yield new propositions. For example, in connection with the operation of a company, let p denote the proposition that the volume of the monthly sales is less than \$200,000 and let q denote the proposition that the monthly expenditure exceeds \$200,000. We might want to have a proposition to describe the situation where the volume of the monthly sales is less than \$200,000 and the monthly expenditure exceeds \$200,000. We might also want to have a proposition to describe the situation where either the volume of the monthly sales is less than \$200,000 or the monthly expenditure exceeds \$200,000. Clearly, these propositions are examples of "combinations" of the propositions p and q.

Let p and q be two propositions. We define the *disjunction* of p and q, denoted $p \vee q$, to be a proposition which is true when either one or both of p and q are true, and which is false when both p and q are false. In the above example on the operation of a company, the disjunction of the proposition that the monthly volume of sales is less than \$200,000 and the proposition that the monthly expenditure is more than \$200,000 is the proposition that either the monthly volume of sales is less than \$200,000 or the monthly expenditure exceeds \$200,000 or both.

Let p and q be two propositions. We define the *conjunction* of p and q, denoted by $p \wedge q$, to be a proposition which is true when both p and q are true, and is false when either one or both of p and q are false. In the foregoing example concerning the operation of a company, the conjunction of the proposition that the monthly volume of sales is less than \$200,000 and the proposition that the monthly expenditure is more than \$200,000 is the proposition that both the monthly volume of sales is less than \$200,000 and the monthly expenditure exceeds \$200,000.

Let p be a proposition. We define the *negation* of p, denoted $\bar{p}$,† to be a proposition which is true when p is false, and is false when p is true. Thus, the negation of the proposition that the monthly volume of sales is less than \$200,000 is the proposition that the monthly volume of sales exceeds or equals \$200,000.

A proposition obtained from the combination of other propositions is

---

†In the literature, the notation $\neg p$ is also used.

referred to as a *compound* proposition. A proposition that is not a combination of other propositions is referred to as an *atomic* proposition. In other words, a compound proposition is made up of atomic propositions. A convenient and precise way to describe the definition of a compound proposition is a table such as those shown in Fig. 1.6 where the values of a compound proposition are specified for all possible choices of the values of the atomic propositions of the compound proposition. Specifically, the tables in Fig. 1.6 show the definitions of the disjunction and conjunction of two propositions and that of the negation of a proposition. Such tables are called *truth tables* for the compound propositions.

p	q	$p \lor q$		p	q	$p \land q$		p	$\bar{p}$
F	F	F		F	F	F		F	T
F	T	T		F	T	F		T	F
T	F	T		T	F	F			
T	T	T		T	T	T			

Figure 1.6

There are two other important ways to construct compound propositions that we want to present: Consider the propositions, "The temperature exceeds 70°C" and "The alarm will be sounded," which we shall denote p and q respectively. Also, consider the proposition, "If the temperature exceeds 70°C, then the alarm will be sounded," which we shall denote r. It is easy to see that r is true if the alarm is sounded when the temperature exceeds 70°C (both p and q are true), and r is false if the alarm is not sounded when the temperature exceeds 70°C (p is true and q is false). On the other hand, when the temperature is equal to or less than 70°C (p is false), the proposition r *cannot possibly be false*, no matter whether the alarm is sounded or not. Consequently, we say that r is also true when the temperature is equal to or less than 70°C.

Let us now formalize the notion of combining two propositions p and q to form one that reads "If p then q" introduced in the example above. Let p and q be two propositions. We define a proposition "If p then q," denoted $p \rightarrow q$, which is true if both p and q are true or if p is false, and is false if p is true and q is false, as specified in the truth table in Fig. 1.7. The compound proposition "If p then q" also reads "p implies q."

p	q	$p \rightarrow q$
F	F	T
F	T	T
T	F	F
T	T	T

Figure 1.7

Let p denote the proposition, "A new computer will be acquired" and q denote the proposition, "Additional funding is available." Consider the proposition, "A new computer will be acquired if and only if additional funding is available," which we shall denote r. Clearly, r is true if a new computer is indeed acquired when additional funding is available. (Both p and q are true.) The proposition r is also true if no new computer is acquired when additional funding is not available. (Both p and q are false.) On the other hand, r is false if a new computer is acquired although no additional funding is available (p is true and q is false), or if no new computer is acquired although additional funding is available (p is false and q is true).

Let p and q be two propositions. We define a proposition, "p if and only if q," denoted $p \leftrightarrow q$, which is true if both p and q are true or if both p and q are false, and which is false if p is true while q is false and if p is false while q is true. The truth table in Fig. 1.8 shows the definition of $p \leftrightarrow q$.

p	q	$p \leftrightarrow q$
F	F	T
F	T	F
T	F	F
T	T	T

Figure 1.8

We can combine compound propositions to yield new propositions. For example, let p and q be propositions. We have $p \wedge q$, $\bar{p} \wedge \bar{q}$ and, consequently,

$$((p \wedge q) \vee (\bar{p} \wedge \bar{q})) \rightarrow p \tag{1.7}$$

as compound propositions where parentheses are used as delimiters. The value of a compound proposition can always be determined by constructing its truth table in a step-by-step manner. As the table in Fig. 1.9 shows, the entries for the columns corresponding to the compound propositions can be constructed column by column from left to right.

p	q	$p \wedge q$	$\bar{p} \wedge \bar{q}$	$(p \wedge q) \vee (\bar{p} \wedge \bar{q})$	$((p \wedge q) \vee (\bar{p} \wedge \bar{q})) \rightarrow p$
F	F	F	T	T	F
F	T	F	F	F	T
T	F	F	F	F	T
T	T	T	F	T	T

Figure 1.9

We can determine whether two propositions are equivalent by examining their truth tables. For example, according to the truth tables in Figs. 1.6 and 1.9, the proposition in (1.7) is equivalent to the proposition $p \vee q$. As another example, we note that the two propositions $p \leftrightarrow q$ and $(p \to q) \wedge (q \to p)$ are equivalent by comparing the truth tables in Figs. 1.8 and 1.10. Indeed, we realize that the choice of the notation $p \leftrightarrow q$ is not accidental.

p	q	$p \to q$	$q \to p$	$(p \to q) \wedge (q \to p)$
F	F	T	T	T
F	T	T	F	F
T	F	F	T	F
T	T	T	T	T

Figure 1.10

As a final example, we ask the reader to verify that the two propositions

$$((p \wedge q) \vee (p \wedge r)) \to s$$

and

$$((\bar{p} \vee (\bar{q} \wedge \bar{r})) \vee s$$

are equivalent.

We shall discuss the subject of constructing, manipulating, and simplifying compound propositions in Chap. 9, after we have developed the concept of Boolean algebras.

A reader probably will recognize that the combination of propositions to yield new propositions bears a strong similarity to the combination of sets to yield new sets. To be specific, we note the similarity between the notions of the union of two sets and of the disjunction of two propositions, between the notions of the intersection of two sets and of the conjunction of two propositions, and between the notions of the complement of a set and of the negation of a proposition. Such similarities are not accidental, as we shall see in Chap. 9.

1.8 REMARKS AND REFERENCES

A number of books can be used as general references for this book. See, for example, Berztiss[1], Birkhoff and Bartee[2], Gill[3], Kemeny, Snell, and Thompson[9], Knuth[10], Korfhage[11], Liu[12], Prather[14], Preparata and Yeh[15], Stone[18], and Tremblay and Manohar[20]. For further discussion on set theory, see Halmos[6] and Stoll[17]. Pólya[13], Sominskii[16] and Golovina[5] are three delightful books covering the subject of mathematical induction and more. See chap. 4 of Liu[12] for further discussion on the principle of inclusion and exclusion. Suppes[19] is a useful reference for the material in Sec. 1.7 on propositions.

[1] Berztiss, A. T.: "Data Structures: Theory and Practice," 2d ed., Academic Press, New York, 1975.
[2] Birkhoff, G., and T. C. Bartee: "Modern Applied Algebra," McGraw-Hill Book Company, New York, 1970.

[3] Gill, A.: "Applied Algebra for the Computer Sciences," Prentice-Hall, Inc., Englewood Cliffs, N.J., 1976.

[4] Golomb, S. W.: "Polyominoes," Scribner's, New York, 1965.

[5] Golovina, L. I., and I. M. Yaglom: "Induction in Geometry," D. C. Heath and Company, Boston, 1963.

[6] Halmos, P.: "Naive Set Theory," D. Van Nostrand Company, Princeton, N.J., 1960.

[7] Hopcroft, J. E., and J. D. Ullman: "Formal Languages and Their Relation to Automata," Addison-Wesley Publishing Company, Reading, Mass., 1969.

[8] Kain, R. Y.: "Automata Theory: Machines and Languages," McGraw-Hill Book Company, New York, 1972.

[9] Kemeny, J. G., J. L. Snell, and G. L. Thompson: "Introduction to Finite Mathematics," 2d ed., Prentice-Hall, Inc., Englewood Cliffs, N.J., 1966.

[10] Knuth, D. E.: "The Art of Computer Programming, Vol. 1, Fundamental Algorithms," 2d ed., Addison-Wesley Publishing Company, Reading, Mass., 1973.

[11] Korfhage, R. R.: "Discrete Computational Structures," Academic Press, New York, 1974.

[12] Liu, C. L.: "Introduction to Combinatorial Mathematics," McGraw-Hill Book Company, New York, 1968.

[13] Pólya, G.: "Induction and Analogy in Mathematics," Princeton University Press, Princeton, N.J., 1954.

[14] Prather, R. E.: "Discrete Mathematical Structures for Computer Science," Houghton Mifflin Company, Boston, 1976.

[15] Preparata, F. P., and R. T. Yeh: "Introduction to Discrete Structures," Addison-Wesley Publishing Company, Reading, Mass., 1973.

[16] Sominskii, I. S.: "The Method of Mathematical Induction," D. C. Heath and Company, Boston, 1963.

[17] Stoll, R. R.: "Set Theory and Logic," W. H. Freeman and Company, San Francisco, 1963.

[18] Stone, H. S.: "Discrete Mathematical Structures and Their Applications," Science Research Associates, Inc., Palo Alto, Calif., 1973.

[19] Suppes, P.: "Introduction to Logic," D. Van Nostrand Company, Inc., Princeton, N.J., 1957.

[20] Tremblay, J. P., and R. P., Manohar: "Discrete Mathematical Structures with Applications to Computer Science," McGraw-Hill Book Company, New York, 1975.

PROBLEMS

1.1 Determine whether each of the following statements is true or false. Explain briefly your answer.

(a) $\emptyset \subseteq \emptyset$

(b) $\emptyset \in \emptyset$

(c) $\emptyset \subseteq \{\emptyset\}$

(d) $\emptyset \in \{\emptyset\}$

(e) $\{a, b\} \subseteq \{a, b, c, \{a, b, c\}\}$

(f) $\{a, b\} \in \{a, b, c, \{a, b, c\}\}$

(g) $\{a, b\} \subseteq \{a, b, \{\{a, b\}\}\}$

(h) $\{a, b\} \in \{a, b, \{\{a, b\}\}\}$

1.2 Give an example of sets A, B, and C such that $A \in B$, $B \in C$, and $A \notin C$.

1.3 Determine whether each of the following statements is true for arbitrary sets A, B, C. Justify your answer.

(a) If $A \in B$ and $B \subseteq C$, then $A \in C$.

(b) If $A \in B$ and $B \subseteq C$, then $A \subseteq C$.

(c) If $A \subseteq B$ and $B \in C$, then $A \in C$.

(d) If $A \subseteq B$ and $B \in C$, then $A \subseteq C$.

1.4 Let A, B, C be subsets of U. Given that

$$A \cap B = A \cap C$$
$$\bar{A} \cap B = \bar{A} \cap C$$

Is it necessary that $B = C$? Justify your answer.

1.5 Given that

$$(A \cap C) \subseteq (B \cap C)$$
$$(A \cap \bar{C}) \subseteq (B \cap \bar{C})$$

show that $A \subseteq B$.

1.6 What can you say about the sets P and Q if
(a) $P \cap Q = P$?
(b) $P \cup Q = P$?
(c) $P \oplus Q = P$?
(d) $P \cap Q = P \cup Q$?

1.7 (a) Let $A \subseteq B$ and $C \subseteq D$. Is it always the case that $(A \cup C) \subseteq (B \cup D)$? Is it always the case that $(A \cap C) \subseteq (B \cap D)$?
(b) Let $W \subset X$ and $Y \subset Z$. Is it always the case that $(W \cup Y) \subset (X \cup Z)$? Is it always the case that $(W \cap Y) \subset (X \cap Z)$?

1.8 (a) Given that $A \cup B = A \cup C$, is it necessary that $B = C$?
(b) Given that $A \cap B = A \cap C$, is it necessary that $B = C$?
(c) Given that $A \oplus B = A \oplus C$, is it necessary that $B = C$?
Justify your answers.

1.9 For $A = \{a, b, \{a, b\}, \emptyset\}$, determine:
(a) $A - \{a, b\}$
(b) $A - \emptyset$
(c) $A - \{\emptyset\}$
(d) $\{\{a, b\}\} - A$
(e) $\emptyset - A$
(f) $\{\emptyset\} - A$

1.10 Let A, B, C be arbitrary sets.
(a) Show that
$$(A - B) - C = A - (B \cup C)$$
(b) Show that
$$(A - B) - C = (A - C) - B$$
(c) Show that
$$(A - B) - C = (A - C) - (B - C)$$

1.11 Let A, B, C be sets. Under what condition is each of the following statements true?
(a) $(A - B) \cup (A - C) = A$
(b) $(A - B) \cup (A - C) = \emptyset$
(c) $(A - B) \cap (A - C) = \emptyset$
(d) $(A - B) \oplus (A - C) = \emptyset$

1.12 Let A, B be two sets.
(a) Given that $A - B = B$, what can be said about A and B?
(b) Given that $A - B = B - A$, what can be said about A and B?

1.13 Let A denote the set of all freshmen, B denote the set of all sophomores, C denote the set of all Mathematics majors, D denote the set of all Computer Science majors, E denote the set of all students in the course Elements of Discrete Mathematics, F denote the set of all students who went to a rock concert on Monday night, G denote the set of all students who stayed up late Monday night. Express the following statements in set-theoretic notation:
(a) All sophomores in Computer Science are in the course Elements of Discrete Mathematics.

(*b*) Those and only those who are in the course Elements of Discrete Mathematics or who went to the rock concert stayed up late Monday night.

(*c*) No student in the course Elements of Discrete Mathematics went to the rock concert Monday night. (The obvious reason is the long problem sets in the course Elements of Discrete Mathematics.)

(*d*) The rock concert was only for freshmen and sophomores.

(*e*) All sophomores who are neither Mathematics nor Computer Science majors went to the rock concert.

1.14 Determine the power sets of the following sets.

(*a*) $\{a\}$

(*b*) $\{\{a\}\}$

(*c*) $\{\emptyset, \{\emptyset\}\}$

1.15 Let $A = \{\emptyset\}$. Let $B = \mathscr{P}(\mathscr{P}(A))$.

(*a*) Is $\emptyset \in B$? $\emptyset \subseteq B$?

(*b*) Is $\{\emptyset\} \in B$? $\{\emptyset\} \subseteq B$?

(*c*) Is $\{\{\emptyset\}\} \in B$? $\{\{\emptyset\}\} \subseteq B$?

1.16 Let $A = \{a, \{a\}\}$.

(*a*) Is $\{a\} \in \mathscr{P}(A)$? $\{a\} \subseteq \mathscr{P}(A)$? $\{\{a\}\} \in \mathscr{P}(A)$? $\{\{a\}\} \subseteq \mathscr{P}(A)$?

(*b*) Repeat part (*a*) if $A = \{a, \{b\}\}$.

1.17 Determine the cardinalities of the sets:

(*a*) $A = \{n^7 | n \text{ is a positive integer}\}$

(*b*) $B = \{n^{109} | n \text{ is a positive integer}\}$

(*c*) $A \cup B$

(*d*) $A \cap B$

Justify your answers.

1.18 Show that at most a countably infinite number of books can ever be written in English. (We define a book to be a finite sequence of words, divided into sentences, paragraphs, and chapters.)

1.19 (*a*) Show that the set of all positive rational numbers is a countably infinite set. (*Hint*: Consider all the points in the first quadrant of the plane whose x coordinate and y coordinate are integral values.)

(*b*) Show that the union of a countably infinite number of countably infinite sets is a countably infinite set.

1.20 Solve the postage stamp problem in Example 1.1 by showing how postages of $3k$, $3k + 1$, $3k + 2$ cents can be made up with 3-cent and 5-cent stamps.

1.21 Mr. J. E. Roberts claims that he is a one-third Indian. When asked how this is possible, his answer was, "My father was a one-third Indian and my mother was a one-third Indian." Is this a correct proof by induction?

1.22 Show that $n^3 + 2n$ is divisible by 3 for all $n \geqq 1$ by induction.

1.23 Show that $2^n \times 2^n - 1$ is indivisible by 3 for all $n \geqq 1$ by induction.

1.24 Show that

$$1 + 2 + 2^2 + 2^3 + \cdots + 2^n = 2^{n+1} - 1$$

by induction.

1.25 Determine the sum

$$1 + 3 + 5 + \cdots + (2n - 1)$$

by (*a*) guessing a general formula on the basis of the values of the sum for $n = 1, 2, 3, 4, 5$; (*b*) proving that the general formula is valid by induction.

1.26 Show that

$$1^2 - 2^2 + 3^2 - 4^2 + \cdots (-1)^{n-1}n^2 = (-1)^{n-1}\frac{n(n+1)}{2}$$

(*a*) By induction.

(*b*) By using the result in Example 1.3.

1.27 Show that
$$1^2 + 3^2 + 5^2 + \cdots + (2n-1)^2 = \frac{n(2n-1)(2n+1)}{3}$$

1.28 Show that
$$1^3 + 2^3 + \cdots + n^3 = (1+2+3+\cdots+n)^2$$

1.29 Show that
$$\frac{1^2}{1 \cdot 3} + \frac{2^2}{3 \cdot 5} + \cdots + \frac{n^2}{(2n-1)(2n+1)} = \frac{n(n+1)}{2(2n+1)}$$

1.30 (*a*) Show that
$$\frac{1}{1 \cdot 2} + \frac{1}{2 \cdot 3} + \frac{1}{3 \cdot 4} + \cdots + \frac{1}{n(n+1)} = \frac{n}{n+1}$$

(*b*) Show that
$$\frac{1}{1 \cdot 3} + \frac{1}{3 \cdot 5} + \cdots + \frac{1}{(2n-1)(2n+1)} = \frac{n}{2n+1}$$

(*c*) Show that
$$\frac{1}{1 \cdot 4} + \frac{1}{4 \cdot 7} + \frac{1}{7 \cdot 10} + \cdots + \frac{1}{(3n-2)(3n+1)} = \frac{n}{3n+1}$$

(*d*) Determine and prove a general formula that includes the results in (*a*), (*b*), and (*c*) as special cases.

1.31 Show that
$$1 \cdot 2 \cdot 3 + 2 \cdot 3 \cdot 4 + 3 \cdot 4 \cdot 5 + \cdots + n(n+1)(n+2) = \frac{n(n+1)(n+2)(n+3)}{4}$$

1.32 Prove by induction that the sum of the cubes of three consecutive integers is divisible by 9.

1.33 Show that for any integer n
$$(11)^{n+2} + (12)^{2n+1}$$
is divisible by 133.

1.34 It is known that for any positive integer $n \geqq 2$
$$\frac{1}{n+1} + \frac{1}{n+2} + \cdots + \frac{1}{2n} - A > 0$$
where A is a constant. How large can A be?

1.35 Show that for any positive integer $n > 1$
$$\frac{1}{\sqrt{1}} + \frac{1}{\sqrt{2}} + \cdots + \frac{1}{\sqrt{n}} > \sqrt{n}$$

1.36 Among the integers $1 - 300$, how many of them are not divisible by 3, nor by 5, nor by 7? How many of them are divisible by 3, but not by 5 nor by 7?

1.37 A survey was conducted among 1000 people. Of these 595 are Democrats, 595 wear glasses, and 550 like ice cream; 395 of them are Democrats who wear glasses, 350 of them are Democrats who like ice cream, and 400 of them wear glasses and like ice cream; 250 of them are Democrats who wear glasses and like ice cream. How many of them who are not Democrats, do not wear glasses, and do not like ice cream? How many of them are Democrats who do not wear glasses, and do not like ice cream?

1.38 Seventy-five children went to an amusement park where they can ride on the merry-go-round, roller coaster, and ferris wheel. It is known that 20 of them have taken all three rides, and 55 of them have taken at least two of the three rides. Each ride costs $0.50, and the total receipt of the amusement park was $70.00. Determine the number of children who did not try any of the rides.

1.39 (*a*) Among 50 students in a class, 26 of them got A in the first examination, 21 got A in the second examination. If 17 students did not get A in either examination, how many students got A in both examinations?

(*b*) If the number of students who got A in the first examination is equal to that in the second examination, if the total number of students who got A in exactly one examination is 40, and if 4 students did not get A in either examination, determine the number of students who got A in the first examination only, who got A in the second examination only, and who got A in both examinations.

1.40 A possible way to define the symmetric difference of two multisets P and Q, denoted $P \oplus Q$, is to let the multiplicity of an element in $P \oplus Q$ be equal to the absolute value of the difference between the multiplicities of the element in P and Q. What is a possible inconsistency with such a definition? (*Hint*: Consider the multisets $(P \oplus Q) \oplus R$ and $P \oplus (Q \oplus R)$.)

1.41 To describe the various restaurants in the city, we let p denote the statement, "The food is good," q denote the statement, "The service is good," and r denote the statement, "The rating is three-star." Write the following statements in symbolic form.

(*a*) Either the food is good, or the service is good, or both.

(*b*) Either the food is good, or the service is good, but not both.

(*c*) The food is good while the service is poor.

(*d*) It is not the case that both the food is good and the rating is three-star.

(*e*) If both the food and services are good, then the rating will be three-star.

(*f*) It is not true that a three-star rating always means good food and good service.

1.42 Let p denote the statement, "The material is interesting," and q denote the statement, "The exercises are challenging," and r denote the statement, "The course is enjoyable." Write the following statements in symbolic form:

(*a*) The material is interesting and the exercises are challenging.

(*b*) The material is uninteresting, the exercises are not challenging, and the course is not enjoyable.

(*c*) If the material is not interesting and the exercises are not challenging, then the course is not enjoyable.

(*d*) The material is interesting means the exercises are challenging, and conversely.

(*e*) Either the material is interesting or the exercises are not challenging, but not both.

1.43 Write the following statements in symbolic form:

(*a*) The sun is bright and the humidity is not high.

(*b*) If I finish my homework before dinner and it does not rain, then I will go to the ball game.

(*c*) If you do not see me tomorrow, it means that I have gone to Chicago.

(*d*) If the utility cost goes up or the request for additional funding is denied, then a new computer will be purchased if and only if we can show that the current computing facilities are indeed not adequate.

1.44 Let p denote the statement, "The weather is nice" and q denote the statement, "We have a picnic." Translate the following in English and simplify if possible:

(*a*) $p \wedge \bar{q}$

(*b*) $p \leftrightarrow q$

(*c*) $\bar{q} \rightarrow \bar{p}$

(*d*) $\overline{(\bar{p} \vee q)} \vee (p \wedge \bar{q})$

1.45 (*a*) Write a compound statement that is true when exactly two of the three statements p, q, r is true.

(*b*) Write a compound statement that is true when none, or one, or two of the three statements p, q, r are true.

1.46 Construct the truth tables for the following statements:

(a) $p \rightarrow p$

(b) $(p \rightarrow p) \vee (p \rightarrow \bar{p})$

(c) $(p \rightarrow p) \rightarrow (p \rightarrow \bar{p})$

(d) $(p \vee \bar{q}) \vee \bar{p}$

(e) $(p \vee \bar{q}) \rightarrow \bar{p}$

(f) $p \leftrightarrow (\bar{p} \vee \bar{q})$

(g) $(p \rightarrow (q \rightarrow r)) \rightarrow ((p \rightarrow q) \rightarrow (p \rightarrow r))$

(h) $(\bar{q} \rightarrow \bar{p}) \rightarrow (p \rightarrow q)$

1.47 (a) Given that the value of $p \rightarrow q$ in false, determine the value of $(\bar{p} \vee \bar{q}) \rightarrow q$.

(b) Given that the value of $p \rightarrow q$ is true, can you determine the value of $\bar{p} \vee (p \leftrightarrow q)$?

1.48 *Phrase structure grammars* Let U be a finite set, called an *alphabet*. The elements of U are called *letters*. Let U^+ denote the set of all possible sequences of letters. For example, let $U = \{a, b\}$ $U^+ = \{a, b, aa, ab, ba, bb, aaa, aab. \ldots\}$. A *language* is a subset of U^+, and the sequences in a language are referred to as *sentences*. Languages can be specified in the usual way we specify sets. For example, we might have

$$L_1 = \{aa, aaaab, bbaaab\}$$

$$L_2 = \{\alpha | \alpha \in U^+, \alpha \text{ contains more } a\text{'s than } b\text{'s}\}$$

We present now another way to specify languages. Let T be a set, called a set of nonterminals, such that U and T are disjoint. We shall use lower-case English letters to denote the elements of U, upper-case English letters to denote the elements of T, and lower-case Greek letters to denote sequences in $(U \cup T)^+$. A *transformation rule*, or a *production*, is of the form $\alpha \rightarrow \beta$. There is a special element in T, called the *sentence symbol*, and it is denoted S.

A *phrase structure grammar* is a set of productions with at least one of them of the form $S \rightarrow \alpha$. For example, $\{S \rightarrow AB, A \rightarrow aaB, B \rightarrow b, B \rightarrow c\}$ is a grammar. A transformation rule can be applied to transform a sequence in $(U \cup T)^+$ in the following way: Let $\gamma\alpha\delta$ be a sequence in $(U \cup T)^+$. If there is a production $\alpha \rightarrow \beta$, then the sequence $\gamma\alpha\delta$ can be transformed into $\gamma\beta\delta$. We now present a way to use a grammar to describe a language: Starting with the sentence symbol S, we shall apply the transformation rules in all possible ways. The sequences in U^+ obtained this way are exactly the sentences in the language. For example, for the grammar shown above, we may have

$$S \rightarrow AB \rightarrow aaBB \rightarrow aabB \rightarrow aabb$$

$$S \rightarrow AB \rightarrow aaBB \rightarrow aacB \rightarrow aacb$$

Indeed, the language is $\{aabb, aabc, aacb, aacc\}$.

(a) Give a verbal description of the sentences in the language specified by the grammar:

$$\{S \rightarrow aSb, S \rightarrow ab\}$$

(b) Give a verbal description of the sentences in the language specified by the grammar:

$$\{S \rightarrow aB, S \rightarrow bA, A \rightarrow a, A \rightarrow aS, A \rightarrow bAA, B \rightarrow b, B \rightarrow bS, B \rightarrow aBB\}$$

(c) Give a verbal description of the sentences in the language specified by the grammar:

$$\{S \rightarrow ABx, A \rightarrow Ay, yB \rightarrow zB, A \rightarrow w, xB \rightarrow u\}$$

(d) Construct a grammar for a language whose sentences are of the form $\{a^n b^{2n} | n \geq 1\}$.

(e) Construct a grammar for a language whose sentences are of the form $\{\alpha abab | \alpha \in \{a, b\}^+\}$.

An interesting question is whether all languages can be described by phrase structure grammars. The answer is negative. However, a discussion of this subject is beyond the scope of this book. The interested reader can consult a book on formal languages such as Hopcroft and Ullman[7] and Kain[8].

1.49 *Russell's paradox* (a) We saw in Sec. 1.1 that membership of a set can be described by stating the properties which uniquely characterize the objects in the set. Such a way of specifying a set can lead to unexpected difficulty as we shall see. Let C be a set of sets such that

$$C = \{x \mid x \text{ is a set and } x \notin x\}$$

Is $C \in C$ or is $C \notin C$? Why? (Such an inconsistency is known as *Russell's paradox*.)

In axiomatic set theory, where all arguments must follow from a chosen set of axioms, Russell's paradox can be avoided by requiring that a set can be constructed only as a subset of a set that has already been constructed. In other words, we can define a set C as

$$C = \{x \mid x \in A, x \text{ possesses some properties}\}$$

where A is a set that has already been constructed. Thus, Russell's paradox is avoided because we cannot construct a set that contains all sets. For further details on axiomatic theory, see, for example, Stoll [17].

(*b*) There is a barber who will shave those and only those who do not shave themselves. Who will shave the barber? Why is this a paradox?

(*c*) An adjective is called *heterological* if it does not have the property it denotes. For example, "monosyllabic" is a heterological adjective, while "polysyllabic" is not. Is "heterological" a heterological adjective? Why is this a paradox?

1.50 *Strong mathematical induction.* We show in this problem the following result which is known as the principle of strong mathematical induction. For a given statement involving a natural number n if we can show that: (1) The statement is true for $n = n_0$; and (2) the statement is true for $n = k + 1$, assuming that the statement is true for $n = n_0, n_0 + 1, n_0 + 2, \ldots, k, (k \geq n_0)$, then we can conclude that the statement is true for all natural numbers $n \geq n_0$.

(*a*) Let S be a set of natural numbers such that: (1) The natural number n_0 is in S. (2) If the natural numbers $n_0, n_0 + 1, n_0 + 2, \ldots, k$ are in S, then the natural number $k + 1$ is also in S.

Show that S is the set of all natural numbers larger than or equal to n_0. (*Hint*: Assume that n_1 is the smallest natural number not in S.)

(*b*) Use the result in part (*a*) to show that the principle of strong mathematical induction is indeed valid.

(*c*) Use the principle of strong mathematical induction to show that any natural number larger than or equal to 2 is either a prime or can be expressed as a product of primes.

PERMUTATIONS AND COMBINATIONS

2.1 INTRODUCTION

In Sec. 1.5 we discussed some results on the size of finite sets. We shall present in this chapter some further results along this line. For example, let A be a finite set of size n. We might wish to know the number of distinct subsets of the set A, that is, the size of the power set of A, $\mathcal{P}(A)$. Furthermore, among all subsets of A, we might want to know the number of subsets that are of size k. For example, let A be a set of ten senators. The number of subsets in $\mathcal{P}(A)$, which is equal to 2^{10}, is the number of different committees the senators can form [including a committee with no members, corresponding to the empty set in $\mathcal{P}(A)$]. Moreover, the number of subsets of size 6 in $\mathcal{P}(A)$, which is equal to 210, is the number of different 6-member committees they can form. In this chapter, we shall discuss these and related problems in the context of permutations and combinations of objects.

2.2 THE RULES OF SUM AND PRODUCT

By an *event* we shall mean something that takes place. Placing a ball in a box, selecting a representative among a group of students, assigning offices to professors, and placing bets in a horse race are all examples of events. An event can occur in one or more ways. For example, in the case of placing a ball in one of several boxes, the number of ways of placing the ball is equal to the number of boxes available. When we consider the occurrence of several events, we shall follow the rules stated below:

Rule of product. If one event can occur in m ways and another event can occur in n ways, then there are $m \times n$ ways in which these two events can occur together.

Rule of sum. If one event can occur in m ways and another event can occur in n ways, then there are $m + n$ ways in which exactly one of these two events can occur.

For example, if there are 52 ways to select a representative for the junior class and 49 ways to select a representative for the senior class, then according to the rule of product, there will be 52×49 ways to select the representatives for both the junior and senior classes. On the other hand, according to the rule of sum, there will be $52 + 49$ ways to select a representative for either the junior or the senior class. As another example, suppose there are seven different courses offered in the morning and five different courses offered in the afternoon. There will be 7×5 choices for students who want to enroll in one course in the morning and one in the afternoon. On the other hand, they will have $7 + 5$ choices if they want to enroll in only one course.

2.3 PERMUTATIONS

Consider the simple problem of placing three balls colored red, blue, and white in ten boxes numbered 1, 2, 3, ..., 10. We want to know the number of distinct ways in which the balls can be placed in the boxes, if each box can hold only one ball. Let us place the balls one at a time, beginning with the red ball, then the blue ball, and then the white ball. Since the red ball can be placed in any one of the ten boxes, the blue ball can be placed in any one of the nine remaining boxes, and the white ball can be placed in any one of the eight remaining boxes, the total number of distinct ways to place these balls is $10 \times 9 \times 8 = 720$.

The result of this numerical example can be generalized immediately: Suppose we are to place r distinctly colored balls in n distinctly numbered boxes with the condition that a box can hold only one ball. Since the first ball can be placed in any one of the n boxes, the second ball can be placed in any one of the remaining $(n - 1)$ boxes, ..., and the rth ball can be placed in any one of the remaining $(n - r + 1)$ boxes, the total number of distinct ways to place the balls is

$$n(n - 1)(n - 2) \cdots (n - r + 1)$$

which can also be written as†

$$\frac{n!}{(n - r)!}$$

We use the notation $P(n, r)$ for the quantity $n(n - 1)(n - 2) \cdots (n - r + 1)$.

---

† $n!$ reads "n factorial" and is defined to be $n(n - 1)(n - 2) \cdots 2 \times 1$. We also have the convention that $0!$ is equal to 1.

The following examples show that the problem of placing balls in boxes is not as uninteresting as it might seem.

Example 2.1 In how many ways can three examinations be scheduled within a five-day period so that no two examinations are scheduled in the same day? Considering the three examinations as distinctly colored balls and the five days as distinctly numbered boxes, we obtain the result $5 \times 4 \times 3 = 60$. □

Example 2.2 Suppose that we have seven rooms and want to assign four of them to four programmers as offices and use the remaining three rooms for computer terminals. The assignment can be made in $7 \times 6 \times 5 \times 4 = 840$ different ways because we can view the problem as that of placing four distinct balls (the programmers) into seven distinct boxes (the rooms), with the three boxes that are left empty being the rooms for computer terminals. (We assume that the programmers are distinct but that all computer terminals are identical.) □

A problem equivalent to placing balls in boxes is that of arranging or permuting distinct objects. By permuting r of n distinct objects, we mean to arrange r of these n objects in some order. For example, there are six ways to permute two of the three objects a, b, c. They are ab, ba, ac, ca, bc, and cb. Since to arrange r of n objects amounts to filling r positions with r of the n objects, there are n choices of an object for the first position, $n - 1$ choices of an object (from the $n - 1$ remaining objects) for the second position, ..., and $n - r + 1$ choices of an object (from the $n - r + 1$ remaining objects) for the rth position. Consequently, there are $n(n - 1) \cdots (n - r + 1)$ ways to arrange r of n objects in order.†

Consider the following examples:

Example 2.3 Let us determine the number of four-digit decimal numbers that contain no repeated digits. Since this is a problem of arranging four of the ten digits 0, 1, 2, ..., 9, the answer is $P(10, 4) = 5040$. Among these 5040 numbers, $9 \times 8 \times 7 = 504$ of them have a leading 0. Consequently, $5040 - 504 = 4536$ of them do not have a leading 0. The result can also be computed as

$$9 \times 9 \times 8 \times 7 = 4536$$

using the argument that the first digit can be any one of the nine digits 1, 2, ..., 9, the second digit can be any of the *nine* remaining digits, and so on. □

---

† A slightly different point of view, which might cause some initial confusion but will prove to be useful eventually, is to consider the problem as that of placing balls in boxes. Consider n boxes corresponding to the n objects, and r balls corresponding to the r positions in the arrangement. The placement of a ball in a certain box is equivalent to putting the object corresponding to the box in a position corresponding to the ball in an arrangement. Consequently, the number of ways to permute r of n objects is $P(n, r)$.

Example 2.4 We note that the number of ways in which we can make up strings of four distinct letters followed by three distinct digits is

$$P(26, 4) \times P(10, 3) = 258,336,000 \qquad \Box$$

Let us return to the problem of placing three distinctly colored balls into ten distinctly numbered boxes. Suppose a box can hold as many balls as we wish. Since the red ball can be placed in any one of the ten boxes, as can the blue ball, and as can the white ball, the total number of ways of placement is

$$10 \times 10 \times 10 = 1000$$

In general, there are n^r ways to place r colored balls into n numbered boxes if a box can hold as many balls as we wish.

We consider now some other examples:

Example 2.5 If we are to schedule three examinations within a five-day period with no restriction on the number of examinations scheduled each day, the total number of ways is $5^3 = 125$. $\qquad \Box$

Example 2.6 Let us determine the number of subsets of a set A whose size is r. Consider the problem of placing the r elements of A in two boxes. Corresponding to each placement, we can define a subset of A by taking the elements placed in box 1 and discarding the elements placed in box 2. Since there are 2^r ways to place the r elements, there are 2^r subsets in $\mathcal{P}(A)$. $\qquad \Box$

Similarly, in terms of permutation of objects, we say that, if there are n distinct kinds of objects with an infinite supply of each kind, then there are n^r ways to arrange r of these n kinds of objects, because there are n choices of an object for the first position, n choices of an object for the second position, ..., and n choices of an object for the rth position. For example, there are 10^4 four-digit decimal sequences. Consequently, $10^4 - P(10, 4) = 4960$ of them contain one or more repeated digits.

Example 2.7 We observe first that there are 2^r r-digit binary sequences. We now ask among the 2^r r-digit binary sequences how many of them have an even number of zeros? We can pair off these binary sequences such that two sequences in a pair differ only in the rth digits. Clearly, one of the two sequences in a pair has an even number of zeros and the other has an odd number of zeros. It follows that there are $\frac{1}{2} \cdot 2^r$ r-digit binary sequences that contain an even number of zeros. We now ask for the number of r-digit quintary sequences (sequences made up of the digits 0, 1, 2, 3, 4) that contain an even number of zeros. We note that among the 5^r r-digit quintary sequences, there are 3^r of them that contain only the digits 2, 3, and 4. These sequences are, of course, counted as sequences containing an even number of zeros. The remaining $5^r - 3^r$ sequences can be divided into groups according to the patterns of 2s, 3s, and 4s in the sequences. (For instance, all sequences of the form $23xx\,344xx\,2xxx$ will be in one group, where each x is either a 0 or a 1.) Since half of the sequences in each group has an even number of zeros, the total number of r-digit quintary sequences with an even number of zeros is $3^r + \frac{1}{2}(5^r - 3^r)$. ☐

Example 2.8 We now give an example of the application of the principle of inclusion and exclusion. Suppose we want to know the number of r-digit quintary sequences that contain at least a 0, a 1, and a 2. Let A_1 be the set of r-digit quintary sequences in which the digit 0 is missing, A_2 be the set of r-digit quintary sequences in which the digit 1 is missing, and A_3 be the set of r-digit quintary sequences in which the digit 2 is missing. Then $A_1 \cup A_2 \cup A_3$ is the set of r-digit quintary sequences in which one or more of the three digits 0, 1, 2 is missing. Since

$$|A_1| = |A_2| = |A_3| = 4^r$$

$$|A_1 \cap A_2| = |A_1 \cap A_3| = |A_2 \cap A_3| = 3^r$$

$$|A_1 \cap A_2 \cap A_3| = 2^r$$

we obtain

$$|A_1 \cup A_2 \cup A_3| = 4^r + 4^r + 4^r - 3^r - 3^r - 3^r + 2^r$$

$$= 3 \cdot 4^r - 3 \cdot 3^r + 2^r$$

Consequently, the number of r-digit quintary sequences that contain at least a 0, a 1, and a 2 is

$$5^r - 3 \cdot 4^r + 3 \cdot 3^r - 2^r$$ ☐

Let us consider the problem of placing four balls, two red, one blue, and one white, in ten numbered boxes. Suppose we repaint the two red balls with two shades of red, light and dark, so that they become distinguishable. The number of ways to place these balls in the ten boxes is then $P(10, 4) = 5040$. Among these 5040 placements, let us consider the placement in which the light red ball is in the first box, the dark red ball is in the second box, the blue ball is in the third box, and the white ball is in the fourth box, and the placement in which the dark red ball is in the first box, the light red ball is in the second box, the blue ball is in the third box, and the white ball is in the fourth box. If we do not distinguish between the two shades of red, that is, if the two red balls are indistinguishable, these two ways of placement actually become one. Indeed, the 5040 placements can be paired off in a similar way so that every pair of placements becomes one when we do not distinguish the two shades of red. Consequently, there are $\frac{5040}{2} = 2520$ ways to place two red balls, one blue ball, and one white ball in ten numbered boxes. Following the same argument, we see that the number of ways of placing three red balls, one blue ball, and one white ball in ten numbered boxes is

$$\frac{P(10, 5)}{3!} = 5040$$

because each way to place three indistinguishable red balls, one blue ball, and one white ball corresponds to 3! ways of placing three distinguishable red balls, one blue ball, and one white ball.

We derive now a general formula for the number of ways to place r colored balls in n number boxes, where q_1 of these balls are of one color, q_2 of them are of a second color, $\ldots$, and q_t of them are of a tth color. We note that a placement of the r balls is not changed by rearranging the q_1 balls of the same color among the boxes in which they are placed, or rearranging the q_2 balls of the same color among the boxes in which they are placed, $\ldots$, or rearranging the q_t balls of the same color among the boxes in which they are placed. On the other hand, if the r balls were distinctly colored, any rearrangement will yield a different placement. It follows that each way to place the r not completely distinctly colored balls corresponds to $q_1!q_2! \cdots q_t!$ ways to place r distinctly colored balls. Since there are $P(n, r)$ ways to place r distinctly colored balls in n numbered boxes, the total number of ways to place r colored balls in n numbered boxes, where q_1 of these balls are of one color, q_2 of them are of a second color, $\ldots$, and q_t of them are of a tth color, is

$$\frac{P(n, r)}{q_1!q_2! \cdots q_t!} \tag{2.1}$$

Example 2.9 We observe that the number of ways to paint 12 offices so that 3 of them will be painted green, 2 of them will be painted pink, 2 of them will be painted yellow, and the remaining ones will be painted white is

$$\frac{12!}{3!\,2!\,2!\,5!} = 166{,}320 \qquad \square$$

In terms of arrangement of objects, we say that there are

$$\frac{n!}{q_1!\,q_2! \cdots q_t!} \tag{2.2}$$

ways to arrange n objects, where q_1 of them are of one kind, q_2 of them are of a second kind, $\ldots$, and q_t of them are of a tth kind. Again, if the n objects were all distinct, there are $n!$ ways to arrange them. On the other hand, in an arrangement of objects that are not completely distinct, permuting the objects of the same kind among themselves will not change the arrangement. We thus obtain the formula in (2.2).

Example 2.10 We note that the number of different messages that can be represented by sequences of three dashes and two dots is

$$\frac{5!}{3!\,2!} = 10 \qquad \square$$

2.4 COMBINATIONS

Consider now the problem of placing three balls, all of them colored red, in ten boxes that are numbered $1, 2, 3, \ldots, 10$. We want to know in how many ways can the balls be placed, if each box can hold only one ball. According to (2.1) the answer is

$$\frac{10 \times 9 \times 8}{3!}$$

In general, the number of ways of placing r balls of the same color in n numbered boxes is

$$\frac{n(n-1)(n-2) \cdots (n-r+1)}{r!} = \frac{n!}{r!(n-r)!}$$

The quantity $\dfrac{n!}{r!(n-r)!}$ is also denoted $C(n, r)$.

Let us consider some examples:

Example 2.11 Suppose a housekeeper wants to schedule spaghetti dinners three times each week. Imagine the spaghetti dinners as three balls and the seven days in the week as seven boxes; then the number of ways of scheduling is

$$\frac{7!}{3! \, 4!} = 35 \qquad \qquad \square$$

Example 2.12 We note that there are $C(32, 7)$ binary sequences of length 32 in each of which there are exactly seven 1s, because we can view the problem as placing seven 1s in 32 numbered boxes (and then fill the empty boxes with 0s). $\qquad \square$

A problem equivalent to placing r indistinguishable balls in n numbered boxes is that of selection of r objects from n distinct objects. If we are to select r objects from n distinct objects, we can imagine the n objects as boxes and mark the selected objects with r identical markers, the balls. Consequently, the number of ways to select r objects from n distinct objects is also $C(n, r)$.

According to the definition of $C(n, r)$ it is clear that $C(n, r) = C(n, n - r)$. There is also a simple combinatorial argument that confirms this result: since to select r objects from n objects is the same as to pick out the $n - r$ objects that are not to be selected, we have $C(n, r) = C(n, n - r)$.

Example 2.13 Among 11 senators there are $C(11, 5) = 462$ ways to select a committee of 5 members. Moreover, there are $C(10, 4) = 210$ ways to select a committee of five members so that a particular senator, senator A, is always included, and there are $C(10, 5) = 252$ ways to select a committee of five members so that senator A is always excluded. We now ask in how many ways we can select a committee of five members so that at least one of senator A and senator B will be included. The number of selections including both senator A and senator B is $C(9,3) = 84$. The number of selections including senator A but excluding senator B is $C(9,4) = 126$, as is the number of selections including senator B but excluding senator A. Consequently, the total number of ways of selection is

$$84 + 126 + 126 = 336$$

Alternatively, since the total number of committees excluding both A and B is $C(9,5)$, the total number of ways of selection is

$$C(11,5) - C(9,5) = 462 - 126 = 336$$

The problem can also be solved by applying the principle of inclusion and exclusion. Among the 462 ways of selecting 5 senators, let A_1 and A_2 be the set of ways of selection that include senator A and senator B, respectively. Since

$$|A_1| = C(10,4) = 210$$

$$|A_2| = C(10,4) = 210$$

$$|A_1 \cap A_2| = C(9,3) = 84$$

it follows that

$$|A_1 \cup A_2| = 210 + 210 - 84 = 336 \qquad \square$$

Example 2.14   If no three diagonals of a convex decagon meet at the same point inside the decagon, into how many line segments are the diagonals divided by their intersections? First of all, the number of diagonals is equal to

$$C(10,2) - 10 = 45 - 10 = 35$$

as there are $C(10,2)$ straight lines joining the $C(10,2)$ pairs of vertices, but 10 of these 45 lines are the sides of the decagon. Since for every four vertices we can count exactly one intersection between the diagonals, as Fig. 2.1 shows (the decagon is convex), there are a total of $C(10,4) = 210$ intersections between the diagonals. Since a diagonal is divided into $k + 1$

Figure 2.1

straight-line segments when there are k intersecting points lying on it, and since each intersecting point lies on two diagonals, the total number of straight-line segments into which the diagonals are divided is $35 + 2 \times 210 = 455$. □

Suppose we are to place r balls of the same color in n numbered boxes, allowing as many balls in a box as we wish. The number of ways to place the balls is

$$\frac{(n + r - 1)!}{r!(n - 1)!} = C(n + r - 1, r)$$

An easy way to see this result is to consider the problem of arranging $n + 1$ 1s and r 0s with a 1 at the beginning and a 1 at the end of each arrangement. If we consider the 1s as interbox partitions and the 0s as balls, then every such arrangement corresponds to a way of placing r balls of the same color in n numbered boxes. For example, let $n = 5$ and $r = 4$, the sequence

$$1011001101$$

can be viewed as a placement of four balls in five boxes having one ball in the first box, no ball in the second box, two balls in the third box, no ball in the fourth box, and one ball in the fifth box. According to (2.1), the number of ways of arranging $n + 1$ 1s and r 0s with 1s at both ends of an arrangement is

$$\frac{(n + r - 1)!}{(n - 1)! \, r!}$$

We note that the problem of selecting r objects from n distinct objects, allowing repeated selections, can be viewed as that of using r identical markers to mark the n distinct objects while each object can be marked with arbitrarily many markers. Therefore, the number of ways to select r objects from n distinct objects, allowing repeated selections, is

$$\frac{(n + r - 1)!}{r! \, (n - 1)!} = C(n + r - 1, r) \tag{2.3}$$

Consider the following examples:

Example 2.15   A domino is made up of two squares each of which is marked with one, two, three, ..., or six spots, or is left blank. We note that there are 28 different dominoes in a set, because the number of distinct dominoes is the same as the number of ways of selecting two subjects from the seven objects "one," "two," "three," ..., "six," and "blank," with repetitions allowed. Thus, according to (2.3), the number of distinct dominoes is

$$C(7 + 2 - 1, 2) = C(8, 2) = 28$$ □

Example 2.16 We note that when three dice are rolled, the number of different outcomes is

$$C(6+3-1, 3) = C(8, 3) = 56$$

because rolling three dice is equivalent to selecting three numbers from the six numbers $1, 2, 3, 4, 5, 6$ with repetitions allowed. ☐

Example 2.17 We ask for the number of different paths for a rook† to move from the southwest corner of a chessboard to the northeast corner by moving eastward and northward only. If we let a 0 denote an eastward step and a 1 denote a northward step, then the number of paths is equal to the number of ways of arranging seven 0s and seven 1s, which is

$$\frac{14!}{7!\,7!} = 3432$$

We ask now how many of these paths consist of four eastward moves and three northward moves. (By an eastward move we mean a certain number of consecutive eastward steps. A northward move is defined similarly.) The number of ways of making up a path with four eastward moves is the same as that of placing seven indistinguishable balls in four distinct boxes with no box left empty. Let us place a ball in each of the four boxes and then distribute the three remaining balls. Since the number of ways to distribute three indistinguishable balls in four distinct boxes with each box holding as many balls as we wish is

$$C(4+3-1, 3) = 20$$

the number of ways to place the seven balls in four boxes with no box left empty is also equal to 20. Similarly, the number of ways of making up a path with three northward moves is

$$C(3+4-1, 4) = 15$$

Therefore, the answer to our question is

$$20 \times 15 = 300$$ ☐

Example 2.18 We want to determine the number of ways to place $2t + 1$ indistinguishable balls in three distinct boxes so that any two boxes together will contain more balls than the other one. The total number of ways to place the balls disregarding the constraint is

$$C(3+2t+1-1, 2t+1) = C(2t+3, 2t+1)$$

The total number of ways to place the balls so that the first box will have more balls than the second and third boxes combined is

$$C(3+t-1, t) = C(t+2, t)$$

† A rook is a chesspiece that can move horizontally and vertically on a chessboard.

(We place $t + 1$ balls in the first box and then place the t remaining balls in the three boxes arbitrarily.) The same result applies to the case that the second box has more balls than the first and third boxes combined, and to the case that the third box has more balls than the first and second boxes combined. Thus, the answer to our question is

$$C(2t + 3, 2t + 1) - 3C(t + 2, t) = C(2t + 3, 2) - 3C(t + 2, 2)$$
$$= \tfrac{1}{2}(2t + 3)(2t + 2) - \tfrac{3}{2}(t + 2)(t + 1)$$
$$= \frac{t(t + 1)}{2} \qquad \square$$

*2.5 GENERATION OF PERMUTATIONS AND COMBINATIONS

Suppose we want to write down the $n!$ permutations of n distinct objects. For $n = 3$, there are only six permutations, so there clearly is little difficulty. For $n = 4$, there are 24 permutations, and it is no longer a totally trivial task to keep track of what we have written down and to make sure we shall write down all the permutations with no omissions or repetitions. An interesting problem is to find systematic procedures that exhaustively generate all the permutations of n objects. As an illustration, we shall present one such procedure. Before we describe our procedure, let us point out an important question one might raise: If we are to design a procedure or to examine a procedure presented to us, how can we ascertain that the procedure will indeed do what it purports to do, namely, to generate all permutations of n objects? Note that there is no way for us to test exhaustively the procedure, since it is supposed to perform correctly for all positive integers n. One way to make sure that a procedure will generate all the permutations exhaustively with no repetition is to introduce an ordering of the $n!$ permutations. If we have a procedure that generates the permutations one by one according to this order, then we can be assured that the procedure will correctly generate all the permutations when it starts with the first permutation in the order and stops at the last permutation in the order. One order we can use is the *lexicographic order*. Without loss of generality, let $\{1, 2, 3, \ldots, n\}$ be the n objects to be permuted. For two permutations $a_1 a_2 \cdots a_n$ and $b_1 b_2 \cdots b_n$, we shall say $a_1 a_2 \cdots a_n$ comes before $b_1 b_2 \cdots b_n$ in the lexicographic order if, for some $1 \leq m < n$, $a_1 = b_1, a_2 = b_2, \ldots, a_{m-1} = b_{m-1}$, and $a_m < b_m$. For example, the permutation 124635 comes before the permutation 125643, and the permutation 125463 comes after the permutation 125346.

Suppose we were given a permutation $a_1 a_2 \cdots a_n$. Our question is what the next permutation is according to the lexicographic order. It is not difficult to show that the next permutation $b_1 b_2 \cdots b_n$ must be such that:

1. $a_i = b_i$, $1 \leq i \leq m - 1$, and $a_m < b_m$, for the largest possible m;
2. b_m is the smallest element among $a_{m+1}, a_{m+2}, \ldots, a_n$ that is larger than a_m;
3. $b_{m+1} < b_{m+2} < \cdots < b_n$.

For example, the permutation following 124653 in the lexicographic order is 125346. For a given permutation $a_1 a_2 \cdots a_n$, we note that the largest possible m for which (1) is satisfied is the largest possible m for which a_m is less than at least one of $a_{m+1}, a_{m+2}, \ldots, a_n$. A moment's reflection shows that this is also the largest possible m for which $a_m < a_{m+1}$.† Therefore, if we examine the permutation $a_1 a_2 \cdots a_n$ element by element from *right to left*, the first time we observe a decrement, we know the value of m and can determine $b_m b_{m+1} \cdots b_m$ according to (2) and (3). For example, suppose we were given the permutation 124653. When we scan the permutation from right to left element by element, according to (1), we determine that the next permutation is of the form 12*xxxx*. In other words, the subscript m is equal to 3. According to (2), we can further determine that the next permutation is of the form 125*xxx*. Finally, according to (3), we determine that the next permutation is 125346. (Actually, we can make use of the fact that $a_{m+1} > a_{m+2} > \cdots > a_n$ to carry out steps 2 and 3 in a rather simple manner. The interested reader is referred to Prob. 2.35.)

Our observation leads immediately to a systematic procedure for generating the $n!$ permutations of n objects by starting with the permutation $1234 \cdots n$ and stopping at the permutation $n \cdots 4321$. Moreover, we know that the procedure is indeed a correct one. We ask the reader to convince herself that our procedure will indeed generate the permutations of the four objects 1, 2, 3, 4 in the following order:

$$1234 \rightarrow 1243 \rightarrow 1324 \rightarrow 1342 \rightarrow 1423 \rightarrow 1432 \rightarrow 2134 \rightarrow 2143 \rightarrow$$
$$2314 \rightarrow 2341 \rightarrow 2413 \rightarrow 2431 \rightarrow 3124 \rightarrow 3142 \rightarrow 3214 \rightarrow 3241 \rightarrow$$
$$3412 \rightarrow 3421 \rightarrow 4123 \rightarrow 4132 \rightarrow 4213 \rightarrow 4231 \rightarrow 4312 \rightarrow 4321$$

Suppose we want to generate all k-subsets‡ of the set $\{1, 2, 3, \ldots, n\}$. So that we can introduce a lexicographic order of the subsets, let us agree first that each subset will be represented by a sequence with the elements in the subset arranged in increasing order. We can then arrange the sequences according to the lexicographic order.§ For example, the 4-subsets of $\{1, 2, 3, 4, 5, 6\}$ are represented and ordered as:

$$1234$$
$$1235$$
$$1236$$
$$1245$$
$$1246$$
$$1256$$
$$1345$$

---

† Suppose that m is the largest possible subscript for which a_m is less than one of a_{m+1}, $a_{m+2}, \ldots, a_n$. Let us assume, however, that $a_m > a_{m+1}$ and $a_m < a_{m+j}$ for some $j > 1$. We then have $a_{m+1} < a_{m+j}$ contradicting the assumption that m is the largest possible subscript for which a_m is less than one of $a_{m+1}, a_{m+2}, \ldots, a_n$. The converse argument is left to the reader.

‡ A subset of size k is abbreviated as a k-subset.

§ For two k-subsets $a_1 a_2 \cdots a_k$ and $b_1 b_2 \cdots b_k$, we shall say $a_1 a_2 \cdots a_k$ comes before $b_1 b_2 \cdots b_k$ in the lexicographic order if for some $1 \leqq m \leqq k$, $a_1 = b_1$, $a_2 = b_2, \ldots, a_{m-1} = b_{m-1}$, and $a_m < b_m$.

1346
1356
1456
2345
2346
2356
2456
3456

Along exactly the same line as our procedure to generate permutations, let us observe how we can design a procedure to generate all k-subsets of the set $\{1, 2, \ldots, n\}$. Let $a_1 a_2 \cdots a_k$ be a k-subset. It can be shown that the next k-subset $b_1 b_2 \cdots b_k$ according to the lexicographic order must be such that:

1. $a_i = b_i$, $1 \leq i \leq m - 1$, and $a_m < b_m$ for the largest possible m;
2. $b_m = a_m + 1$
3. $b_{j+1} = b_j + 1$ for $m \leq j \leq k - 1$.

In the sequence $a_1 a_2 \cdots a_k$, we define the maximum possible value of a_j to be $n - k + j$. Thus, the maximum possible value of a_k is n, the maximum possible value of a_{k-1} is $n - 1$, the maximum possible value of a_{k-2} is $n - 2$, ..., and the maximum possible value of a_1 is $n - k + 1$. Since in $a_1 a_2 \cdots a_k$ the largest m for which a_m is not equal to its maximum possible value is the largest m that satisfies (1), we can determine m by examining $a_1 a_2 \cdots a_k$ from *right to left*, element by element. Once the value of m is determined we can determine $b_m b_{m+1} \cdots b_k$ according to (2) and (3). We ask the reader to check the example on the 4-subsets of $\{1, 2, 3, 4, 5, 6\}$ shown above which were generated by our procedure.

2.6 REMARKS AND REFERENCES

As general references on the subject of combinatorics see Berge[1], Berman and Fryer[2], Even[3], Liu[4], Riordan[5], and Ryser[7]. See Whitworth[8, 9], for extensions and further examples on the subject of permutations and combinations. See chaps. 1 and 2 of Even[3], and chap. 5 of Reingold, Nievergelt, and Deo[6] on algorithms for generating permutations and combinations of a given set of objects.

[1] Berge, C.: "Principles of Combinatorics," Academic Press, New York, 1971.
[2] Berman, G., and K. D. Fryer: "Introduction to Combinatorics," Academic Press, New York, 1972.
[3] Even, S.: "Algorithmic Combinatorics," The Macmillan Company, New York, 1973.
[4] Liu, C. L.: "Introduction to Combinatorial Mathematics," McGraw-Hill Book Company, New York, 1968.
[5] Riordan, J.: "An Introduction to Combinatorial Analysis," John Wiley & Sons, Inc., New York, 1958.

[6] Reingold, E. M., J. Nievergelt, and N. Deo: "Combinatorial Algorithms: Theory and Practice," Prentice-Hall, Inc., Englewood Cliffs, N.J., 1977.

[7] Ryser, H. J.: "Combinatorial Mathematics," published by the Mathematical Association of America, distributed by John Wiley & Sons, Inc., New York, 1963.

[8] Whitworth, W. A.: "Choice and Chance," reprint of the 5th ed. (1901), Hafner Publishing Company, Inc., New York, 1965.

[9] Whitworth, W. A.: "DCC Exercises in Choice and Chance," reprint of the edition of 1897, Hafner Publishing Company, Inc., New York, 1965.

PROBLEMS

2.1 A menu in a restaurant reads like the following:

Group A: Wonton Soup
     Shark's Fin Soup
     Egg Rolls
     Rumayki

Group B: Almond Duck
     Chicken Chow Mein
     Moo Goo Gai Pan

Group C: Sweet and Sour Pork
     Pepper Steak
     Dragon Beef
     Butterfly Shrimp
     Shrimp with Lobster Sauce
     Egg Foo Young

Group D: Coffee
     Tea
     Milk

(*a*) Suppose you are to select one course from each group without omission and substitution. How many different "complete four-course dinners" can you make out of this menu?

(*b*) Suppose the waiter will not force you to make a selection if it is your desire to omit a group completely. (After all, you are going to pay for it.) How many different dinners can you make out of this menu?

(*c*) Suppose you are to select two courses from group *C* without omission and substitution. How many different dinners can you make out? Suppose you are to select either one or two from group *C* without any other omission and substitution. How many dinners can you make out?

2.2 (*a*) In how many ways can two integers be selected from the integers 1, 2, ..., 100 so that their difference is exactly seven?

(*b*) Repeat part (*a*) if the difference is to be seven or less.

2.3 In how many ways can two adjacent squares be selected from an 8×8 chessboard?

2.4 (*a*) In how many ways can 10 boys and 5 girls stand in a line so that no 2 girls are next to each other?

(*b*) Repeat part (*a*) if they are to stand around a circle.

2.5 In how many ways can the letters in the English alphabet be arranged so that there are exactly seven letters between the letters *a* and *b*?

2.6 (*a*) In how many ways can the letters *a, a, a, a, a, b, c, d, e* be permuted such that no two *a*'s are adjacent?

(*b*) Repeat part (*a*) if no two of *b, c, d, e* can be adjacent.

2.7 (*a*) In how many ways can the letters *a, b, c, d, e, f* be arranged so that the letter *b* is always to the immediate left of the letter *e*?

(*b*) Repeat part (*a*) if the letter *b* is always to the left of the letter *e*.

2.8 There are 15 "true or false" questions in an examination. How many different ways can a student do the examination if he or she can also choose not to answer some of the questions?

2.9 A palindrome is a word that reads the same forward or backward. How many seven-letter palindromes can be made out of the English alphabet?

2.10 (*a*) How many different automobile license plates made up of two letters followed by four digits are there?

(*b*) Repeat part (*a*) if the two letters must be distinct.

2.11 (*a*) In how many ways can we make up the pattern

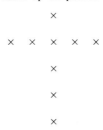

with 0s and 1s?

(*b*) How many of these patterns are not symmetrical with respect to the vertical axis?

2.12 (*a*) Cards are drawn from a deck of 52 cards with replacements. In how many ways can 10 cards be drawn so that the 10th card is the first repetition?

(*b*) Repeat part (*a*) if the 10th card is a repetition.

2.13 From a total of 20 seats, in how many ways can 3 blocks of consecutive seats with 5 seats in each block be selected?

2.14 (*a*) Show that the total number of permutations of p red balls and 0, or 1, or 2, ..., or q white balls is

$$\frac{p!}{p!} + \frac{(p+1)!}{p!1!} + \frac{(p+2)!}{p!2!} + \cdots + \frac{(p+q)!}{p!q!}$$

(*b*) Show that the sum in part (*a*) is

$$\frac{(p+q+1)!}{(p+1)!q!}$$

(*c*) Show that the total number of permutations of 0, or 1, or 2, ..., or p red balls with 0, or 1, or 2, ..., or q white balls is

$$\frac{(p+q+2)!}{(p+1)!(q+1)!} - 2$$

2.15 (*a*) From 200 automobiles, 30 are selected to test whether they meet the safety requirements. Also, 30 (from the same 200 automobiles) are selected to test whether they meet the antipollution requirements. In how many ways can the selection be made?

(*b*) In how many ways can the selection be made so that there are exactly five automobiles that undergo both tests?

2.16 (*a*) Fifteen basketball players are to be drafted by the three professional teams in Boston, Chicago, and New York such that each team will draft five players. In how many ways can this be done?

(*b*) Fifteen basketball players are to be divided into three teams of five players each. In how many ways can this be done?

2.17 (*a*) In how many ways can two numbers be selected from the integers $1, 2, \ldots, 100$ so that their sum is an even number? An odd number?

(*b*) Use a combinatorial argument to show that

$$C(2n, 2) = 2C(n, 2) + n^2$$

2.18 There are 50 students in each of the junior and the senior classes. Each class has 25 male and 25 female students. In how many ways can eight representatives be selected so that there are four females and three juniors?

2.19 Three integers are selected from the integers $1, 2, \ldots, 1000$. In how many ways can these integers be selected such that their sum is divisible by 4?

2.20 In how many ways can a group of eight people be divided into committees, subject to the constraint that each person must belong to exactly one committee, and each committee must contain at least two people. (Note that a division into committees of three, three, and two people is considered the same as a division into committees of two, three, and three people.)

2.21 From 100 students two groups of 10 students each are selected. In how many ways can the selection be made so that the tallest student in the first group is shorter than the shortest student in the second group? (Assume all 100 students are of different heights.)

2.22 In how many ways can 22 different books be given to 5 students so that 2 of them will have 5 books and the other 3 will have 4 books?

2.23 In how many ways can $2n$ people be divided into n pairs?

2.24 (a) In how many ways can two squares be selected from an 8×8 chessboard so that they are not in the same row or the same column?

(b) In how many ways can four squares, not all in the same row or column, be selected from an 8×8 chessboard to form a rectangle?

2.25 Show that the product of k successive integers is divisible by $k!$ (*Hint:* Consider the number of ways of selecting k objects from $n + k$ objects.)

2.26 (a) Show that

$$C(2n + 2, n + 1) = C(2n, n + 1) + 2C(2n, n) + C(2n, n - 1)$$

(b) Give a combinatorial interpretation to the equality in part (a).

2.27 (a) Among $2n$ objects, n of them are identical. Find the number of ways to select n objects out of these $2n$ objects.

(b) Among $3n + 1$ objects, n of them are identical. Find the number of ways to select n objects out of these $3n + 1$ objects.

2.28 In how many ways can two integers be selected from $1, 2, \ldots, n - 1$ so that their sum is larger than n?

2.29 Out of a large number of pennies, nickels, dimes, and quarters, in how many ways can five coins be selected?

2.30 (a) Show that the number of ways to place r indistinguishable balls in n distinct boxes, $n \leq r$, with no box left empty is $C(r - 1, n - 1)$. (A box can hold arbitrarily many balls.)

(b) In how many ways can a student schedule 15 hours of study in a 5-day period so that she will study at least an hour each day?

(c) In how many ways can r indistinguishable balls be placed in n distinct boxes with each box holding at least q balls?

(d) Repeat part (b) if she must study at least two hours each day.

2.31 (a) r distinct balls are to be placed in n distinct boxes with balls in the same box arranged in order. Show that there are

$$(n + r - 1)(n + r - 2) \cdots (n + 1)n$$

ways to do so.

(b) In how many ways can the letters a, b, c, d, e, f, g, h be arranged so that a is to the left of b and b is to the left of c?

2.32 Find the number of permutations of the letters a, b, c, d, e, f, g so that neither the pattern *beg* nor the pattern *cad* appears. (*Hint:* Use the principle of inclusion and exclusion.)

2.33 Ten books were arranged on a shelf in alphabetical order of the names of the authors. In how many ways can a monkey rearrange the books so that no book is put in its original place? (*Hint:* Use the principle of inclusion and exclusion.)

2.34 Among all n-digit numbers, how many of them contain the digits 2 and 7 but not the digits 0, 1, 8, 9?

2.35 Steps 2 and 3 in the procedure for generating all permutations of a set in Sec. 2.5 can be carried out as follows:

(a) Examine the permutation $a_1 a_2 \ldots a_n$ element by element from right to left. Let a_m be the rightmost element such that $a_m < a_{m+1}$.

(b) Examine the permutation element by element from right to left again. Let a_p denote the rightmost element such that $a_m < a_p$.

(c) Interchange a_m and a_p.

(d) Interchange a_{m+1} and a_n, a_{m+2} and a_{n-1}, a_{m+3} and a_{n-2} and so on. [Note that after step (c), the original a_m becomes a_p.]

Convince yourself that these steps indeed will yield the next permutation in the lexicographic order.

RELATIONS AND FUNCTIONS

3.1 INTRODUCTION

In many problems concerning discrete objects, it is often the case that there is some kind of relationship among the objects. Among a set of computer programs, we might say that two of the programs are related if they share some common data and are not related otherwise. Among a group of students, we might say two students are related if the first letters of their last names are the same. On the other hand, in a different situation we might want to say that two students are related if the first letters of their last names are different. Also, consider the set of integers {1, 2, 3, ..., 15}. We might say that three integers in the set are related if their sum is divisible by 5. Thus, the integers 2, 3, 5 are related and integers 5, 10, 15 are related, but the integers 1, 2, 4 are not. We study in this chapter relations among discrete objects.

As was mentioned in Chap. 1, we are going to introduce the notion of *ordered sets* of objects. To this end, we define first the notion of *ordered pairs* of objects. An *ordered pair* of objects is a pair of objects arranged in a fixed order. We use the notation (a, b) to denote the ordered pair in which the first object is a and the second object is b. Note that an ordered pair differs from a set of two objects in two ways: First, the order of the objects in an ordered pair is important. Thus, (a, b) and (b, a) are two different ordered pairs. Second, the two objects in an ordered pair do not have to be distinct. Thus, (a, a) is a well-defined ordered pair. We often encounter the notion of ordered pairs of objects. For example, among all players in a tennis tournament, an ordered pair (a, b) might denote the champion and the runner-up of the tournament. Thus, the ordered pair (a, b) is not the same as the ordered pair (b, a). As another example, among all students in a class, an ordered pair (a, b) might denote the two students who obtained the highest scores in two examinations. Thus, the pair (a, a) means that student a obtained the highest scores in both examinations.

Let A and B be two sets. The *cartesian product* of A and B, denoted $A \times B$, is the set of all ordered pairs of the form (a, b) where $a \in A$ and $b \in B$. For example,

$$\{a, b\} \times \{a, c, d\} = \{(a, a), (a, c), (a, d), (b, a), (b, c), (b, d)\}$$

A *binary relation from A to B* is a subset of $A \times B$. A binary relation is indeed only a formalization of the intuitive notion that some of the elements in A are related to some of the elements in B. As a matter of fact, if R is a binary relation from A to B and if the ordered pair (a, b) is in R, we would say that the element a *is related to* the element b. For example, let $A = \{a, b, c, d\}$ be a set of four students, let $B = \{CS121, CS221, CS257, CS264, CS273, CS281\}$ be a set of six courses. The cartesian product $A \times B$ gives all the possible pairings of students and courses. On the other hand, a relation $R = \{(a, CS121), (b, CS221), (b, CS264), (c, CS221), (c, CS257), (c, CS273), (d, CS257), (d, CS281)\}$ might describe the courses the students are taking, and a relation $T = \{(a, CS121), (c, CS257), (c, CS273)\}$ might describe the courses the students are having difficulty with.

Besides a list of the ordered pairs, a binary relation can also be represented in tabular form or graphical form. For example, let $A = \{a, b, c, d\}$ and $B = \{\alpha, \beta, \gamma\}$, and let $R = \{(a, \alpha), (b, \gamma), (c, \alpha), (c, \gamma), (d, \beta)\}$ be a binary relation from A to B. R can be represented in tabular form, as shown in Fig. 3.1a, where the rows of the table correspond to the elements in A and the columns of the table correspond to the elements in B, and a check mark in a cell means the element in the row containing the cell is related to the element in the column containing the cell. R can also be represented in graphical form as shown in Fig. 3.1b, where the points in the left-hand column are the elements in A, the points in the

(a)   (b)

Figure 3.1

right-hand column are the elements in B, and an arrow from a point in the left-hand column to a point in the right-hand column indicates that the corresponding element in A is related to the corresponding element in B.

Since binary relations are sets of ordered pairs, the notions of the intersection of two relations, the union of two relations, the symmetric difference of two relations, and the difference of two relations follow directly from that of sets. To be specific, let R_1 and R_2 be two binary relations from A to B, then $R_1 \cap R_2$, $R_1 \cup R_2$, $R_1 \oplus R_2$, and $R_1 - R_2$ are also binary relations from A to B which are known as the intersection, the union, the symmetric difference, and the difference of R_1 and R_2. For example, let $A = \{a, b, c, d\}$ be a set of students and $B = \{CS121, CS221, CS257, CS264, CS273, CS281\}$ be a set of courses. We might have a binary relation R_1 from A to B describing the courses the students are taking, and a binary relation R_2 from A to B describing the courses the students are interested in, as shown in Fig. 3.2. Then the binary relations $R_1 \cap R_2$, which is $\{(a, CS121), (b, CS221), (d, CS264), (d, CS281)\}$, describes the courses that the students are taking and are also interested in. The binary relation $R_1 \cup R_2$, which is $\{(a, CS121), (a, CS264), (b, CS221), (b, CS257), (b, CS273), (c, CS221), (c, CS273), (c, CS281), (d, CS264), (d, CS273), (d, CS281)\}$, describes the courses the students are either taking or interested in. The binary relation $R_1 \oplus R_2$, which is $\{(a, CS264), (b, CS257), (b, CS273), (c, CS221), (c, CS273), (c, CS281), (d, CS273)\}$, describes the courses the students are interested in but not taking or are taking but not interested in. The binary relation $R_1 - R_2$, which is $\{(b, CS257), (c, CS221), (c, CS273), (c, CS281)\}$, describes the courses the students are taking but not interested in.

	CS121	CS221	CS257	CS264	CS273	CS281
a	✓					
b		✓	✓			
c		✓			✓	✓
d				✓		✓

$$R_1$$

	CS121	CS221	CS257	CS264	CS273	CS281
a	✓			✓		
b		✓			✓	
c						
d				✓	✓	✓

$$R_2$$

Figure 3.2

As binary relations describe relationship between pairs of objects, we would like to define ternary relations to describe relationship among triples of objects, and quaternary relations to describe relationship among quadruples of objects, and so on. Although it is intuitively obvious that we might wish to define ordered triples of objects such as (a, b, c) to mean the three objects a, b, c are related in a ternary relation, we define, instead, an *ordered triple* formally as an ordered pair $((a, b), c)$, where the first component of the ordered pair is in turn an ordered pair.† [Indeed, on many occasions we shall abbreviate $((a, b), c)$ as (a, b, c).] Thus, a *ternary relation* among three sets A, B, and C is defined as a subset of the cartesian product of the two sets $A \times B$ and C, $(A \times B) \times C$. For example, let $A = \{a, b\}$, $B = \{\alpha, \beta\}$, and $C = \{1, 2\}$. We have

$$(A \times B) \times C = \{((a, \alpha), 1), ((a, \alpha), 2), ((a, \beta), 1), ((a, \beta), 2),$$
$$((b, \alpha), 1), ((b, \alpha), 2), ((b, \beta), 1), ((b, \beta), 2)\}$$

Also, let A be a set of students, B be a set of courses, and C be a set of all possible grades. Then a ternary relation among A, B, and C can be defined to describe the grades the students obtained in the courses they took. Similarly, an *ordered quadruple* is defined as an ordered pair $(((a, b), c), d)$ whose first component is the ordered triple $((a, b), c)$ and whose second component is the element d. A *quarternary relation* among four sets A, B, C, and D is defined as a subset of $((A \times B) \times C) \times D$. In general, an *ordered n-tuple* is defined as an ordered pair whose first component is an ordered $(n - 1)$-tuple; and an *n-ary relation* among the sets $A_1, A_2, A_3, \ldots, A_n$ is defined as a subset of $((A_1 \times A_2) \times A_3) \cdots \times A_n$.

Note that the notion of an ordered n-tuple is indeed that of an ordered set. For example, if we have arranged the eight students in a class in some order (such as the alphabetical order of their names), then the ordered 8-tuple (20, 18, 20, 21, 19, 19, 20, 20) describes unambiguously their ages. Also, an ordered triple such as $(15, -21, -9)$ might be used to describe the position of a point in a three-dimensional space according to some coordinate system, and an ordered 7-tuple such as (1, 1, 0, 1, 0, 0, 0) might be used to describe the work record of an employee in a week, where a 1 indicates that he or she was present and a 0 indicates that he or she was absent on a certain day of the week.

3.2 A RELATIONAL MODEL FOR DATA BANKS

As an example of relations among discrete objects, we present a brief introduction to a relational model for data banks. Modern large-scale computing systems are capable of handling large amounts of data such as the records of charge account transactions of customers in a department store, the employment history and personal data of employees in a company, and the

---

† The reason such a formal definition is used is that no new concept needs to be introduced in defining ordered triples, quadruples, ..., and k-tuples.

records of parts ordered and received in a factory. So that large amounts of data can be handled effectively, they must be organized in some form suitable for the more frequent operations such as inserting new data, deleting old data, updating existing data, and searching for items with special attributes. One general way to view the organization of large varieties of data is with a *relational data model*. Let $A_1, A_2, \ldots, A_n$ be n (not necessarily distinct) sets. An n-ary relation among $A_1, A_2, \ldots, A_n$ is known as a *table*† on $A_1, A_2, \ldots, A_n$ in the language of relational data model. The sets $A_1, A_2, \ldots, A_n$ are called the *domains* of the table, and n is called the *degree* of the table. For example, let SUPPLIER $= \{s_1, s_2, s_3, s_4\}$ be the set of suppliers of parts, PART $= \{p_1, p_2, p_3, p_4, p_5, p_6, p_7\}$ be the set of parts, PROJECT $= \{j_1, j_2, j_3, j_4, j_5\}$ be the set of projects, and QUANTITY be the set of positive integers. We may have a table named SUPPLY on the sets SUPPLIER, PART, PROJECT, and QUANTITY describing the names of suppliers who supply parts to the various projects and the quantities they supply. Thus,

SUPPLY $= \{(s_1, p_2, j_5, 5), (s_1, p_3, j_5, 17), (s_2, p_3, j_3, 9), (s_2, p_1, j_5, 5), (s_4, p_1, j_1, 4)\}$

is an example of a table. We shall also represent a table in tabular form as shown in Fig. 3.3.

SUPPLY

SUPPLIER	PART	PROJECT	QUANTITY
s_1	p_2	j_5	5
s_1	p_3	j_5	17
s_2	p_3	j_3	9
s_2	p_1	j_5	5
s_4	p_1	j_1	4

Figure 3.3

As another example, consider a table ASSEMBLE on PART, PART, and QUANTITY, where PART $= \{p_1, p_2, \ldots, p_7\}$ is the set of parts and QUANTITY is the set of integers. An ordered triple (p_i, p_j, c) in the table ASSEMBLE means that part p_i is a subcomponent of part p_j; moreover, it takes c units of part p_i to assemble each unit of part p_j. Therefore, we might have the table shown in Fig. 3.4.

A domain of a table is called a *primary key*‡ if its value in an ordered n-tuple uniquely identifies the ordered n-tuple in the table. In the table EMPLOYEES in Fig. 3.5, EMPLOYEE NO. is a primary key; so is NAME. On the other hand, DEPARTMENT is not a primary key and neither is HOURLY WAGE. If a table does not have a domain that can serve as a primary key, we might wish

---

† The term *relation* is also used. Here, we choose to use a more intuitive term.

‡ Since, in general, ordered n-tuples are added to and deleted from a table from time to time, being a primary key could be a *time-varying* property of a domain.

ASSEMBLE

PART	PART	QUANTITY
p_1	p_5	9
p_2	p_5	7
p_3	p_5	2
p_2	p_6	12
p_3	p_6	3
p_4	p_7	1
p_6	p_7	1

Figure 3.4

to use a combination of domains to identify the ordered n-tuples in the table. We define a *composite primary key* as the cartesian product of two or more domains such that its value† in an ordered n-tuple uniquely identifies the ordered n-tuple in the table. For example, in the table SUPPLY in Fig. 3.3, SUPPLIER $\times$ PART is a composite primary key.

EMPLOYEES

EMPLOYEE NO.	NAME	DEPARTMENT	HOURLY WAGE
20835	Bernstein, E. M.	2	4.50
11273	Jones, D. J.	1	4.00
10004	Smith, C. W.	1	5.25
21524	Vögeli, W. J.	2	7.50
17734	Wong, J. W. S.	1	4.50
30219	Yamamoto, S.	3	5.00

Figure 3.5

Given a collection of tables, we might wish to manipulate the tables in various ways. As an illustration, we describe two important operations on tables—*projection* and *join*. We frequently want to abstract subtables from a table. The operation *projection* enables us to do so. Let R be a table of degree n. A *projection* of R is an m-ary relation, $m \le n$, obtained from R by deleting $n - m$ of the components in each ordered n-tuple in R. We use the notation $\pi_{i_1 i_2 \cdots i_m}(R)$, $1 \le i_1 < i_2 < \cdots < i_m \le n$, to denote a projection of R, that is, a table of degree m obtained from R such that for each ordered n-tuple in R there is a corresponding ordered m-tuple in $\pi_{i_1 i_2 \cdots i_m}(R)$ with the kth component of the ordered m-tuple being the i_kth component of the ordered n-tuple. For example, for the table SUPPLY in Fig. 3.3, the projection $\pi_{1,3}$ (SUPPLY) is shown in Fig. 3.6.

† By the values of a composite primary key, we mean the ordered tuples in the cartesian product.

$\pi_{1,3}$ (SUPPLY)

SUPPLIER	PROJECT
s_1	j_5
s_2	j_3
s_2	j_5
s_4	j_1

Figure 3.6

Note that, as illustrated in this example, there might be fewer ordered m-tuples in a projection of a table than there are ordered n-tuples in the table, because several distinct ordered n-tuples in the table might yield the same ordered m-tuple in the projection.

Tables can also be combined to yield bigger tables. The operation *join* combines two tables into one. Let R be a table of degree n and S be a table of degree m. For p less than n and m, we can construct a *join* of R and S, that is, a table denoted $\tau_p(R * S)$ such that

$$\tau_p(R * S) = \{(a_1, a_2, \ldots, a_{n-p}, b_1, b_2, \ldots, b_p, c_1, c_2, \ldots, c_{m-p})|$$
$$(a_1, a_2, \ldots, a_{n-p}, b_1, b_2, \ldots, b_p) \in R,$$
$$(b_1, b_2, \ldots, b_p, c_1, c_2, \ldots, c_{m-p}) \in S\}$$

As an example, for the tables SUPPLY and COLOR of Fig. 3.7*a*, we have the join τ_2 (SUPPLY * COLOR) shown in Fig. 3.7*b*.

SUPPLY

SUPPLIER	PART	PROJECT
s_1	p_1	j_1
s_2	p_1	j_1
s_2	p_2	j_2

COLOR

PART	PROJECT	COLOR
p_1	j_1	c_1
p_2	j_2	c_2
p_2	j_2	c_3

(*a*)

τ_2 (SUPPLY * COLOR)

SUPPLIER	PART	PROJECT	COLOR
s_1	p_1	j_1	c_1
s_2	p_1	j_1	c_1
s_2	p_2	j_2	c_2
s_2	p_2	j_2	c_3

(*b*)

Figure 3.7

A great deal more can be said about the relational data model, especially concerning other operations on tables, its implementation on computer systems, and so on; many of the details can be found in Codd[1] and Date[3].

3.3 PROPERTIES OF BINARY RELATIONS

A binary relation from a set A to A is said to be a *binary relation on A*. For example, let A be a set of positive integers. We may define a binary relation R on A such that (a, b) is in R if and only if $a - b \geq 10$. Thus, $(12, 1)$ is in R, but $(12, 3)$ is not; neither is $(1, 12)$. As another example, let $B = \{CS121, CS221, CS257, CS264, CS273, CS281\}$. A binary relation on $B = \{(CS121, CS221), (CS121, CS257), (CS257, CS281)\}$ might describe the prerequisite structure of these courses in that an ordered pair in the binary relation means the first course in the pair is a prerequisite of the second course in the pair. For the rest of this chapter, we shall study binary relations on a set because this is the case that we shall encounter most often.

Let R be a binary relation on A. R is said to be a *reflexive relation* if (a, a) is in R for every a in A. In other words, in a reflexive relation every element in A is related to itself. For example, let A be a set of courses and R be a binary relation on A so that for two courses a and b in A, (a, b) is in R if and only if their final examinations are scheduled in the same time period. Clearly, for any course a, (a, a) is in R. Thus, R is a reflexive relation. As another example, let A be a set of positive integers, and let us define a binary relation R on A such that (a, b) is in R if and only if a divides b. Since an integer always divides itself, R is a reflexive relation. On the other hand, let us define a binary relation T on a set of integers A such that (a, b) is in T if and only if $a > b$. Clearly, T is not a reflexive relation. When a binary relation on a set is represented in tabular form, it is very simple to determine whether the binary relation is a reflexive relation. To be specific, a binary relation on a set is reflexive if and only if all the cells on the main diagonal of the table contain check marks. For example, the binary relation in Fig. 3.8a is reflexive while that in Fig. 3.8b is not.

Let R be a binary relation on A. R is said to be a *symmetric relation* if (a, b) in R implies that (b, a) is also in R. For example, let A be a set of students and let R be a binary relation on A such that (a, b) is in R if and only if a is in a class that b is in. If a is in a class that b is in, then, clearly, b is also in a class that a is in. Thus, the relation R is a symmetric relation. Let A be a set of positive integers and T be a binary relation on A such that (a, b) is in T if

(a)

(b)

Figure 3.8

and only if $a \geq b$. Since, for instance, $(10, 9)$ is in T but $(9, 10)$ is not in T, T is not a symmetric relation. As another example, let $A = \{a, b, c\}$. Let $U = \{ \ \}$, $V = \{(a, a), (b, b)\}$, $W = \{(a, b), (b, a)\}$, and $X = A \times A$ be four binary relations on A. We note that all four of these are symmetric relations. When a binary relation on a set is represented in tabular form, we can determine whether it is symmetric by observing whether the check marks are in cells that are symmetrical with respect to the main diagonal. For example, the binary relation in Fig. 3.9a is symmetric while that in Fig. 3.9b is not.

Let R be a binary relation on A. R is said to be an *antisymmetric relation* if (a, b) in R implies that (b, a) is not in R unless $a = b$. In other words, if both (a, b) and (b, a) are in R, then it must be the case that $a = b$. For example, let A be a set of tests to be performed on a patient in the hospital and let R be a binary relation on A such that if (a, b) is in R, then test a must be performed before test b. Clearly, if test a must be performed before test b, then test b must not be performed before test a for two distinct tests a and b. Thus, R is an antisymmetric relation. As another example, let A be a set of positive integers and R be a binary relation on A such that (a, b) is in R if and only if $a \geq b$. We note that R is an antisymmetric relation. Let $A = \{a, b, c\}$. Let $S = \{(a, a), (b, b)\}$ and $N = \{(a, b), (a, c), (c, a)\}$ be binary relations on A. Note that S is both symmetric and antisymmetric, yet N is neither symmetric nor antisymmetric.

Let R be a binary relation on A. R is said to be a *transitive relation* if (a, c) is in R whenever both (a, b) and (b, c) are in R. For example, let $A = \{a, b, c\}$ and $X = \{(a, a), (a, b), (a, c), (b, c)\}$. We note that X is a transitive relation. We also note that $Y = \{(a, b)\}$ is a transitive relation, yet $Z = \{(a, b), (b, c)\}$ is not.† As another example, let A be a set of people, and let R be a binary relation on A such that (a, b) is in R if and only if a is an ancestor of b. Clearly, R is a transitive relation. On the other hand, if we let T be a binary relation on A such that (a, b) is in T if and only if a is the father of b, then T is not a transitive relation.

(a)

(b)   **Figure 3.9**

† At this point, a reader probably becomes aware of the problem of checking whether a given binary relation is transitive or not. This clearly can be done by an exhaustive search, but looking for efficient procedures for checking transitivity is still a topic of current research.

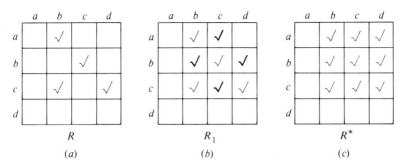

Figure 3.10

Let R be a binary relation on A. The *transitive extension of R*, denoted R_1, is a binary relation on A such that R_1 contains R, and moreover, if (a, b) and (b, c) are in R, then (a, c) is in R_1. For example, let $A = \{a, b, c, d\}$ and R be the binary relation shown in Fig. 3.10a. The transitive extension of R, R_1, is shown in Fig. 3.10b, where the ordered pairs in R_1 but not in R are marked with heavy check marks. Note that if R is a transitive relation, R is equal to R_1. Let R_2 denote the transitive extension of R_1, and, in general, let R_{i+1} denote the transitive extension of R_i. We define the *transitive closure* of R, denoted R^*, to be the set union of R, R_1, R_2, ... For example, the transitive closure of the binary relation R in Fig. 3.10a is shown in Fig. 3.10c. As another example, let A be a set of cities and R be a binary relation on A so that the ordered pair (a, b) is in R if there is a communication link from city a to city b for the transmission of messages. Thus, the transitive extension of R, R_1, describes how messages can be transmitted from one city to another, either on a direct communication link or through one intermediate city. Similarly, the transitive extension of R_1, R_2, describes how messages can be transmitted from one city to another, either on a direct communication link or through at most two intermediate cities. Finally, the transitive closure of R, R^*, describes how messages can be transmitted from one city to another, either through a direct communication link or through as many intermediate cities as we wish. We note that for any binary relation R, R^* is always a transitive relation.

3.4 EQUIVALENCE RELATIONS AND PARTITIONS

A binary relation might have one or more of the following properties: reflexivity, symmetry, antisymmetry, and transitivity. For example, the binary relation in Fig. 3.11a is a reflexive and transitive relation, and the binary relation in Fig. 3.11b is a reflexive and symmetric relation. We shall study in this and the next sections two important classes of binary relations, namely, equivalence relations and partial ordering relations.

A binary relation on a set is said to be an *equivalence relation* if it is reflexive, symmetric, and transitive. For example, the binary relation on the set

(a)   (b)

Figure 3.11

$\{a, b, c, d, e, f\}$ shown in Fig. 3.12 is an equivalence relation. Let A be a set of students and R be a binary relation on A such that (a, b) is in R if and only if a lives in the same dormitory as b does. We leave it to the reader to check that R is an equivalence relation. Also, let A be a set of strings of 0s and 1s the lengths of which are at least three. Let R be a binary relation on A such that for two strings a and b, (a, b) is in R if and only if the last three digits in a are the same as the last three digits in b. Again, R is an equivalence relation.

We define now the notion of a *partition* of a set. A *partition* of a set A is a set of nonempty subsets of A denoted $\{A_1, A_2, \ldots, A_k\}$ such that the union of A_i's is equal to A and the intersection of A_i and A_j is empty for any distinct A_i and A_j. In other words, a partition of a set is a division of the elements in the set into disjoint subsets. These subsets are also called *blocks* of the partition. For example, let $A = \{a, b, c, d, e, f, g\}$, then $\{\{a\}, \{b, c, d\}, \{e, f\}, \{g\}\}$ is a partition of A. As another example, we note that a deck of playing cards is partitioned into the four suits. It is also partitioned into the 13 ranks. We also introduce the notation $\{\overline{a}, \overline{bcd}, \overline{ef}, \overline{g}\}$ for a partition where we place an overbar above the elements that are in the same block. We see now a connection between an equivalence relation on a set A and a partition of the set A. From an equivalence relation on A, we can define a partition of A so that every two elements in a block are related and any two elements in different blocks are not.† This partition is said to be the partition induced by the equivalence

	a	b	c	d	e	f
a	√	√				
b	√	√				
c			√			
d				√	√	√
e				√	√	√
f				√	√	√

Figure 3.12

† See Prob. 3.18, part (a).

relation, and the blocks in the partition are called the *equivalence classes.* Conversely, from a partition of a set A we can define an equivalence relation on A so that every two elements in the same block of the partition are related, and any two elements in different blocks are not related.† For example, let A be a set of people and R be a binary relation on A such that (a, b) is in R if and only if a and b have the same family name. We note that R is an equivalence relation which induces a partition of A where the equivalence classes are families.‡ As another example, let A be the set of all natural numbers. Let n be a fixed integer. Let R be a binary relation on A so that (a, b) is in R if the remainders of a divided by n and b divided by n are the same. R is an equivalence relation which divides A into n equivalence classes. These equivalence classes are those numbers that are divisible by n; those numbers that leave a remainder of 1 when divided by n; those numbers that leave a remainder of 2 when divided by n ...; and those numbers that leave a remainder of $n - 1$ when divided by n. Two numbers a and b that are in the same equivalence class are said to be *equal modulo n*, and the notation $a \equiv b$ (mod n) is frequently used.

Let π_1 and π_2 be two partitions of a set A. Let R_1 and R_2 be the corresponding equivalence relations. We define the *product* of π_1 and π_2, denoted $\pi_1 \cdot \pi_2$, to be the partition corresponding to the equivalence relation $R_1 \cap R_2$. (We leave it to the reader to show that the intersection of two equivalence relations is always an equivalence relation.) In other words, the product of π_1 and π_2 is a partition of A such that two elements a and b are in the same block of $\pi_1 \cdot \pi_2$ if a and b are in the same block of π_1 and also in the same block of π_2. We define the *sum* of π_1 and π_2, denoted $\pi_1 + \pi_2$, to be the partition corresponding to the equivalence relation $(R_1 \cup R_2)^*$. (We leave it to the reader to show that the union of two equivalence relations is always a reflexive and symmetric relation.) In other words, the sum of π_1 and π_2 is a partition of A such that two elements a and b are in the same block of $\pi_1 + \pi_2$ if there exist elements $c_1, c_2, c_3, \ldots, c_k$ such that a and c_1 are in the same block of π_1 or π_2, c_1 and c_2 are in the same block of π_1 or π_2, c_2 and c_3 are in the same block of π_1 or π_2, $\ldots$, and c_k and b are in the same block of π_1 or π_2. That is, two elements a and b are in the same block of $\pi_1 + \pi_2$ if they are *chain-connected*, by which we mean there exists a sequence of elements $a, c_1, c_2, \ldots, c_k, b$ such that each pair of successive elements in the sequence is in the same block of π_1 or π_2.

For example, let $A = \{a, b, c, d, e, f, g, h, i, j, k\}$. Let

$$\pi_1 = \{\overline{abcd}, \overline{efg}, \overline{hi}, \overline{jk}\} \qquad \pi_2 = \{\overline{abch}, \overline{di}, \overline{efjk}, \overline{g}\}$$

be two partitions of A. We then have

$$\pi_1 \cdot \pi_2 = \{\overline{abc}, \overline{d}, \overline{ef}, \overline{g}, \overline{h}, \overline{i}, \overline{jk}\}$$

and

$$\pi_1 + \pi_2 = \{\overline{abcdhi}, \overline{efgjk}\}$$

---

† See Prob. 3.18, part (b).

‡ We assume that no two families have the same family name.

An interesting physical interpretation can be given to the product and sum of partitions. Suppose that A is a set of people, and for the partitions π_1 and π_2 given in the foregoing, π_1 is a partition of them into age groups and π_2 is a partition of them into height groups. Suppose we wish to identify a certain person in A. If we are told what age group she is in, then we can identify her up to the blocks of π_1. In other words, we shall be able to identify her as one of a, b, c, d, as one of e, f, g, as one of h, i, or as one of j, k, depending on the age group she is in. Similarly, if we are told what height group she is in, we can identify her up to the blocks of π_2. If we are told the age group and the height group she is in, then we can identify her up to the blocks of $\pi_1 \cdot \pi_2$. In other words, we shall be able to identify her as one of a, b, c, as d, as one of e, f, as g, as h, as i, or as one of j, k. If we are told either the age group or the height group she is in, but not both, we can identify her up to the blocks of $\pi_1 + \pi_2$. In other words, whether we are provided with the information on the age group or with the information on the height group the person to be identified is in, we never have any ambiguity on the identity of the person beyond the blocks of $\pi_1 + \pi_2$. That is, we are assured that we can distinguish someone in the a, b, c, d, h, i group from someone in the e, f, g, j, k group, no matter whether we are given the age group information or the height group information. Indeed, a partition on a set can be viewed either as possessing some *information* on the identification of one of the objects in the set (two objects in different blocks can always be distinguished) or as possessing some *ambiguity* on the identification of one of the objects in the set (two objects in the same block can never be distinguished). It follows that the product of two partitions $\pi_1 \cdot \pi_2$ represents the total information we have on the identification of one of the objects when we have the information from both π_1 and π_2, and the sum of two partitions $\pi_1 + \pi_2$ represent the *most*† ambiguity we might have when we are only sure that either the information from π_1 or that from π_2 will be available to us.

3.5 PARTIAL ORDERING RELATIONS AND LATTICES

A binary relation is said to be a *partial ordering relation* if it is reflexive, antisymmetric, and transitive. For example, the binary relation on the set $\{a, b, c, d, e\}$ shown in Fig. 3.13a is a partial ordering relation. As another example, let A be a set of positive integers, and let R be a binary relation on A such that (a, b) is in R if a divides b. We ask the reader to check that R is a partial ordering relation. Consider also the example of a set of books, A, each of which possesses a certain number of attributes. Let R be a binary relation on A such that (a, b) is in R if and only if every attribute of book a is also an attribute of book b. Note that R is a partial ordering relation. As another example, let A be a set of food items of different prices. Let R be a binary relation on A such that

---

† Note the word *most*. In many cases, the ambiguity will be less than that in $\pi_1 + \pi_2$.

	a	b	c	d	e
a	✓	✓	✓	✓	✓
b		✓	✓		✓
c			✓		✓
d				✓	✓
e					✓

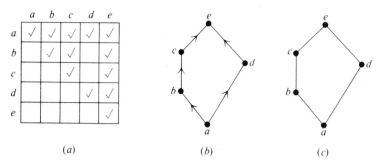

(a) (b) (c)

Figure 3.13

(a, b) is in R if a is not inferior to b, both in terms of nutritional value and of price. Again, R is a partial ordering relation.

It was pointed out in Sec. 3.1 that a binary relation from a set A to a set B can be represented graphically as illustrated in Fig. 3.1b. For a binary relation R on a set A, we can have a slightly simpler graphical representation. Instead of having two columns of points as in Fig. 3.1b, we represent the elements in A by points and use arrows to represent the ordered pairs in R. For example, the binary relation on the set $\{a, b, c, d\}$ in Fig. 3.14a is represented graphically in Fig. 3.14b. When the binary relation is a partial ordering relation, the graphical representation can be further simplified. Since the relation is understood to be reflexive, we can omit arrows from points back to themselves. Since the relation is understood to be transitive, we can omit arrows between points that are connected by sequences of arrows. For example, such a simplified representation for the partial ordering relation in Fig. 3.13a is shown in Fig. 3.13b. In many cases, when the graphical representation is so oriented that all arrowheads point in one direction (upward, downward, left to right, or right to left), we can even omit the arrowheads as the example in Fig. 3.13c shows. Such a graphical representation of a partial ordering relation in which all

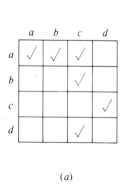

	a	b	c	d
a	✓	✓	✓	
b			✓	
c				✓
d			✓	

(a)

(b) **Figure 3.14**

arrowheads are understood to be pointing upward is also known as the *Hasse diagram* of the relation.

A set A, together with a partial ordering relation R on A, is called a *partially ordered set* and is denoted by (A, R). In the literature, a partially ordered set is also abbreviated as a *poset*. There is also an alternative notation for specifying a partial ordering relation: For each ordered pair (a, b) in R, we write $a \le b$ instead of $(a, b) \in R$, where $\le$ is a generic symbol corresponding to the set of ordered pairs R and is commonly read "less than or equal to." (We frequently say a is less than or equal to b to mean $a \le b$, and say a is less than b to mean $a \le b$ and $a \ne b$. We also say b is larger than or equal to a and write $b \ge a$ to mean $a \le b$.) Indeed, a partially ordered set is usually denoted $(A, \le)$.

Let $(A, \le)$ be a partially ordered set. A subset of A is called a *chain* if every two elements in the subset are related. Note that, because of antisymmetry and transitivity, in any chain with finite number of elements $\{a_1, a_2, \ldots, a_k\}$ there is an element a_{i_1} that is less than every other element in the chain, there is an element a_{i_2} that is less than every other element except a_{i_1}, there is an element a_{i_3} that is less than every other element except a_{i_1} and $a_{i_2}, \ldots$, and so on. We shall use the notation $a_{i_1} \le a_{i_2} \le a_{i_3} \le \cdots \le a_{i_k}$ as an abbreviation for the list of ordered pairs $a_{i_1} \le a_{i_2}, a_{i_1} \le a_{i_3}, \ldots, a_{i_1} \le a_{i_k}, a_{i_2} \le a_{i_3}, a_{i_2} \le a_{i_4}, \ldots, a_{i_2} \le a_{i_k}, \ldots$. We frequently refer to the number of elements in a chain as the *length* of the chain. A subset of A is called an *antichain* if no two distinct elements in the subset are related. For example, for the partially ordered set in Fig. 3.13a, $\{a, b, c, e\}$, $\{a, b, c\}$, $\{a, d, e\}$, and $\{a\}$ are chains, and $\{b, d\}$, $\{c, d\}$, $\{a\}$ are antichains. Consider a partially ordered set $(A, \le)$, where A is the set of all employees in a company, and for a and b in A, $a \le b$ if and only if b is a or is a superior of a. In this case, a chain is a subset of employees in which there, indeed, exists a chain of command. On the other hand, an antichain is a subset of employees in which no one has command over another. A partially ordered set $(A, \le)$ is called a *totally ordered set* if A is a chain. In this case, the binary relation $\le$ is called a *total ordering relation*.

Let $(A, \le)$ be a partially ordered set. An element a in A is called a *maximal element* if for no b in A, $a \ne b$, $a \le b$. An element a in A is called a *minimal element* if for no b in A, $a \ne b$, $b \le a$. For example, in the partially ordered set of Fig. 3.15a, j and k are maximal elements, a, b, e are minimal elements. An element a is said to *cover* another element b if $b \le a$ and for no other element c, $b \le c \le a$. In the partially ordered set of Fig. 3.15a, f covers b, f also covers c, but f does not cover a.

Let a and b be two elements in a partially ordered set $(A, \le)$. An element c is said to be an *upper bound* of a and b if $a \le c$ and $b \le c$. For example, in the partially ordered set of Fig. 3.15a, h is an upper bound of f and g; so are i, j, and k. An element c is said to be a *least upper bound* of a and b if c is an upper bound of a and b, and if there is no other upper bound d of a and b such that $d \le c$. For example, in the partially ordered set of Fig. 3.15a, h is a least upper bound of f and g; so is i. Similarly, an element c is said to be a *lower bound* of a and b if $c \le a$ and $c \le b$, and an element c is said to be a *greatest lower bound* of

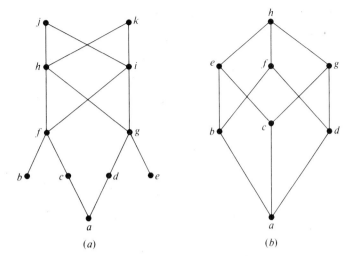

Figure 3.15

a and *b* if *c* is a lower bound of *a* and *b*, and if there is no other lower bound *d* of *a* and *b* such that $c \le d$. In the partially ordered set of Fig. 3.15*a*, *a*, *b*, *c*, *d*, *e*, *f*, and *g* are all lower bounds of *h* and *i*, while *f* and *g* are greatest lower bounds of *h* and *i*. As another example, let *A* be the set of all positive integers and *R* be a binary relation on *A* such that (a, b) is in *R* if *a* divides *b*. We can readily check that *R* is a partial ordering relation. For two integers *a* and *b*, a common multiple of *a* and *b* is an upper bound of *a* and *b*, and the least common multiple of *a* and *b* is a least upper bound (and the only one) of *a* and *b*. Similarly, a common divisor of *a* and *b* is a lower bound of *a* and *b*, and the greatest common divisor of *a* and *b* is a greatest lower bound (and the only one) of *a* and *b*.

A partially ordered set is said to be a *lattice* if every two elements in the set have a unique least upper bound and a unique greatest lower bound. The partially ordered set in Fig. 3.15*a* is not a lattice, whereas the one in Fig. 3.15*b* is. We shall study the many properties of lattices in Chap. 9.

Let (A, R_1) and (B, R_2) be two partially ordered sets. We define a binary relation R_3 over the set $A \times B$ such that for a_1 and a_2 in *A*, and b_1 and b_2 in *B* $((a_1, b_1), (a_2, b_2))$ is in R_3 if and only if (a_1, a_2) is in R_1 and (b_1, b_2) is in R_2. We ask the reader to show that R_3 is a partial ordering relation. Consequently, $(A \times B, R_3)$ is a partially ordered set which is usually referred to as the *cartesian product* of the two partially ordered sets (A, R_1) and (B, R_2). Figure 3.16*b* shows the cartesian product of the two partially ordered sets in Fig. 3.16*a*. Let *A* denote the set of all positive divisors of an integer *n*. Let *R* be a binary relation on *A* such that (a, b) is in *R* if *a* divides *b*. Expressing *n* as a product of powers of primes $p_1^{\alpha_1} p_2^{\alpha_2} \cdots p_t^{\alpha_t}$, we note that a positive divisor of *n* can then be expressed as $p_1^{\beta_1} p_2^{\beta_2} \cdots p_t^{\beta_t}$, where $0 \le \beta_i \le \alpha_i$, for $i = 1, 2, \ldots, t$. Indeed, the integers in *A* can all be represented by ordered *t*-tuples of the form $(\beta_1, \beta_2, \ldots, \beta_t)$, where $0 \le \beta_i \le \alpha_i$, $i = 1, 2, \ldots, t$. Furthermore, let the ordered

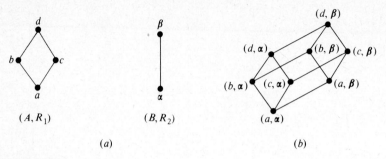

Figure 3.16

t-tuples corresponding to the integers a and b be $(\beta_1, \beta_2, \ldots, \beta_t)$ and $(\beta'_1, \beta'_2, \ldots, \beta'_t)$. That a divides b means $\beta_i \leq \beta'_i$, $i = 1, 2, \ldots, t$. We now ask the reader to check that the partially ordered set (A, R) can be expressed as a cartesian product $(((A_1 \times A_2) \times A_3) \times \cdots) \times A_t$ where, for $i = 1, 2, \ldots, t$, $(A_i, \leq)$ is a totally ordered set such that $A_i = \{0, 1, 2, \ldots, \alpha_i\}$ and $0 \leq 1 \leq 2 \leq \cdots \leq \alpha_i$.

3.6 CHAINS AND ANTICHAINS

As an illustration of the concepts of chains and antichains in a partially ordered set, we present a theorem that shows a close relationship between chains and antichains.†

Theorem 3.1 Let $(P, \leq)$ be a partially ordered set. Suppose the length of the longest chains in P is n. Then the elements in P can be partitioned into n disjoint antichains.

PROOF We shall prove the theorem by induction on n.
Basis of induction. For $n = 1$, no two elements in P are related. Clearly, they constitute an antichain.
Induction step. We assume that the theorem holds when the length of the longest chains in a partially ordered set is $n - 1$. Let P be a partially ordered set with the length of its longest chains being n. Let M denote the set of maximal elements in P. Clearly, M is a nonempty antichain. Consider now the partially ordered set $(P - M, \leq)$. Since there is no chain of length n in $P - M$, the length of the longest chains is at most $n - 1$. On the other hand, if the length of the longest chains in $P - M$ is less than $n - 1$, M must contain two or more elements that are members of the same chain, which is certainly an impossibility. Consequently, we conclude that the length of the

---

† Theorem 3.1 is a dual of what is known as Dilworth's theorem, which states that if the size of the largest antichains in P is n, then the elements in P can be partitioned into n disjoint chains. For a proof of Dilworth's theorem, see Mirsky[6].

longest chain in $P - M$ is $n - 1$, and, according to the induction hypothesis, $P - M$ can be partitioned into $n - 1$ disjoint antichains. ☐

A direct consequence of Theorem 3.1 can be stated as:

Corollary 3.1.1 Let $(P, \leq)$ be a partially ordered set consisting of $mn + 1$ elements. Either there is an antichain consisting of $m + 1$ elements or there is a chain of length $n + 1$ in P.

PROOF Suppose the length of the longest chains in P is n. According to Theorem 3.1, P can be partitioned into n disjoint antichains. If each of these antichains consists of m of fewer elements, the total number of elements in P is at most nm, which is a contradiction to the assumption of the corollary. ☐

Example 3.1 Show that among $ab + 1$ white mice either there is a sequence of $a + 1$ mice, each a descendant of the next, or there is a group of $b + 1$ mice no one of which is a descendant of another. Let us order the mice according to the descendant relationship. Clearly, such an ordering relation is a partial ordering relation. If there is an antichain of size $b + 1$ or larger in this partially ordered set, there is a group of $b + 1$ or more mice, not one of which is a descendant of another. On the other hand, if there is a chain of length $a + 1$ or larger, there is a sequence of $a + 1$ or more mice, each a descendant of the next. ☐

Example 3.2 A point (x, y) in the first quadrant of the xy plane defines a rectangle with the points $(0, 0)$, $(x, 0)$, $(0, y)$, (x, y) as its vertices. (See Fig. 3.17.) We want to show that for the rectangles defined by any five distinct points in the first quadrant either there are three rectangles $R_{i_1}, R_{i_2}, R_{i_3}$ such that R_{i_1} and R_{i_2} are inside of R_{i_3}, and R_{i_1} is inside of R_{i_2}, or there are three rectangles such that no one is inside of another. Let $P = \{(x_1, y_1), (x_2, y_2), (x_3, y_3), (x_4, y_4), (x_5, y_5)\}$ denote the set of five points in the first quadrant.

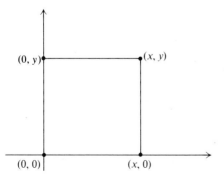

Figure 3.17

We define a partial ordering relation $\leq$ on P such that $(x_i, y_i) \leq (x_j, y_j)$ if and only if $x_i \leq x_j$ and $y_i \leq y_j$. According to Corollary 3.1.1, either there is a chain of length 3 in P or there is an antichain of size 3 in P. Clearly, a chain of length 3 corresponds to a set of three rectangles $R_{i_1}, R_{i_2}, R_{i_3}$ such that R_{i_1} and R_{i_2} are inside of R_{i_3}, and R_{i_1} is inside of R_{i_2}. On the other hand, an antichain of size 3 corresponds to a set of three rectangles such that no one is inside of another. □

3.7 A JOB-SCHEDULING PROBLEM

Consider the problem of scheduling the execution of a set of tasks on a multiprocessor computing system that has a set of identical processors. (The problem can also be phrased as that of making up a schedule for a certain number of workers to complete a given set of tasks.) Let $P_1, P_2, \ldots, P_n$ denote the n identical processors in a multiprocessor computing system. Let $\mathcal{T} = \{T_1, T_2, \ldots, T_r\}$ denote a set of tasks to be executed on the computing system. We assume that the execution of a task occupies one and only one processor. Moreover, since the processors are identical, a task can be executed on any one of the processors. Let $\mu(T_i)$ denote the *execution time* of task T_i, that is, the amount of time it takes to execute T_i on a processor. There is also a partial ordering $\leq$ specified over $\mathcal{T}$ such that for $T_i \neq T_j$, $T_i \leq T_j$ if and only if the execution of task T_j cannot begin until the execution of task T_i has been completed. (T_i is said to be a predecessor of T_j, and T_j is said to be a successor of T_i.) A partially ordered set of tasks can be described graphically as illustrated in Fig. 3.18a, where the execution time of each task is written next to the name of the task. An obvious interpretation can be given to our model of a set of tasks. Consider the tasks $T_1, T_2, \ldots, T_r$ to be subprograms of a larger program. Then $T_i \leq T_j$ might mean that subprogram T_j uses some of the data generated by subprogram T_i, so that the execution of T_j must await the completion of T_i. For example, if a computer system is used in some space mission, task T_i might be a subprogram that determines the course of the spacecraft and task T_j might be a subprogram that estimates the total fuel

Figure 3.18

consumption for midcourse adjustment. Clearly, we should complete T_i before executing T_j.†

By *scheduling* a set of tasks on a multiprocessor computing system, we mean to specify for each task T_i both the time interval within which it will be executed and the processor P_k on which execution will take place. (Without loss of generality, we assume that execution of the set begins at time $t = 0$.) An explicit way to describe a schedule is a *timing diagram*. For example, the timing diagram of a schedule for execution of the set of tasks in Fig. 3.18a on a three-processor computing system is shown in Fig. 3.18b, where ϕ_1, ϕ_2, ϕ_3, ϕ_4 denote periods within which a processor is left idle. For a given schedule, an *idle period* of a processor is defined to be a time interval within which the processor is not executing a task. We use ϕ_1, ϕ_2, ... to denote idle periods of the processors, and $\mu(\phi_1)$, $\mu(\phi_2)$, ... to denote the lengths of the idle periods. Notice that in a given schedule, a processor might be left idle either because there is no executable task‡ at that time or because it is an intentional choice. Clearly, it is never necessary nor beneficial in a given schedule to leave all processors idle at the same time. On the other hand, as we shall see later, it might be beneficial to leave some of the processors idle even though there are tasks that are executable at that time.

The total *elapsed time* of a schedule is the total time it takes to complete the execution of all tasks according to the schedule. Clearly, it is desirable to obtain a schedule that has the minimum total elapsed time. Unfortunately, there is no known procedure (short of exhaustive trial and error) for constructing schedules with minimum total elapsed time. Consequently, an alternative way to approach the problem is to look for schedules that are good but not necessarily the best possible ones. As a matter of fact, there is a very simple and intuitive way to schedule a given set of tasks, namely, never leave a processor intentionally idle. That is, a processor is left idle for a period of time only if no task is executable within that period. Conversely, whenever a processor becomes available at any time, we shall execute on this processor any one of the tasks that are executable at that time. For example, the schedule in Fig. 3.18b for the set of tasks in Fig. 3.18a was obtained this way. At the beginning, since only T_1 and T_2 are executable at that time, they are executed on processors P_1 and P_2, respectively. The completion of T_1 and T_2 lead to the execution of T_3. After the execution of T_3 is completed, T_4, T_5, T_6, T_7 all become executable. An arbitrary choice of executing T_4, T_5, T_6 on P_1, P_2, P_3, respectively, yields the schedule in Fig. 3.18b. We hasten to point out that such a simple way to schedule tasks does not always give us a schedule with minimum total elapsed time. As a matter of fact, Fig. 3.19b shows a schedule for the set of tasks in Fig. 3.19a in which between $t = 9$ and $t = 10$, processor

---

† Outside of the context of computing, we may also have interpretations such as, "One cannot put on one's shoes before putting on one's socks," or "One should not mop the floor before dusting the furniture," or "The machine cannot be assembled until all its subparts have been built."

‡ A task is said to be *executable* at a certain instant of time if the execution of its predecessors has been completed at that time.

Figure 3.19

P_2 is left idle although T_3 is executable at $t = 9$. The reader can convince himself that the schedule in Fig. 3.19b is better than any schedule in which processors are not left idle intentionally.

We want to determine how good or how bad such a simple scheduling procedure is. As it turns out, the result is rather surprising:

> **Theorem 3.2** For a given set of tasks, let ω denote the total elapsed time when the tasks are executed according to a schedule that contains no intentional idle periods, and let ω_0 denote the minimum possible total elapsed time. Then,
>
> $$\frac{\omega}{\omega_0} \leq 2 - \frac{1}{n}$$
>
> where n is the number of processors in the computing system. Moreover, the bound is best possible.

PROOF To simplify the presentation, we shall prove the result for $n = 2$. The proof for the general case is analogous. Consider the schedule that contains no intentional idle periods, as illustrated in Fig. 3.20. We observe first that the termination of an idle period in one processor coincides with the completion of the execution of a task in the other processor. (Otherwise, the idle period would not be terminated.) Let ϕ_i be an idle

Termination of an idle period coincides with
the completion of the execution of a task

Figure 3.20

period for a processor. A task T_{ij} is said to overlap ϕ_i if the execution of T_{ij} in the other processor overlaps ϕ_i. For example, in Fig. 3.20, T_{11}, T_{12}, T_{13} overlap ϕ_1. Let T_{i1}, T_{i2}, T_{i3}, ..., T_{il} be the tasks that overlap ϕ_i. We claim that

$$T_{i1} \le T_{i2} \le T_{i3} \le \cdots \le T_{il}$$

If this was not the case, the tasks T_{i1}, T_{i2}, ..., T_{il} would not have to be executed sequentially in P_2, and some of them could be executed in P_1 in place of some portions of the idle period ϕ_i in P_1. Similarly, if ϕ_j is another idle period, and T_{j1}, T_{j2}, T_{j3}, ..., T_{jm} are the tasks that overlap ϕ_j, then by repeating the argument, we have

$$T_{j1} \le T_{j2} \le T_{j3} \le \cdots \le T_{jm}$$

It is clear that every task executed after the completion of T_{il} must be a successor of T_{il} (or else it would be executed in ϕ_i). Consequently, there is a subset of tasks $\mathscr{C}$ such that:

1. $\mathscr{C}$ is a chain,
2. $\Sigma_{T_k \in \mathscr{C}} \mu(T_k) \ge \Sigma_{\phi_i \in \Phi} \mu(\phi_i)$,

where Φ is the set of all idle periods in the schedule. We note that

$$\omega = \frac{1}{2}\left[\sum_{T_j \in \mathscr{T}} \mu(T_j) + \sum_{\phi_i \in \Phi} \mu(\phi_i) \right]$$

$$\le \frac{1}{2}\left[\sum_{T_j \in \mathscr{T}} \mu(T_j) + \sum_{T_k \in \mathscr{C}} \mu(T_k) \right] \qquad (3.1)$$

Certainly,

$$\omega_0 \ge \frac{1}{2} \sum_{T_j \in \mathscr{T}} \mu(T_j)$$

and

$$\omega_0 \ge \sum_{T_k \in \mathscr{C}} \mu(T_k)$$

Thus, (3.1) becomes

$$\omega \le \omega_0 + \tfrac{1}{2}\omega_0$$

or

$$\frac{\omega}{\omega_0} \le \frac{3}{2}$$

That this bound is best possible can be demonstrated by the example in Fig. 3.21, where Fig. 3.21*b* shows two schedules for the set of tasks in Fig. 3.21*a*. □

It is interesting to observe that, according to the result we have just proved, the schedule obtained by following the simple rule of not leaving a processor intentionally idle is never worse than a best possible schedule by more than 50 percent in a two-processor computing system, and by more than 100 percent in an *n*-processor computing system. Such a result indeed supports very strongly

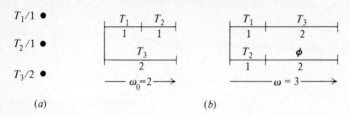

Figure 3.21

the idea of employing simple procedures to obtain good, but not necessarily best possible, results in many optimization problems.

3.8 FUNCTIONS AND THE PIGEONHOLE PRINCIPLE

A binary relation R from A to B is said to be a *function* if for every element a in A, there is a unique element b in B so that (a, b) is in R. For a function R from A to B, instead of writing $(a, b) \in R$, we also use the notation $R(a) = b$, where b is called the *image* of a. The set A is called the *domain* of the function R, and the set B is called the *range* of the function R. The notion of a function is but a formalization of the notion of associating or assigning an element in the range to each of the elements in the domain. For example, let A be a set of houses and B be a set of colors. Then a function from A to B is an assignment of colors for painting the houses. A function can often be represented in graphical form. Figure 3.22a shows a function R from $A = \{a, b, c, d, e\}$ to $B = \{\alpha, \beta, \gamma, \delta\}$. Following the convention of representing a binary relation in tabular form that was introduced earlier, we can represent the function in Fig. 3.22a as that in Fig. 3.22b. However, a more convenient tabular form for representing functions is that shown in Fig. 3.22c, where the left column contains all the elements in the domain and the right column contains their corresponding images.

A function from A to B is said to be an *onto* function if every element of B is the image of one or more elements of A. Figure 3.23a shows an example of

Figure 3.22

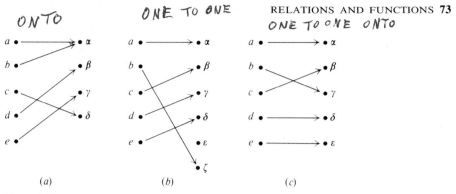

Figure 3.23

an onto function. A function from A to B is said to be a *one-to-one* function if no two elements of A have the same image. Figure 3.23b shows an example of a one-to-one function. A function from A to B is said to be a *one-to-one onto* function if it is both an onto and a one-to-one function.† Figure 3.23c shows a one-to-one onto function. Let A be a set of workers and B_1, B_2, and B_3 be sets of jobs. An onto function from A to B_1 is an assignment of the workers to the jobs so that every job has at least one worker assigned to it; a one-to-one function from A to B_2 is an assignment such that no two workers will have the same job; and a one-to-one onto function from A to B_3 is an assignment such that every job has a worker assigned to it, and no two workers are assigned to the same job.

In the literature, an onto function is also called a *surjection*, a one-to-one function is also called an *injection*, and a one-to-one onto function is also called a *bijection*.

A well-known proof technique in mathematics is the so-called *pigeonhole principle*, also known as the *shoe box argument* or *Dirichlet drawer principle*. In an informal way the pigeonhole principle says that if there are "many" pigeons and "a few" pigeonholes, then there must be some pigeonhole occupied by two or more pigeons. Formally, let D and R be finite sets. If $|D| > |R|$, then for any function f from D to R, there exist d_1, $d_2 \in D$ such that $f(d_1) = f(d_2)$. Some trivial applications of the pigeonhole principle are: Among 13 people, there are at least 2 of them who were born in the same month. Here the 13 people are the pigeons, and the 12 months are the pigeonholes. Also, if 11 shoes are selected from 10 pairs of shoes there must be a pair of matched shoes among the selection. Here the 11 shoes are the pigeons and the 10 pairs are the pigeonholes. The pigeonhole principle can be stated in a slightly more general form: For any function f from D to R, there exist i elements $d_1, d_2, \ldots, d_i$ in D, $i = \lceil |D|/|R| \rceil$, such that $f(d_1) = f(d_2) = \cdots = f(d_i)$.‡ (See Prob. 3.26.)

---

† We recall that we introduced the notion of a one-to-one correspondence between the elements in two sets. Formally, we say that there is a one-to-one correspondence between the elements of two sets if there exists a one-to-one onto function from one set to the other.

‡ We use $\lceil x \rceil$ to denote the smallest integer not less than x.

In the following we present some more examples, and we ask the reader to observe how the pigeonhole principle is applied, since we shall not make an explicit statement every time we use it.

Example 3.3 A chess player wants to prepare for a championship match by playing some practice games in 77 days. She wants to play at least one game a day but no more than 132 games altogether. We show now that no matter how she schedules the games there is a period of consecutive days within which she plays *exactly* 21 games. Let a_i denote the total number of games she plays up through the ith day. Clearly, the sequence $a_1, a_2, \ldots, a_{77}$ is a monotonically increasing sequence, with $a_1 \geqq 1$ and $a_{77} \leqq 132$. Let us compute the sequence $a_1 + 21, a_2 + 21, \ldots, a_{77} + 21$, which again is a monotonically increasing sequence with $a_{77} + 21 \leqq 153$. Since the values of the 154 numbers $a_1, a_2, \ldots, a_{77}, a_1 + 21, \ldots, a_{77} + 21$ range from 1 to 153, two of them must be the same. Moreover, because both the sequence $a_1, a_2, \ldots, a_{77}$ and the sequence $a_1 + 21, a_2 + 21, \ldots, a_{77} + 21$ are monotonically increasing, we have $a_i = a_j + 21$ for some a_i and a_j. □

Example 3.4 We want to show that in a sequence of $n^2 + 1$ distinct integers, there is either an increasing subsequence of length $n + 1$ or a decreasing subsequence of length $n + 1$. Let $a_1, a_2, \ldots, a_{n^2+1}$ denote the sequence of integers. Let us label the integer a_k with an ordered pair (x_k, y_k), where x_k is the length of a longest increasing subsequence starting at a_k and y_k is the length of a longest decreasing subsequence starting at a_k. Suppose there is no increasing subsequence or decreasing subsequence of length $n + 1$ in the sequence $a_1, a_2, \ldots, a_{n^2+1}$. That is, the values of x_k and y_k lie between 1 and n for $k = 1, 2, \ldots, n^2 + 1$. With only n^2 distinct ordered pairs as possible labels for the $n^2 + 1$ integers, there must exist a_i and a_j in the sequence that are labeled with the same ordered pair. However, this is impossible because if $a_i < a_j$, we must have $x_i > x_j$; and if $a_i > a_j$, we must have $y_i > y_j$. Consequently, we conclude that there is either an increasing subsequence or a decreasing subsequence of length $n + 1$ in the sequence $a_1, a_2, \ldots, a_{n^2+1}$.† □

Example 3.5 We want to show that among six persons, either there are three persons who are mutual friends or there are three persons who are complete strangers to each other. Let A be a person in the group. According to the pigeonhole principle, either there are three (or more) persons who are friends of A or there are three (or more) persons who are strangers to A. Let us consider the former case only, since the latter case can be resolved by a similar argument. Let B, C, D denote the friends of A. If any two of B, C, D know each other, then these two, together with A, form a friendly threesome. On the other hand, if no two of B, C, D know

---

† See Prob. 3.27 for a slightly different argument. See also Prob. 3.28.

each other, then they are three persons who are complete strangers to each other. □

Example 3.6 A rooming house has 90 rooms and 100 guests. Keys are to be issued to the guests so that any 90 guests can have access to the 90 rooms in the sense that each guest will have a key to an unoccupied room. (In other words, we can assign the 90 rooms to any 90 guests so that every guest has a key to the room he is assigned.) We want to use a scheme that will minimize the total number of keys. Although the following scheme seems to be wasteful, it turns out to be a minimal scheme. Ninety of the guests will be given one key each so that together they will have access to the 90 rooms. Each of the remaining 10 guests will be given 90 keys, one key to each of the 90 rooms. Clearly such a scheme, which uses a total of 990 keys, works. To show that this is a minimal scheme, we observe that, if 989 or fewer keys were issued, there is a room that has at most 10 outstanding keys. Clearly, the scheme fails when none of the holders of the keys to this room are among the 90 guests. □

Example 3.7 The circumference of the two concentric disks are divided into 200 sections each, as shown in Fig. 3.24. For the outer disk, 100 of the sections are painted red and 100 of the sections are painted blue. For the inner disk the sections are painted red or blue in an arbitrary manner. Show that it is possible to align the two disks so that 100 or more of the sections on the inner disk have their colors matched with the corresponding sections on the outer disk. Let us hold the outer disk fixed and rotate the inner disk through the 200 possible alignments. For each alignment, let us count the

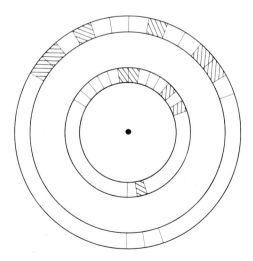

Figure 3.24

number of matches. We note that the sum of the counts for the 200 possible alignments is 20,000, because each of the 200 sections on the inner disk will match its corresponding section on the outer disk in exactly 100 of the alignments. Therefore, there must be an alignment in which there are 100 or more matches. ☐

3.9 REMARKS AND REFERENCES

For further discussion on the subject of data base management, see Codd[1] and Date[3]. See Mirsky[6] for Dilworth's theorem and its application. The notions of partitions and partition pairs are very useful in studying the structural properties of finite-state machines. See, for example, chap. 12 of Kohavi[5]. Theorem 3.2 is due to Graham[4]. Coffman[2] covers the many aspects of the job-scheduling problem introduced in Section 3.7. See chap. 4 of Ryser[7] for Ramsey's theorem which is a nontrivial generalization of the pigeonhole principle.

[1] Codd, E. F.: A Relational Model of Data for Large Shared Data Banks, *Communications of ACM*, **13**: 377–387 (1970).
[2] Coffman, E. G., Jr. (Ed.): "Computer and Job-Shop Scheduling Theory," John Wiley & Sons, Inc., New York, 1976.
[3] Date, C. J.: "An Introduction to Database Systems," Addison-Wesley Publishing Company, Reading, Mass., 1975.
[4] Graham, R. L.: Bounds for Certain Multiprocessing Anomalies, *Bell System Technical Journal*, **45**: 1563–1581 (1966).
[5] Kohavi, Z.: "Switching and Finite Automata Theory," McGraw-Hill Book Company, New York, 1970.
[6] Mirsky, L.: "Transversal Theory," Academic Press, New York, 1971.
[7] Ryser, H. J.: "Combinatorial Mathematics," published by the Mathematical Association of America, distributed by John Wiley & Sons, Inc., New York, 1963.

PROBLEMS

3.1 Let A denote the set of shirts and B denote the set of slacks a man owns. What possible interpretation can be given to the cartesian product $A \times B$? to a binary relation from A to B?

3.2 Let $A = \{1, 2\}$. Construct the set $\mathscr{P}(A) \times A$.

3.3 (*a*) Given that $A \subseteq C$ and $B \subseteq D$, show that $A \times B \subseteq C \times D$.
(*b*) Given that $A \times B \subseteq C \times D$, does it necessarily follow that $A \subseteq C$ and $B \subseteq D$?

3.4 (*a*) Let A be an arbitrary set. Is the set $A \times \emptyset$ well defined?
(*b*) Given that $A \times B = \emptyset$, what can one say about the sets A and B?
(*c*) Is it possible that $A \subseteq A \times A$ for some set A?

3.5 Let A, B, C, D be arbitrary sets.
(*a*) Show that
$$(A \cap B) \times (C \cap D) = (A \times C) \cap (B \times D)$$

(b) Confirm or disprove the following identities:

$$(A \cup B) \times (C \cup D) = (A \times C) \cup (B \times D)$$

$$(A - B) \times (C - D) = (A \times C) - (B \times D)$$

$$(A \oplus B) \times (C \oplus D) = (A \times C) \oplus (B \times D)$$

3.6 Let A, B, C be arbitrary sets.
(a) Show that

$$(A \cap B) \times C = (A \times C) \cap (B \times C)$$

(b) Confirm or disprove the following identities:

$$(A \cup B) \times C = (A \times C) \cup (B \times C)$$

$$(A - B) \times C = (A \times C) - (B \times C)$$

$$(A \oplus B) \times C = (A \times C) \oplus (B \times C)$$

3.7 Let $A = \{6{:}00, 6{:}30, 7{:}00, \ldots, 9{:}30, 10{:}00\}$ denote the set of nine half-hour periods in the evening. Let $B = \{3, 12, 15, 17\}$ denote the set of the four local television channels. Let R_1 and R_2 be two binary relations from A to B. What possible interpretations can be given to the binary relations R_1, R_2, $R_1 \cup R_2$, $R_1 \cap R_2$, $R_1 \oplus R_2$, and $R_1 - R_2$?

3.8 Let I be the set of all integers.
(a) Is there a natural way to interpret the ordered pairs in $I \times I$ as geometric points in the plane?
(b) Let R_1 be a binary relation on $I \times I$ such that the ordered pair (of ordered pairs) $((a, b), (c, d))$ is in R_1 if and only if $a - c = b - d$. What is a geometric interpretation of the binary relation R_1?
(c) Let R_2 be a binary relation on $I \times I$ such that $((a, b), (c, d))$ is in R_2 if and only if $\sqrt{(a - c)^2 + (b - d)^2} \le 10$. What is a geometric interpretation of R_2? of $R_1 \cup R_2$? of $R_1 \cap R_2$? of $R_1 - R_2$? of $R_1 \oplus R_2$?

3.9 Let A be a set of workers and B be a set of jobs. Let R_1 be a binary relation from A to B such that (a, b) is in R_1 if worker a is assigned to job b. (We assume that a worker might be assigned to more than one job and more than one worker might be assigned to the same job.) Let R_2 be a binary relation on A such that (a_1, a_2) is in R_2 if a_1 and a_2 can get along with each other if they were assigned to the same job. State a condition in terms of R_1, R_2 and (possibly) binary relations derived from R_1 and R_2 such that an assignment of the workers to the jobs according to R_1 will not put workers that cannot get along with one another on the same job.

3.10 Let R be a symmetric and transitive relation on a set A. Show that if for every a in A there exists b in A such that (a, b) is in R, then R is an equivalence relation.

3.11 Let R be a transitive and reflexive relation on A. Let T be a relation on A such that (a, b) is in T if and only if both (a, b) and (b, a) are in R. Show that T is an equivalence relation.

3.12 Let R be a binary relation. Let

$$S = \{(a, b) | (a, c) \in R \text{ and } (c, b) \in R \text{ for some } c\}$$

Show that if R is an equivalence relation, then S is also an equivalence relation.

3.13 Let A be a set of books.
(a) Let R_1 be a binary relation on A such that (a, b) is in R_1 if book a costs more and contains fewer pages than book b. In general, is R_1 reflexive? Symmetric? Antisymmetric? Transitive?
(b) Let R_2 be a binary relation on A such that (a, b) is in R_2 if book a costs more or contains fewer pages than book b. In general, is R_2 reflexive? Symmetric? Antisymmetric? Transitive?

3.14 Let R be a reflexive relation on a set A. Show that R is an equivalence relation if and only if (a, b) and (a, c) are in R implies that (b, c) is in R.

3.15 A binary relation on a set that is reflexive and symmetric is called a *compatible relation*.
(a) Let A be a set of people and R be a binary relation on A such that (a, b) is in R if a is a friend of b. Show that R is a compatible relation.

(b) Let A be a set of English words and R be a binary relation on A such that two words in A are related if they have one or more letters in common. Show that R is a compatible relation.

(c) Give more examples of compatible relations.

(d) Let R_1 and R_2 be two compatible relations on A. Is $R_1 \cap R_2$ a compatible relation? Is $R_1 \cup R_2$ a compatible relation?

(e) Let A be a set. A *cover* of A is a set of nonempty subsets of A, $\{A_1, A_2, \ldots, A_k\}$, such that the union of the A_i's is equal to A. Suggest a way to define a compatible relation on A from a cover of A. Give an interpretation of the notion of a cover in terms of the example in part (a).

(f) Suggest a way to define a cover of A from a compatible relation on A. Does your suggested way define *uniquely* a cover of A?

3.16 (a) Show that the transitive closure of a symmetric relation is symmetric.

(b) Is the transitive closure of an antisymmetric relation always antisymmetric?

(c) Show that the transitive closure of a compatible relation (see Prob. 3.15) is an equivalence relation.

3.17 Let R be a binary relation from A to B. The *converse* of R, denoted R^{-1}, is a binary relation from B to A such that

$$R^{-1} = \{(b, a) | (a, b) \in R\}$$

(a) Let R_1 and R_2 be binary relations from A to B. Is it true that $(R_1 \cup R_2)^{-1} = R_1^{-1} \cup R_2^{-1}$?

(b) Let R be a binary relation on A. If R is reflexive, is R^{-1} necessarily reflexive? If R is symmetric, is R^{-1} necessarily symmetric? If R is transitive, is R^{-1} necessarily transitive?

3.18 (a) Let R be an equivalence relation on a set A. Let $\{A_1, A_2, \ldots, A_k\}$ be a set of subsets of A such that $A_i \not\subseteq A_j$ for $i \neq j$ and such that a and b are contained in one of the subsets if and only if the ordered pair (a, b) is in R. Show that $\{A_1, A_2, \ldots, A_k\}$ is a partition of A.

(b) Let $\{A_1, A_2, \ldots, A_k\}$ be a partition of a set A. We define a binary relation R on A such that an ordered pair (a, b) is in R if and only if a and b are in the same block of the partition. Show that R is an equivalence relation.

3.19 Let A be a set and f be a function from A to A. A partition π of A is said to have the *substitution property* with respect to f if for any two elements a and b that are together in one block of π the two elements $f(a)$ and $f(b)$ are also together in one block of π. Let $A = \{1, 2, 3, 4, 5, 6\}$. Let f be a function from A to A such that $f(1) = 3, f(2) = 3, f(3) = 2, f(4) = 5, f(5) = 4, f(6) = 4$.

(a) Does $\pi_1 = \{\overline{123}, \overline{456}\}$ have the substitution property with respect to f? How about $\pi_2 = \{\overline{16}, \overline{25}, \overline{34}\}$? $\pi_3 = \{\overline{12}, \overline{34}, \overline{56}\}$?

(b) Let A be the set of integers and π be a partition of A into even and odd integers. Let $f(a) = a + 1$ for every a in A. Does π have the substitution property with respect to f? Let

$$g(a) = \begin{cases} \dfrac{a}{2} & \text{if } a \text{ is even} \\ \dfrac{a-1}{2} & \text{if } a \text{ is odd} \end{cases}$$

Does π have the substitution property with respect to g?

(c) Let π_1 and π_2 be two partitions that have the substitution property with respect to f. Show that both $\pi_1 \cdot \pi_2$ and $\pi_1 + \pi_2$ also have the substitution property with respect to f.

3.20 Let $(A, \leq)$ be a partially ordered set. Let $\leq_R$ be a binary relation on A such that for a and b in A, $a \leq_R b$ if and only if $b \leq a$.

(a) Show that $\leq_R$ is a partial ordering relation.

(b) Show that if $(A, \leq)$ is a lattice, then $(A, \leq_R)$ is also a lattice.

3.21 Is the cartesian product of two lattices always a lattice? Prove your claim.

3.22 Let P be an arbitrary partially ordered set, and L be a chain of two elements. Let Q denote the cartesian product $P \times L$. Let A be an antichain in Q. Let B be a largest possible subset of P such that B does not contain a chain of length exceeding 2. Show that $|A| \leq |B|$.

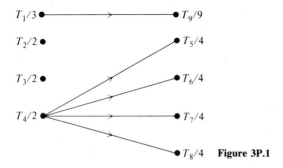

Figure 3P.1

3.23 The procedure of scheduling a set of tasks according to the rule of never leaving a processor idle intentionally discussed in Sec. 3.7 did not specify how ties can be broken when several executable tasks compete for a free processor. One way to break ties is to assign distinct priorities to the tasks, and schedule a task that has the highest priority among all executable tasks at any time instant on a processor that is free at that time instant. For example, the set of tasks shown in Fig. 3P.1 is to be executed on a computing system with three processors. Let us assign priorities in decreasing order to the tasks T_1, T_2, T_3, T_4, T_5, T_6, T_7, T_8, T_9.

(a) Construct the corresponding schedule.

(b) Suppose we remove the arrows between T_4 and T_5 and between T_4 and T_6 in Fig. 3P.1. Construct the corresponding schedule.

(c) Suppose we reduce the execution time of each task by 1. Construct the corresponding schedule.

(d) Suppose we execute the set of tasks on a computing system with four processors. Construct the corresponding schedule.

(This problem illustrates some of the *anomalies* when tasks are scheduled according to the rule of never leaving a processor intentionally idle.)

3.24 Let f be a function from A to B and g be a function from B to C. We can define a function h from A to C such that for every a in A, $h(a) = g(f(a))$. h is called the *composition* of f and g and is denoted $g \circ f$.

(a) For the functions f and g in Fig. 3P.2, determine $g \circ f$.

(b) For the functions f, g, and h in Fig. 3P.2, determine $h \circ (g \circ f)$ and $(h \circ g) \circ f$.

(c) Show that for any functions f, g, and h, $h \circ (g \circ f) = (h \circ g) \circ f$.

(d) State a necessary and sufficient condition on f and g so that:

    1. $g \circ f$ is an onto function;

    2. $g \circ f$ is a one-to-one function;

    3. $g \circ f$ is a one-to-one onto function.

(e) Construct an example to show that for two functions f and g from A to A, in general, $f \circ g \neq g \circ f$.

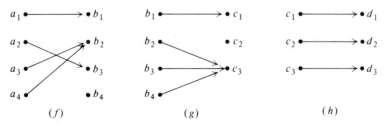

Figure 3P.2

3.25 Let f, g, h be functions from N to N, where N is the set of natural numbers so that

$$f(n) = n + 1$$
$$g(n) = 2n$$
$$h(n) = \begin{cases} 0 & n \text{ is even} \\ 1 & n \text{ is odd} \end{cases}$$

Determine $f \circ f$, $f \circ g$, $g \circ f$, $g \circ h$, $h \circ g$, $(f \circ g) \circ h$.

3.26 Let f be a function from D to R. Suppose that $|D| > |R|$.

(*a*) Let i be the quotient and r be the remainder when $|D|$ is divided by $|R|$. Show that

$$\left\lceil \frac{|D|}{|R|} \right\rceil = \begin{cases} i + 1 & \text{if } r \neq 0 \\ i & \text{if } r = 0 \end{cases}$$

(*b*) Show that there exist i elements $d_1, d_2, \ldots, d_i$ in D such that $f(d_1) = f(d_2) = \cdots = f(d_i)$.

3.27 In this problem, we prove the result in Example 3.4 using a slightly different argument. Let $a_1, a_2, \ldots, a_{n^2+1}$ be a sequence of $n^2 + 1$ distinct integers. Suppose there is no increasing subsequence of length $n + 1$ in the sequence. Let us label the integer a_k with a label x_k, where x_k is the length of a longest increasing subsequence starting at a_k, $1 \leq k \leq n^2 + 1$.

(*a*) Show that there are at least $n + 1$ integers in the sequence that have the same label.

(*b*) Show that these integers form a decreasing subsequence.

3.28 In this problem we show how we can apply Corollary 3.1.1 to prove the result in Example 3.4. Let $a_1, a_2, \ldots, a_{n^2+1}$ be a sequence of $n^2 + 1$ distinct integers. We define a partial ordering relation $\leq$ over the $n^2 + 1$ ordered pairs (a_i, i) such that $(a_i, i) \leq (a_j, j)$ if and only if $a_i \leq a_j$ and $i \leq j$.

(*a*) What is the significance of a chain in the partially ordered set? of an antichain?

(*b*) Apply Corollary 3.1.1 to prove the result in Example 3.4.

3.29 Use the pigeonhole principle to show that one of any m consecutive integers is divisible by m.

3.30 Use the pigeonhole principle to show that the decimal expansion of a rational number must, after some point, become periodic.

3.31 A man hiked for 10 hours and covered a total distance of 45 miles. It is known that he hiked 6 miles in the first hour and only 3 miles in the last hour. Show that he must have hiked at least 9 miles within a certain period of two consecutive hours.

3.32 The circumference of a "roulette wheel" is divided into 36 sectors to which the numbers $1, 2, \ldots, 36$ are assigned in some arbitrary manner. Show that there are three consecutive sectors such that the sum of their assigned numbers is at least 56.

3.33 Let $x_1, x_2, \ldots, x_n$ be n arbitrary integers. Show that $x_i + x_{i+1} + x_{i+2} + \cdots x_{i+k}$ is divisible by n for some i and k, $i \geq 1$, $k \geq 0$.

3.34 From the integers 1–200, 101 of them are chosen arbitrarily. Show that, among the chosen numbers, there exist two such that one divides another.

3.35 Show that for an arbitrary integer N, there exists a multiple of N that contains only the digits 0 and 7. (For example, for $N = 3$, we have $259 \times 3 = 777$; for $N = 4$, we have $1925 \times 4 = 7700$; for $N = 5$, we have $14 \times 5 = 70$; for $N = 6$, we have $1295 \times 6 = 7770$.) (*Hint:* Consider the remainders when the following integers are divided by N: $7, 77, 777, 7777, \ldots, \underbrace{7777 \cdots 77}_{N \ 7s}$.)

3.36 (*a*) Show that among $n + 1$ arbitrarily chosen integers, there are two whose difference is divisible by n.

(*b*) Show that among $n + 2$ arbitrarily chosen integers, either there are two whose difference is divisible by $2n$ or there are two whose sum is divisible by $2n$.

3.37 Consider a sequence of N positive integers containing precisely n distinct integers. If $N \geq 2^n$, show that there is a consecutive block of integers in the sequence whose product is a perfect square. (For example, in the sequence $3, 7, 5, 3, 7, 3, 5, 7$; $7 \cdot 5 \cdot 3 \cdot 7 \cdot 3 \cdot 5$ is a perfect square.)

3.38 Show that among $n + 1$ positive integers less than or equal to $2n$ there are two of them that are relatively prime.

3.39 The following procedure can be used to select the larger n of $2n$ distinct numbers:

1. Arrange n of the $2n$ numbers in ascending order. Denote these numbers $a_1, a_2, \ldots, a_n$ such that $a_1 < a_2 < a_3 < \cdots < a_n$.
2. Arrange the remaining n numbers in descending order. Denote these numbers $b_1, b_2, \ldots, b_n$ such that $b_1 > b_2 > b_3 > \cdots > b_n$.
3. Compare a_i and b_i and select the larger one of the two for $i = 1, 2, \ldots, n$.

For example, given eight numbers 4, 2, 7, 6, 5, 3, 1, 8:

1. Arrange 4, 2, 7, 6 in ascending order: $2 < 4 < 6 < 7$.
2. Arrange 5, 3, 1, 8 in descending order: $8 > 5 > 3 > 1$.
3. Compare 2 and 8; select 8. Compare 4 and 5; select 5. Compare 6 and 3; select 6. Compare 7 and 1; select 7.

Prove the procedure is correct. (*Hint:* Show that exactly one of a_i and b_i is among the larger n of the $2n$ numbers.)

GRAPHS AND PLANAR GRAPHS

4.1 INTRODUCTION

As was pointed out in Chap. 3, there are many real life problems that can be abstracted as problems concerning sets of discrete objects and binary relations on them. For example, consider a series of public-opinion polls conducted to determine the popularity of the presidential candidates. In each poll, voters' opinions were sought on two of the candidates, and a favorite between the two was determined. The results of the polls will be interpreted as follows: Candidate a is considered to be running ahead of candidate b if:

1. Candidate a was ahead of candidate b in a poll conducted between them; or
2. candidate a was ahead of candidate c in a poll, and candidate c was ahead of candidate b in another poll; or
3. candidate a was ahead of candidate c, and candidate c was ahead of candidate d, and candidate d was ahead of candidate b in three separate polls, and so on.

Given two candidates, we might want to know whether one of them is running ahead of the other.† Let $S = \{a, b, c, \ldots\}$ be the set of candidates and R be a binary relation on S such that (a, b) is in R if a poll between a and b was conducted and a was chosen the favorite candidate. We recall that a binary relation on a set can be represented in tabular form, as in Fig. 4.1a, or in graphical

---

† Note that, according to our interpretation of the results of the polls, it is possible that candidate a is running ahead of candidate b, and, at the same time, candidate b is also running ahead of candidate a. (See how politics works!)

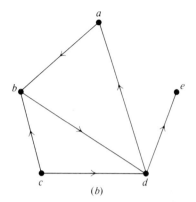

Figure 4.1

form, as in Fig. 4.1*b*. Suppose that the binary relation in Fig. 4.1*a* and Fig. 4.1*b* represents the results of the polls conducted. We observe that candidate *a* is more popular than candidate *e* because of the ordered pairs $(a, b), (b, d), (d, e)$ in *R*. One probably would agree that the graphical representation of the binary relation *R* in Fig. 4.1*b* is quite useful in comparing the popularity of two candidates, since there must be "a sequence of arrows" leading from the point corresponding to the more popular candidate to that corresponding to the less popular one.

As another example, consider a number of cities connected by highways. Given a map of the highways, we might want to determine whether there is a highway route between two cities on the map. Also, consider all the board positions† in a chess game. We might want to know whether a given board position can be reached from some other board position through a sequence of legal moves. It is clear that both of these examples are again concerned with discrete objects and binary relations on them. In the example of the highway map, let $S = \{a, b, c, \ldots\}$ be the set of cities and *R* be a binary relation on *S* such that (a, b) is in *R* if there is a highway from city *a* to city *b*. In the chess game example, let $S = \{a, b, c, \ldots\}$ be the set of all board positions and *R* be a binary relation on *S* such that (a, b) is in *R* if board position *a* can be transformed into board position *b* in one legal move. Furthermore, in both of these cases as in the example on the popularity of the presidential candidates, we want to know whether, for given *a* and *b* in *S*, there exist *c, d, e, …, h* in *S* such that $\{(a, c), (c, d), (d, e), \ldots, (h, b)\} \subseteq R$.

Indeed, in many problems dealing with discrete objects and binary relations, a graphical representation of the objects and the binary relations on them is a very convenient form of representation. This leads us naturally to a study of the theory of graphs.

† To be precise, a board position means the position of all pieces together with a specification of whose turn it is (i.e., either black or white).

4.2 BASIC TERMINOLOGY

A *directed graph G* is defined abstractly as an ordered pair (V, E), where V is a set and E is a binary relation on V. As was pointed out before, a directed graph can be represented geometrically as a set of marked points V with a set of arrows E between pairs of points.† For example, Fig. 4.2 shows a directed graph. The elements in V are called the *vertices*, and the ordered pairs in E are called the *edges* of the directed graph. An edge is said to be *incident* with the vertices it joins. For example, the edge (a, b) is incident with the vertices a and b. Sometimes, when we wish to be more specific, we say that the edge (a, b) is incident *from a* and is incident *into b*. The vertex a is called the *initial vertex*, and the vertex b is called the *terminal vertex* of the edge (a, b). An edge that is incident from and into the same vertex, like (c, c) in Fig. 4.2, is called a *loop*. Two vertices are said to be *adjacent* if they are joined by an edge. Moreover, corresponding to an edge (a, b), the vertex a is said to be *adjacent to* the vertex b, and the vertex b is said to be *adjacent from* the vertex a. A vertex is said to be an *isolated* vertex if there is no edge incident with it.

An *undirected graph G* is defined abstractly as an ordered pair (V, E), where V is a set and E is a set of multisets of two elements from V. For example, $G = (\{a, b, c, d\}, \{\{a, b\}, \{a, d\}, \{b, c\}, \{b, d\}, \{c, c\}\})$ is an undirected graph. An undirected graph can be represented geometrically as a set of marked points V with a set of lines E between the points. The undirected graph G above is shown in Fig. 4.3. As another example, let $V = \{a, b, c, d, e\}$ be a set of computer programs. Figure 4.4 shows an undirected graph in which there is an edge between two vertices if the corresponding programs share some common data. From now on, when it is clear from the context, we shall use the term "graph" to refer to "directed graph" or "undirected graph," or both.

Let $V = \{a, b, c, d\}$ be the four players in a round-robin tennis tournament. Let $E = \{(a, b), (a, d), (b, d), (c, a), (c, b), (d, c)\}$ be a binary relation on V so

Figure 4.2

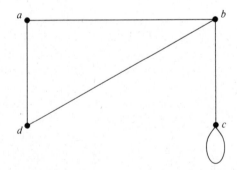

Figure 4.3

† In fact, we could choose to *define* a directed graph to be a set of marked points V with a set of arrows between the points so that there is at most one arrow from one point to another point.

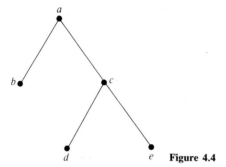

Figure 4.4

that (x, y) in E means that x beats y in the match between them. The graph $G = (V, E)$ is shown in Fig. 4.5a. Let $V' = \{1, 2, 3, 4\}$ be the four chapters in a book. Let $E' = \{(1, 2), (2, 3), (3, 1), (3, 4), (4, 1), (4, 2)\}$ be a binary relation on V' such that $(1, 2)$ in E' means that the material in chapter 1 refers to that in chapter 2, and so on. The graph $G' = (V', E')$ is shown in Fig. 4.5b. A careful reader might have recognized that the graph in Fig. 4.5b "resembles" the graph in Fig. 4.5a. Indeed, such a "resemblance" becomes even more evident if we

(a)

(b)

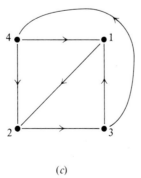

(c)

Figure 4.5

redraw the graph in Fig. 4.5b to that shown in Fig. 4.5c. Comparing the graphs
in Fig. 4.5a and Fig. 4.5c, we realize that both the tennis tournament problem
and the problem of cross reference among chapters can be represented
abstractly by the same graph. Consequently, many of the results concerning the
tennis players can be rephrased immediately as results concerning the chapters
of the book. For example, according to Fig. 4.5a, player b beats player d who,
in turn, beats player c, and according to Fig. 4.5c, chapter 1 refers to chapter 2
which, in turn, refers to chapter 3. The concept of resemblance of two graphs
can be made precise: Two graphs are said to be *isomorphic* if there is a
one-to-one correspondence between their vertices and between their edges
such that incidences are preserved. In other words, there is an edge between
two vertices in one graph if and only if there is a corresponding edge between
the corresponding vertices in the other graph. For example, Fig. 4.6a shows a
pair of isomorphic undirected graphs, and Fig. 4.6b shows a pair of isomorphic
directed graphs. In these two figures, corresponding vertices in the two
isomorphic graphs are labeled with the same letter, primed and unprimed. The
reader can convince himself that the graphs are isomorphic by checking the
incidence relations.

Let $G = (V, E)$ be a graph. A graph $G' = (V', E')$ is said to be a *subgraph* of
G if E' is a subset of E and V' is a subset of V such that the edges in E' are
incident only with the vertices in V'. For example, Fig. 4.7b shows a subgraph
of the graph in Fig. 4.7a. A subgraph of G is said to be a *spanning subgraph* if it

(a)

(b)

Figure 4.6

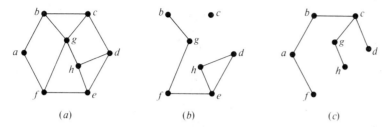

Figure 4.7

contains all the vertices of G. The *complement of a subgraph $G' = (V', E')$* with respect to the graph G is another subgraph $G'' = (V'', E'')$ such that E'' is equal to $E - E'$ and V'' contains only the vertices with which the edges in E'' are incident.† For example, Fig. 4.7c shows the complement of the subgraph in Fig. 4.7b. The *undirected complete graph* of n vertices, denoted K_n, is a graph with n vertices in which there is an edge between each pair of distinct vertices. The *complement of a graph G* of n vertices is defined to be its complement with respect to K_n and is denoted $\bar{G}$. For example, let G be a graph of n vertices. Let the n vertices in G represent n people, and let the set of edges in G represent a compatible relationship such that an edge between two vertices means that the two corresponding persons can work cooperatively as a team. Clearly, the set of edges in $\bar{G}$ will represent the incompatibility relationship among the n people. We also define a *directed complete* graph of n vertices to be a graph with n vertices in which there is exactly one arrow between each pair of distinct vertices.

4.3 MULTIGRAPHS AND WEIGHTED GRAPHS

The definition of a graph can be extended in several ways. Let $G = (V, E)$, where V is a set and E is a multiset of ordered pairs from $V \times V$. G is called a *directed multigraph*. Geometrically, a directed multigraph can be represented as a set of marked points V with a set of arrows E between the points where there is no restriction on the number of arrows from one point to another point. (Indeed, the multiplicity of an ordered pair of vertices in the multiset E is the number of arrows between the corresponding marked points.) For example, Fig. 4.8 shows a multigraph. Also, consider a graphical representation of a highway map in which an edge between two cities corresponds to a lane in a highway between the cities. Since there are often multilane highways between pairs of cities, this representation yields a multigraph. The notion of an *undirected multigraph* can be defined in a similar manner. From now on, when

---

† According to our definition, some isolated vertices in G that are not included in G' will not be included in G'' either. However, one can always modify the definition if it is desirable to include such isolated vertices in G''.

Figure 4.8

it is clear from the context, we shall use the term *graph* to mean either "graph" or "multigraph" or both. On the other hand, when it is necessary to emphasize that we are referring to a graph (instead of a multigraph) we shall use the term *linear graph.*

In modeling a physical situation as an abstract graph, there are many occasions in which we wish to attach additional information to the vertices and/or the edges of the graph. For example, in a graph that represents the highway connection among cities, we might wish to assign a number to each edge to indicate the distance between the two cities connected by the edge. We might also wish to assign a number to each vertex to indicate the population of the city. In a graph that represents the outcomes of the matches in a tennis tournament we might wish to label each edge with the score and the date of the match between the players connected by the edge. In a formal and general way we define a *weighted graph* as either an ordered quadruple (V, E, f, g), or an ordered triple (V, E, f), or an ordered triple (V, E, g), where V is the set of vertices, E is the set of edges, f is a function whose domain is V, and g is a function whose domain is E. The function f is an assignment of weights to the vertices, and the function g is an assignment of weights to the edges. The weights can be numbers, symbols, or whatever quantities that we wish to assign to the vertices and edges.

We present some examples:

Example 4.1 Let us consider the problem of modeling the behavior of a vending machine that sells candy bars for 15¢ apiece. For simplicity, we assume that the machine accepts only nickels and dimes and will not return any change when more than 15¢ is deposited. The weighted graph in Fig. 4.9 is a description of the behavior of the machine where the vertices correspond to the amounts that have already been deposited for the current sale, namely, 0, 5, 10, and 15¢ or more. At any moment, a customer can do one of three things: deposit a nickel, deposit a dime, and press a button for a candy bar of her choice. Correspondingly, in the graph in Fig. 4.9, there are three outgoing edges from each vertex labeled 5, 10, and P. An edge with weight 5 updates the total amount deposited in the machine when the customer puts in a nickel, and an edge with weight 10 updates the total amount deposited in the machine when the customer puts in a dime. Clearly, when we are at vertices a, b, and c, nothing will happen when we press a button to select a candy bar; the machine will release a candy bar only when vertex d is reached. □

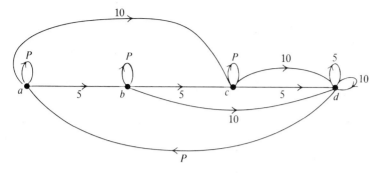

a: 0¢ deposited

b: 5¢ deposited

c: 10¢ deposited

d: 15¢ or more deposited

Figure 4.9

Example 4.2 We consider the problem of recognizing sentences consisting of an article, followed by at most three adjectives, followed by a noun, and then followed by a verb, as shown in the following.

The train stops.
A little girl laughs.
The large fluffy white clouds appear.

When we examine a sentence word by word, we can determine whether it is in this special form by following the weighted graph in Fig. 4.10, starting at vertex *a*. If vertex *g* is reached, the sentence is in the special form. To simplify the drawing of the graph, we use dotted arrows to indicate the discovery of words that are out of the normal order. In that case we reach vertex *h* which signifies the detection of an "illegal" sentence. □

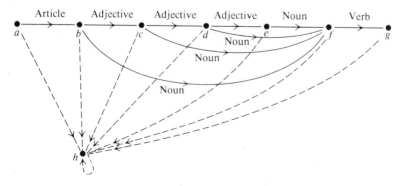

Figure 4.10

These two examples illustrate a general model that is useful in describing many different physical problems ranging from automatic machines and electronic hardware for digital computers to grammatical structures of languages. Such a model is known as a *finite-state model*, referring to the fact that there is a finite number of vertices in the graphs which describe the problems. We shall not go into any details here on this subject since it is beyond our scope of study.

4.4 PATHS AND CIRCUITS

In a directed graph, a *path* is a sequence of edges $(e_{i_1}, e_{i_2}, \ldots, e_{i_k})$† such that the terminal vertex of e_{i_j} coincides with the initial vertex of $e_{i_{(j+1)}}$ for $1 \leq j \leq k - 1$. A path is said to be *simple* if it does not include the same edge twice. A path is said to be *elementary* if it does not meet the same vertex twice.‡ In Fig. 4.11, (e_1, e_2, e_3, e_4) is a path; $(e_1, e_2, e_3, e_5, e_8, e_3, e_4)$ is a path, but not a simple one; $(e_1, e_2, e_3, e_5, e_9, e_{10}, e_{11}, e_4)$ is a simple path, but not an elementary one. In the example on the popularity of the presidential candidates in Sec. 4.1, that candidate a is more popular than candidate b means the existence of a path from vertex a to vertex b in the graph representing the results of the polls. In the example of the highway map in Sec. 4.1, a path from one vertex to another vertex in the graph representing the highway connections is exactly a highway route between the corresponding cities.

A *circuit* is a path $(e_{i_1}, e_{i_2}, \ldots, e_{i_k})$ in which the terminal vertex of e_{i_k} coincides with the initial vertex of e_{i_1}. A circuit is said to be *simple* if it does not include the same edge twice. A circuit is said to be *elementary* if it does not meet the same vertex twice. In Fig. 4.11 $(e_1, e_2, e_3, e_5, e_9, e_{10}, e_{12}, e_6, e_7)$ is a simple circuit, but not an elementary one; $(e_1, e_2, e_3, e_5, e_6, e_7)$ is an elementary circuit.

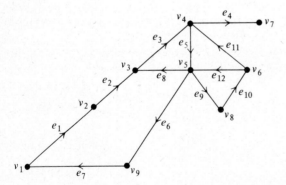

Figure 4.11

---

† To simplify the notation, we identify the edges of a graph by letter names such as $e_1, e_2, \ldots,$ as shown in Fig. 4.11.

‡ In other words, no two edges in the sequence have the same terminal vertex.

On many occasions, we shall also represent a path or a circuit by the sequence of vertices the path or the circuit meets when it is traced. For example, the path (e_1, e_2, e_3, e_4) in the graph in Fig. 4.11 can also be represented as $(v_1, v_2, v_3, v_4, v_7)$, and the circuit $(e_5, e_9, e_{10}, e_{11})$ can be represented as $(v_4, v_5, v_8, v_6, v_4)$.

The notions of paths and circuits in an undirected graph can be defined in a similar way. We shall leave the details to the reader.

As was illustrated in the foregoing several examples, there are many problems in which we want to determine whether there is a path from one vertex to another vertex. We present now a result that is useful in answering the question of existence of such paths. In the next section, we shall see a general procedure for finding such paths.

Theorem 4.1 In a (directed or undirected) graph with n vertices, if there is a path from vertex v_1 to vertex v_2, then there is a path of no more than $n - 1$ edges from vertex v_1 to vertex v_2.

PROOF Suppose there is a path from v_1 to v_2. Let $(v_1, \ldots, v_i, \ldots, v_2)$ be the sequence of vertices that the path meets when it is traced from v_1 and v_2. If there are l edges in the path, there are $l + 1$ vertices in the sequence. For l larger than $n - 1$, there must be a vertex v_k that appears more than once in the sequence, that is, $(v_1, \ldots, v_i, \ldots, v_k, \ldots, v_k, \ldots, v_2)$. Deleting the edges in the path that leads v_k back to v_k, we have a path from v_1 to v_2 which has fewer edges than the original one. This argument can be repeated until we have a path that has $n - 1$ or fewer edges. □

An undirected graph is said to be *connected* if there is a path between every two vertices, and is said to be *disconnected* otherwise. A directed graph is said to be connected if the undirected graph derived from it by ignoring the directions of the edges is connected and is said to be disconnected otherwise. It follows that a disconnected graph consists of two or more components each of which is a connected graph. A directed graph is said to be *strongly connected* if for every two vertices a and b in the graph there is a path from a to b as well as a path from b to a. For example, Fig. 4.12a shows a connected graph which, however, is not strongly connected, while Fig. 4.12b shows a disconnected graph.

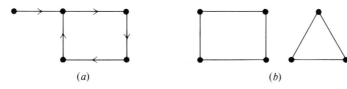

(a) (b)

Figure 4.12

4.5 SHORTEST PATHS IN WEIGHTED GRAPHS

Let $G = (V, E, w)$ be a weighted graph, where w is a function from E to the set of positive real numbers. Consider V as a set of cities and E as a set of highways connecting these cities. The weight of an edge $\{i, j\}$, $w(i, j)$,† is usually referred to as the length of the edge $\{i, j\}$, which has an obvious interpretation as the distance between cities i and j, although other interpretations such as the yearly cost to maintain the highway, or the number of accidents on the highway each month are also possible and meaningful. The *length* of a path in G is defined to be the sum of the lengths of the edges in the path. One problem of significant interest is to determine a shortest path from one vertex to another vertex in V. There are several well-known procedures for solving this problem. We present here one discovered by E. W. Dijkstra[3]. Our presentation is in terms of undirected graphs, although the procedure works also for directed graphs.

Suppose we are to determine a shortest path from vertex a to vertex z in G. In our procedure, a shortest path from a to some other vertex is determined, then a shortest path from a to still another vertex is determined, and so on. Eventually, our procedure terminates when a shortest path from a to z is determined.

Our procedure is based on the following observations: Let T be a subset of vertices in V with $a \notin T$. Let P denote the subset of vertices $V - T$. A shortest path from a to one of the vertices in T can be determined as follows: For each vertex t in T let $l(t)$ denote the length of a shortest path among all paths from a to t *that do not include any other vertex in T.*‡ [Notice that $l(t)$ is not necessarily the shortest distance from a to t since there might be a shorter path from a to t that includes other vertices in T]. We call $l(t)$ the *index of t with respect to P.* Among all the vertices in T, let t_1 be a vertex that has the smallest index. We claim that the shortest distance between a and t_1 is indeed equal to $l(t_1)$. To prove our claim, let us assume that there is a path from a to t_1 whose length is less than $l(t_1)$. In this case such a path must include one or more of the vertices in $T - \{t_1\}$. Let t_2 be the first vertex in $T - \{t_1\}$ we encounter when we trace this path from a to t_1. It follows that $l(t_2)$ is less than $l(t_1)$, which is a contradiction. Consequently, if, when computing $l(t)$, we recorded the sequence of vertices in the path that yielded $l(t)$ for each t in T, we would also have determined a shortest path from a to t_1. For example, for the graph in Fig. 4.13a, let $T = \{c, d, e, z\}$. We ask the reader to check that $l(c) = 3$, $l(d) = 8$, $l(e) = 6$, $l(z) = \infty$. It follows that the shortest distance from a to c is 3.

We must also, however, find an efficient way to compute $l(t)$ for each t in T. Again, let T be a subset of V and let $P = V - T$. Suppose that for every vertex p in P, there is a shortest path from a to p that includes only the

---

† Strictly speaking, we should use the notation $w[\{i, j\}]$. However, the notation $w(i, j)$ is simpler and unambiguous.

‡ We set $l(t)$ to ∞ if no such path exists.

vertices in P. We assume that for each vertex t in T we have already computed its index with respect to P, $l(t)$. Let x be a vertex in T. Let P' be $P \cup \{x\}$ and T' be $T - \{x\}$. Let $l'(t)$ denote the index of a vertex t in T' with respect to P'. We claim that

$$l'(t) = \min[l(t), l(x) + w(x, t)]\dagger \qquad (4.1)$$

To show this, we observe that there are two possible ways to obtain a shortest path from a to t that does not include any vertex in T'. The first way is to have a path that includes neither a vertex in T' nor the vertex x. In that case, the index of t with respect to P' is $l(t)$. The second way is to have a path consisting of a path from a to x that includes no vertex in T, followed by the edge $\{x, t\}$. In that case, the index of t with respect to P' is $l(x) + w(x, t)$. We should point out that we need not consider the possibility of having a path that goes from a to x, then to some p_1 in P, and then to t. Because if this is the case, while there is a shortest path from a to p_1 that includes x, there is also one that does not include x which can replace it. Consequently, this is reduced to the first possibility considered above. For example, for the graph in Fig. 4.13a, let $P = \{a, b\}$ and $T = \{c, d, e, z\}$. Suppose we have already computed $l(c) = 3$, $l(d) = 8$, $l(e) = 6$, $l(z) = \infty$. Let $P' = \{a, b, c\}$, $T' = \{d, e, z\}$. We have

$$l'(d) = \min(8, 3 + \infty) = 8$$

$$l'(e) = \min(6, 3 + 1) = 4$$

$$l'(z) = \min(\infty, 3 + \infty) = \infty$$

These observations lead to the following procedure for computing the shortest distance from a to any vertex in G:

1. Initially, let $P = \{a\}$ and $T = V - \{a\}$. For every vertex t in T, let $l(t) = w(a, t)$.
2. Let x be the vertex in T that has the smallest index with respect to P.
3. If x is the vertex we wish to reach from a, stop. If not, let $P' = P \cup \{x\}$ and $T' = T - \{x\}$. For every vertex t in T', compute its index with respect to P' according to (4.1). Repeat step 2 using P' as P and T' as T.

We ask the reader to incorporate into this procedure the necessary computation for keeping track of the path from a to x whose length is equal to $l(x)$ for each x in T. In that case, we shall be able to determine a shortest path from a to x as well as the shortest distance.

For example, for the weighted graph in Fig. 4.13a, the successive steps for determining a shortest path between a and z are shown in Fig. 4.13b through Fig. 4.13f, where the vertices in P are encircled. Note that the vertices in T are labeled with their indices. Moreover, a vertex whose index is not infinity is also labeled with the sequence of vertices corresponding to a path from a to that vertex whose length is equal to the index of the vertex. We urge the reader's

---

† We set $w(x, t)$ to infinity if there is no edge joining x and t in the graph.

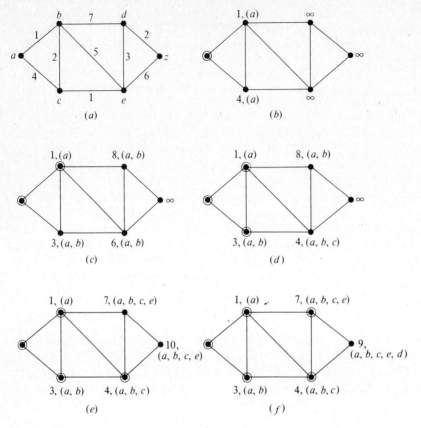

Figure 4.13

effort to understand fully how these labels of sequences of vertices corresponding to the paths from a are constructed and modified. Thus, the minimum distance between a and z is 9, and moreover, the shortest path is (a, b, c, e, d, z).

4.6 EULERIAN PATHS AND CIRCUITS

L. Euler became the father of the theory of graphs when he proved in 1736 that it was not possible to cross each of the seven bridges on the river Pregel in Königsberg, Germany, once and only once in a walking tour. A map of the Königsberg bridges is shown in Fig. 4.14*a*, and it can be represented as the graph shown in Fig. 4.14*b*, where the edges represent the bridges and the vertices represent the islands and the two banks of the river. It is clear that the problem of crossing each of the Königsberg bridges once and only once is equivalent to finding a path in the graph in Fig. 4.14*b* that traverses each of the edges once and only once. As it turns out, instead of using a brute-force trial-

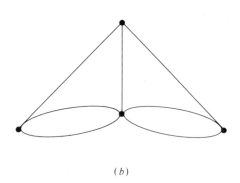

(a)   (b)

Figure 4.14

and-error approach, which is probably what the people at Königsberg did, Euler discovered a very simple criterion for determining whether there is a path in a graph that traverses each of the edges once and only once. We define an *eulerian path* in a graph to be a path that traverses each edge in the graph once and only once. Similarly, we define an *eulerian circuit* in a graph to be a circuit that traverses each edge in the graph once and only once. We restrict our discussion to undirected graphs for the time being. As will be seen later, extension to directed graphs is rather straightforward.

To establish a necessary and sufficient condition for the existence of eulerian paths or circuits in an arbitrary graph, we introduce the notion of the *degree* of a vertex. The degree of a vertex is the number of edges incident with it. We observe first that in any graph there is an even number of vertices of odd degree. Since each edge contributes a count of 1 to the degree of each of the two vertices with which it is incident, the sum of the degrees of the vertices is equal to twice the number of edges in a graph. It follows that there must be an even number of vertices of odd degree.

The existence of eulerian paths or eulerian circuits in a graph is related to the degrees of the vertices. We show now a result due to Euler:

Theorem 4.2 An undirected graph possesses an eulerian path if and only if it is connected and has either zero or two vertices of odd degree.†

PROOF Suppose that the graph possesses an eulerian path. That the graph must be connected is obvious. When the eulerian path is traced, we observe that every time the path meets a vertex, it goes through two edges which are incident with the vertex and have not been traced before. Thus, except for the two vertices at the two ends of the path, the degree of any vertex in the graph must be even. If the two vertices at the two ends of the eulerian path are distinct, they are the only two vertices with odd degree. If they

_____
† We rule out the uninteresting case of graphs containing isolated vertices.

coincide, all vertices have even degree, and the eulerian path becomes an eulerian circuit. Thus, the necessity of the stated condition is proved.

To prove the sufficiency of the stated condition, we construct an eulerian path by starting at one of the two vertices that are of odd degree† and going through the edges of the graph in such a way that no edge will be traced more than once. For a vertex of even degree, whenever the path "enters" the vertex through an edge, it can always "leave" the vertex through another edge that has not been traced before. Therefore, when the construction eventually comes to an end, we must have reached the other vertex of odd degree. If all the edges in the graph were traced this way, clearly, we would have an eulerian path. If not all of the edges in the graph were traced, we shall remove those edges that have been traced and obtain a subgraph formed by the remaining edges. The degrees of the vertices of this subgraph are all even. Moreover, this subgraph must touch the path that we have traced at one or more vertices since the original graph is connected. Starting from one of these vertices, we can again construct a path that passes through the edges. Because the degrees of the vertices are all even, this path must return eventually to the vertex at which it starts. We can combine this path with the path we have constructed to obtain one which starts and ends at the two vertices of odd degree. If necessary, the argument is repeated until we obtain a path that traverses all the edges in the graph. □

Corollary 4.2.1 An undirected graph possesses an eulerian circuit if and only if it is connected and its vertices are all of even degree.

We conclude from Theorem 4.2 and Corollary 4.2.1 that the graph in Fig. 4.15a has an eulerian path but not an eulerian circuit, because the graph is

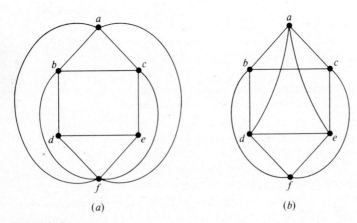

(a)   (b)

Figure 4.15

---

† We start at an arbitrary vertex if there is no vertex of odd degree.

connected and there are exactly two vertices, *d* and *e*, of odd degree. Also, the graph in Fig. 4.15*b* has an eulerian circuit because the graph is connected and all the vertices are of even degree.

We show now some illustrative examples.

Example 4.3 We often encounter the puzzle of determining the possibility of drawing a given figure in a continuous trace with no part of the figure being repeated. For example, the figure in Fig. 4.16*a* can be traced as shown in Fig. 4.16*b*. If we imagine the figure in Fig. 4.16*a* is a graph as shown in Fig. 4.16*c*, the problem of tracing the figure in Fig. 4.16*a* is that of determining the existence of an Euler path in the graph in Fig. 4.16*c*. Since the graph in Fig. 4.16*c* has only two vertices of odd degree, it has an eulerian path. Similarly, the two figures in Fig. 4.16*d* can also be drawn in a continuous trace with no portion of the figures being repeated. ☐

Example 4.4 We want to know whether it is possible to arrange the 28 different dominoes† in a circle so that the adjacent halves of every two adjacent dominoes in the arrangement are the same. We construct a graph

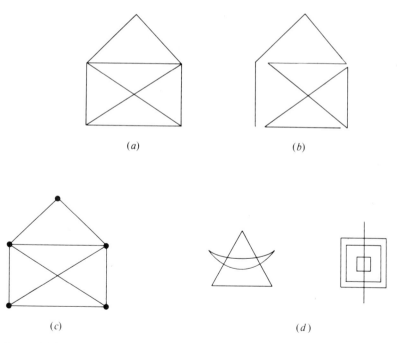

(a) (b)

(c) (d)

Figure 4.16

† See Example 2.15.

with seven vertices corresponding to blank, 1, 2, 3, 4, 5, and 6. There is an edge between every pair of vertices corresponding to a domino with its two halves being the two vertices the edge is incident with. It follows that an eulerian circuit in this graph will correspond to a circular arrangement, as specified in the foregoing. Since the degree of every vertex in this graph is 8, an eulerian circuit indeed exists. $\qquad\qquad\square$

Our results can be extended immediately to directed graphs. In a directed graph, the *incoming degree* of a vertex is the number of edges that are incident *into* it, and the *outgoing degree* of a vertex is the number of edges that are incident *from* it. Similar to the results on undirected graphs, we have the following:

Theorem 4.3 A directed graph possesses an eulerian circuit if and only if it is connected and the incoming degree of every vertex is equal to its outgoing degree. A directed graph possesses an eulerian path if and only if it is connected and the incoming degree of every vertex is equal to its outgoing degree with the possible exception of two vertices. For these two vertices, the incoming degree of one is one larger than its outgoing degree, and the incoming degree of the other is one less than its outgoing degree.

Example 4.5 An interesting example is from an analog-to-digital conversion problem. The surface of a rotating drum is divided into 16 sectors and is shown in Fig. 4.17*a*. The positional information of the drum is to be represented by the digital binary signals, *a*, *b*, *c*, and *d*, as shown in Fig. 4.17*b*, where conducting (lined area) and nonconducting (white area) materials are used to make up the sectors. Depending on the position of the drum, the terminals *a*, *b*, *c*, and *d* will either be connected to ground or be insulated from it. For example, when the position of the drum is that shown in Fig. 4.17*b*, terminals *a*, *c*, and *d* are connected to ground, whereas terminal *b* is not. In order that the 16 different positions of the drum will be distinctly represented by the binary signals at the terminals, the sectors must be constructed in such a way that no two conducting and nonconduct-

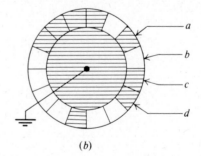

(*a*)　　　　　　　　　　　　　　(*b*)

Figure 4.17

ing patterns of four consecutive sectors are the same. The problem is to determine whether such an arrangement of conducting and nonconducting sectors exists and, if so, to determine what that arrangement is. Letting the binary digit 0 denote a conducting sector and the binary digit 1 denote a nonconducting sector, we can rephrase the problem as follows: Arrange 16 binary digits in a circular array such that the 16 sequences of 4 consecutive digits are all distinct.

The answer to the question of the possibility of such an arrangement is affirmative, and it is actually quite obvious once the right point of view has been taken. We shall construct a directed graph with eight vertices which are labeled with the eight 3-digit binary numbers {000, 001, ..., 111}. From a vertex labeled $\alpha_1\alpha_2\alpha_3$, there is an edge to the vertex labeled $\alpha_2\alpha_30$ and an edge to the vertex labeled $\alpha_2\alpha_31$. The graph so constructed is shown in Fig. 4.18. Moreover, we shall label each edge of the graph with a 4-digit binary number. In particular, the edge from the vertex $\alpha_1\alpha_2\alpha_3$ to the vertex $\alpha_2\alpha_30$ is labeled $\alpha_1\alpha_2\alpha_30$, and the edge from the vertex $\alpha_1\alpha_2\alpha_3$ to the vertex $\alpha_2\alpha_31$ is labeled $\alpha_1\alpha_2\alpha_31$. Since the vertices are labeled with the eight distinct 3-digit binary numbers, the edges will be labeled with the 16 distinct 4-digit binary numbers. In a path of the graph, the labels for any two consecutive edges must be of the form $\alpha_1\alpha_2\alpha_3\alpha_4$ and $\alpha_2\alpha_3\alpha_4\alpha_5$; namely, the three trailing digits of the label of the first edge are identical to the three leading digits of the label of the second edge. Since the 16 edges in the graph are

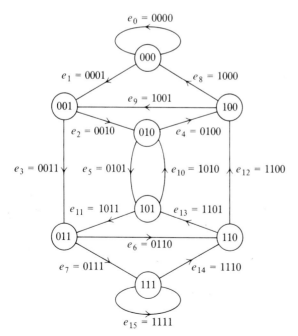

Figure 4.18

labeled with distinct binary numbers, it follows that corresponding to an eulerian circuit of the graph, there is a circular arrangement of 16 binary digits in which all sequences of 4 consecutive digits are distinct. For instance, corresponding to the eulerian circuit (e_0, e_1, e_2, e_5, e_{10}, e_4, e_9, e_3, e_6, e_{13}, e_{11}, e_7, e_{15}, e_{14}, e_{12}, e_8), the sequence of 16 binary digits is 0000101001101111. (The circular arrangement is obtained by closing the two ends of the sequence.) According to our result in the foregoing, the existence of an eulerian circuit in the graph is obvious, because every one of the vertices has an incoming degree equal to 2 and an outgoing degree equal to 2. Moreover, we can find an eulerian circuit in the graph by following the construction procedure suggested in the proof of Theorem 4.2.

Using a similar argument, we can show that it is possible to arrange 2^n binary digits in a circular array such that the 2^n sequences of n consecutive digits in the arrangement are all distinct. To show this, we construct a directed graph with 2^{n-1} vertices which are labeled with the 2^{n-1} $(n-1)$-digit binary numbers. From the vertex $\alpha_1\alpha_2\alpha_3 \cdots \alpha_{n-1}$, there is an edge to the vertex $\alpha_2\alpha_3 \cdots \alpha_{n-1}0$ which is labeled $\alpha_1\alpha_2\alpha_3 \cdots \alpha_{n-1}0$, and an edge to the vertex $\alpha_2\alpha_3 \cdots \alpha_{n-1}1$ which is labeled $\alpha_1\alpha_2\alpha_3 \cdots \alpha_{n-1}1$. Clearly, this graph has an eulerian circuit which corresponds to a circular arrangement of the 2^n binary digits.   □

4.7 HAMILTONIAN PATHS AND CIRCUITS

A problem similar to the determination of an eulerian path or an eulerian circuit is to determine a path or a circuit that passes through each vertex in a graph once and only once. We define a *hamiltonian path* (circuit) to be a path (circuit) that passes through each of the vertices in a graph exactly once. (Sir William Hamilton invented the game "all around the world" in which the player is asked to determine a route along the dodecahedron which will pass through each angular point once and only once.)

As an example, consider the problem of seating a group of people at a round table. If we let the vertices of an undirected graph denote the people and the edges represent the relation that two people are friends, a hamiltonian circuit corresponds to a way of seating them so that everyone has a friend on each side.

Although the problem of determining the existence of hamiltonian paths or circuits has the same flavor as that of determining the existence of eulerian paths or circuits, no simple necessary and sufficient condition is known. To show that a given graph has a hamiltonian path or circuit, we can resort to an explicit construction of such a path or circuit. For example, in the graph in Fig. 4.19 the heavy edges constitute a hamiltonian circuit.

We show now some general results on the existence of hamiltonian paths or circuits. Our first result is a rather general condition which is *sufficient* to

Figure 4.19

guarantee the existence of a hamiltonian path in an undirected graph:

Theorem 4.4 Let G be a linear graph of n vertices. If the sum of the degrees for each pair of vertices in G is $n - 1$ or larger, then there exists a hamiltonian path in G.†

PROOF We show first that G is a connected graph. Suppose G has two or more disconnected components. Let v_1 be a vertex in one component that has n_1 vertices and v_2 be a vertex in another component that has n_2 vertices. Since the degree of v_1 is at most $n_1 - 1$ and the degree of v_2 is at most $n_2 - 1$, the sum of their degrees is at most $n_1 + n_2 - 2$, which is less than $n - 1$, and we have a contradiction.

We show now how a hamiltonian path can be constructed in a step-by-step manner, starting with a path containing a single edge. Let there be a path of $p - 1$ edges, $p < n$, in G which meets the sequence of vertices $(v_1, v_2, \ldots, v_p)$. If either v_1 or v_p is adjacent to a vertex that is not in the path, we can immediately extend the path to include this vertex and obtain a path of p edges. Otherwise, both v_1 and v_p are adjacent only to the vertices that are in the path. We want to show that in that case there is a circuit containing exactly the vertices $v_1, v_2, \ldots, v_p$. If v_1 is adjacent to v_p, then the circuit $(v_1, v_2, \ldots, v_p, v_1)$ will suffice, so let us assume that v_1 is adjacent only to $v_{i_1}, v_{i_2}, \ldots, v_{i_k}$, where $2 \leq i_j \leq p - 1$.‡ If v_p is adjacent to one of $v_{i_1-1}, v_{i_2-1}, \ldots, v_{i_k-1}$, say v_{j-1}, then, as shown in Fig. 4.20, the circuit $(v_1, v_2, v_3, \ldots, v_{j-1}, v_p, v_{p-1}, \ldots, v_j, v_1)$ contains exactly the vertices $v_1, v_2, \ldots, v_p$. If v_p is not adjacent to any one of $v_{i_1-1}, v_{i_2-1}, \ldots, v_{i_k-1}$, then v_p is adjacent to at

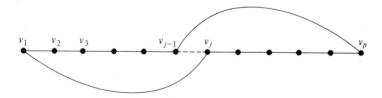

Figure 4.20

† See Prob. 4.25 for a generalization of this result.
‡ As a matter of fact, $i_1 = 2$.

most $p - k - 1$ vertices. Consequently, the sum of the degrees of v_1 and v_p is at most $n - 2$, which is a contradiction.

Now that we have a circuit containing all the vertices $v_1, v_2, \ldots, v_p$, let us pick a vertex v_x that is not in the circuit. Because G is connected, there must be an edge between v_x and v_k for some v_k in $\{v_1, v_2, \ldots, v_p\}$ as shown in Fig. 4.21a. We now have the path $(v_x, v_k, v_{k+1}, \ldots, v_{j-1}, v_p, v_{p-1}, \ldots, v_j, v_1, v_2, \ldots, v_{k-1})$ which contains p edges, as shown in Fig. 4.21b.

We can repeat the foregoing construction until we obtain a path with $n - 1$ edges. $\qquad\square$

It is easy to see that the condition in Theorem 4.4 is a sufficient but not a necessary condition for the existence of a hamiltonian path in a graph. Let G be an n-gon, $n > 5$. It is clear that G has a hamiltonian path although the sum of the degrees of any two vertices is 4.

Example 4.6 Consider the problem of scheduling seven examinations in seven days so that two examinations given by the same instructor are not scheduled on consecutive days. If no instructor gives more than four examinations, show that it is always possible to schedule the examinations. Let G be a graph with seven vertices corresponding to the seven examinations. There is an edge between any two vertices which correspond to two examinations given by different instructors. Since the degree of each vertex is at least 3, the sum of the degrees of any two vertices is at least 6. Consequently, G always contains a hamiltonian path, which corresponds to a suitable schedule for the seven examinations. $\qquad\square$

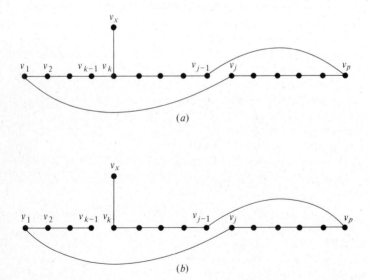

Figure 4.21

Another interesting result is:

Theorem 4.5 There is always a hamiltonian path in a directed complete graph.

PROOF Let there be a path with $p - 1$ edges in a directed complete graph which meets the sequence of vertices $(v_1, v_2, \ldots, v_p)$. Let v_x be a vertex that is not included in this path. If there is an edge (v_x, v_1) in the graph, we can augment the original path by adding the edge (v_x, v_1) to the path so that the vertex v_x will be included in the augmented path. If, on the other hand, there is no edge from v_x to v_1, then there must be an edge (v_1, v_x) in the graph. Suppose that (v_x, v_2) is also an edge in the graph. We can replace the edge (v_1, v_2) in the original path with the two edges (v_1, v_x) and (v_x, v_2) so that the vertex v_x will be included in the augmented path. On the other hand, if there is no edge from v_x to v_2, then there must be an edge (v_2, v_x) in the path and we can repeat the argument. Eventually, if we find that it is not possible to include the vertex v_x in any augmented path by replacing an edge (v_k, v_{k+1}) in the original path with two edges (v_k, v_x) and (v_x, v_{k+1}) with $1 \leq k \leq p - 1$, then we conclude that there must be an edge (v_p, v_x) in the graph. We can, therefore, augment the original path by adding to it the edge (v_p, v_x) so that the vertex v_x will be included in the augmented path. We can repeat the argument until all the vertices in the graph are included in a path. $\square$

Example 4.7 As an illustration of the application of the result in Theorem 4.5, let us consider the problem of ranking players in a round-robin tennis tournament such that player a will be ranked higher than player b if a beats b, or a beats a player who beats b, or a beats a player who beats another player who beats b, and so on. Since the outcomes of the matches can be represented as a directed complete graph, the existence of a hamiltonian path in the graph means that it is always possible to rank the players linearly. (Note that such a ranking is not necessarily unique.) $\square$

Example 4.8 As another illustrative example, we consider the problem of printing and then binding n books. There is one printing machine, and one binding machine. Let p_i and b_i denote the printing time and binding time of book i, respectively. If it is known that for any two books, i and j, either $b_i \geq p_j$ or $b_j \geq p_i$, show that it is possible to specify an order in which the books are printed (and then bound) so that the binding machine will be kept busy until all books are bound, once the first book is printed. (Thus, the total time it takes for completing the whole task is $p_k + \sum_{i=1}^{n} b_i$ for some k.) Let us construct a directed graph of n vertices corresponding to the n books. There is an edge from vertex i to vertex j if and only if $b_i \geq p_j$. We note that this is a directed complete graph, and a hamiltonian path in the directed complete graph will be an ordering of the books satisfying the condition stated above. $\square$

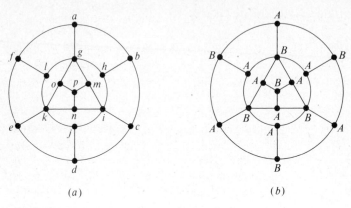

Figure 4.22

There is no general method of solution to the problem of proving the nonexistence of a hamiltonian path or circuit in a graph. We present here, however, an illustrative example:

Example 4.9 We want to show that the graph in Fig. 4.22a has no hamiltonian path. We label the vertex a by A and label all the vertices that are adjacent to it by B. Continuing, we label all the vertices that are adjacent to a B vertex by A and label all the vertices that are adjacent to an A vertex by B, until all vertices are labeled. The labeled graph is shown in Fig. 4.22b. If there is a hamiltonian path in the graph, then it must pass through the A vertices and the B vertices alternately. However, since there are nine A vertices and seven B vertices, the existence of a hamiltonian path is impossible. □

*4.8 THE TRAVELING SALESPERSON PROBLEM

We discuss in this section a natural extension of the problem of finding a hamiltonian circuit in a graph. The *traveling salesperson problem* has long been of great interest. Let $G = (V, E, w)$ be a complete graph of n vertices, where w is a function from E to the set of positive real numbers such that for any three vertices i, j, k in V

$$w(i, j) + w(j, k) \geqq w(i, k)†$$

We shall refer to $w(i, j)$ as the length of edge $\{i, j\}$. The traveling salesperson problem asks for a hamiltonian circuit of minimum length, where, again, the

---

† This condition, known as the *triangle inequality*, is needed to prove the result in (4.2). On the other hand, one might be able to argue that the triangular inequality is satisfied in most real-life cases anyway.

length of a circuit is defined to be the sum of the lengths of the edges in the circuit.

A physical interpretation of the abstract formulation is rather obvious: Consider the graph G as a map of n cities where $w(i, j)$ is the distance between cities i and j. A salesperson wants to have a tour of the n cities which starts and ends at the same city and includes visiting each of the remaining $n - 1$ cities once and only once. Moreover, an itinerary that has a minimum total distance is desired. The traveling salesperson problem turns out to be a difficult one in that we know of no "efficient"† procedure for solving the problem. Recalling our discussion in Sec. 3.7 on the job-scheduling problem, one might wish to look for simple procedures that will give good results to the traveling salesperson problem. To illustrate such a possibility, we present a procedure, known as the *nearest-neighbor method*, which gives reasonably good results for the traveling salesperson problem:

1. Start with an arbitrarily chosen vertex, and find the vertex that is closest to the starting vertex to form an initial path of one edge. We shall augment this path in a vertex-by-vertex manner as described in step 2.
2. Let x denote the latest vertex that was added to the path. Among all vertices that are not in the path, pick the one that is closest to x, and add to the path the edge connecting x and this vertex. Repeat this step until all vertices in G are included in the path.
3. Form a circuit by adding the edge connecting the starting vertex and the last vertex added.

For example, for the graph shown in Fig. 4.23a, if we start from vertex a, a vertex-by-vertex construction of a hamiltonian circuit according to the nearest-neighbor method is shown in Fig. 4.23b, through Fig. 4.23e. Note that the total distance of this circuit is 40 whereas the total distance of a minimum hamiltonian circuit is 37, as shown in Fig. 4.23f.

We shall prove the following result:

Theorem 4.6 For a graph with n vertices, let d be the total distance of a hamiltonian circuit obtained according to the nearest-neighbor method and d_0 be the total distance of a minimum hamiltonian circuit. Then,

$$\frac{d}{d_0} \leq \tfrac{1}{2}\lceil \lg n \rceil + \tfrac{1}{2}‡ \tag{4.2}$$

---

† As the reader would immediately respond, it entirely depends on what we mean by an "efficient procedure." The efficiency of computation procedures is an important subject in the study of theory of computation. It is beyond the scope of our discussion to go into this subject, so let us just say that there is no known procedure that can solve the traveling salesperson problem with more than several hundred cities with the use of a large-scale digital computer for a reasonable amount of time.

‡ We adopt the notation $\lg n$ for $\log_2 n$. We use $\lceil x \rceil$ to denote the smallest integer larger than or equal to x.

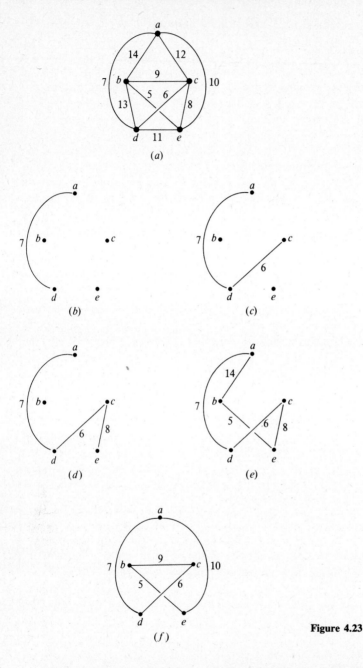

Figure 4.23

PROOF Before proceeding with the proof, let us illustrate the general idea of the proof by considering a specific case. Let D denote the hamiltonian circuit obtained according to the nearest-neighbor method. Let l_1 denote the length of the longest edge in D, l_2 be the length of the next longest edge in D, and, in general, l_i be the length of the ith longest edge in D. Thus,

$$d = \sum_{i=1}^{n} l_i$$

For the sake of illustration, let us assume $n = 14$. *Suppose* we can prove that

$$d_0 \geq 2l_1$$

$$d_0 \geq 2l_2$$

$$d_0 \geq 2(l_3 + l_4) \tag{4.3}$$

$$d_0 \geq 2(l_5 + l_6 + l_7 + l_8)$$

$$d_0 \geq 2(l_9 + l_{10} + l_{11} + l_{12} + l_{13} + l_{14})$$

then we shall have

$$5d_0 \geq 2 \sum_{i=1}^{14} l_i = 2d$$

or

$$\frac{d}{d_0} \leq \frac{5}{2} = \frac{1}{2}\lceil \lg 14\rceil + \frac{1}{2}$$

So that we can have a set of inequalities similar to that in (4.3) for general n, we shall prove

$$d_0 \geq 2l_1 \tag{4.4}$$

$$d_0 \geq 2 \sum_{i=k+1}^{2k} l_i \qquad 1 \leq k \leq \left\lfloor \frac{n}{2} \right\rfloor \tag{4.5}$$

and

$$d_0 \geq 2 \sum_{i=\lceil n/2\rceil+1}^{n} l_i \tag{4.6}$$

Note that if n is even, (4.6) is included in (4.5).

First of all, because of the triangle inequality, (4.4) is obviously true. Suppose the longest edge in D is incident with vertices x and y. Then the triangle inequality implies that the length of any path between x and y is larger than or equal to l_1. Since any hamiltonian circuit of G can be broken up as two paths between x and y, (4.4) follows immediately.

Let a_i denote the vertex *to which* the ith longest edge in D was added in our construction according to the nearest-neighbor method. (For example, according to this convention, the vertices in the graph in Fig. 4.23a are so named in Fig. 4.24a.) For a fixed k, $1 \leq k \leq \lfloor n/2 \rfloor$, let H be the complete subgraph of G containing the vertices a_i, $1 \leq i \leq 2k$. Let T denote the hamiltonian circuit in H that visits the vertices in H in the same circular order as a minimum hamiltonian circuit visits the vertices in G.[†] Let t denote the length of T. (In the illustrative example in Fig. 4.23,

---

† All vertices in G but not in H are ignored in the circular order.

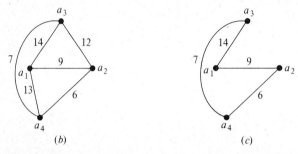

Figure 4.24

consider the case $k = 2$. The subgraph H is that shown in Fig. 4.24b. In the minimum hamiltonian circuit in Fig. 4.23f, the vertices are visited in the circular order a_2, a_1, a_5, a_3, a_4, so the circuit T is that shown in Fig. 4.24c, and $t = 36$.) Because of the triangle inequality, we have

$$d_0 \geqq t \qquad (4.7)$$

Let $\{a_i, a_j\}$ be an edge in T. If in our construction according to the nearest-neighbor method, vertex a_i was added before a_j, we have $w(a_i, a_j) \geqq l_i$. If a_j was added before a_i, we have $w(a_j, a_i) \geqq l_j$, which is the same as $w(a_i, a_j) \geqq l_j$. Thus, we have

$$w(a_i, a_j) \geqq \min (l_i, l_j) \qquad (4.8)$$

Summing (4.8) for all the edges in T, we obtain

$$t \geqq \sum_{(a_i,\, a_j) \in T} \min (l_i, l_j) \qquad (4.9)$$

Note that the smallest possible value of $\min (l_i, l_j)$ in the sum in (4.9) is l_{2k}, the next smallest possible value is l_{2k-1}, and so on. Furthermore, for any $1 \leqq i \leqq 2k$, l_i can appear in the sum at most two times. Since there are $2k$ edges in T, the sum in (4.9) is larger than or equal to two times the sum of the lengths of the k shortest edges. Consequently, we have

$$t \geq \sum_{(a_i, a_j) \in T} \min(l_i, l_j) \geq 2(l_{2k} + l_{2k-1} + \cdots + l_{k+1})$$

$$= 2 \sum_{i=k+1}^{2k} l_i \qquad (4.10)$$

Combining (4.7) and (4.10), we obtain (4.5).

Inequality (4.6) can be proved in a similar manner:[†] Let D_0 denote a minimum hamiltonian circuit. Using the same argument as that of deriving (4.9), we obtain

$$d_0 \geq \sum_{(a_i, a_j) \in D_0} \min(l_i, l_j)$$

Using the same argument as that for deriving (4.10), we obtain

$$d_0 \geq \sum_{(a_i, a_j) \in D_0} \min(l_i, l_j) \geq 2(l_n + l_{n-1} + \cdots + l_{\lceil n/2 \rceil + 1}) + l_{\lceil n/2 \rceil}$$

$$\geq 2(l_n + l_{n-1} + \cdots + l_{\lceil n/2 \rceil + 1})$$

$$= 2 \sum_{i=\lceil n/2 \rceil + 1}^{n} l_i$$

Now, for $k = 1, 2^1, 2^2, \ldots, 2^{\lceil \lg n \rceil - 2}$, (4.5) yields $\lceil \lg n \rceil - 1$ inequalities:

$$d_0 \geq 2l_2$$

$$d_0 \geq 2(l_3 + l_4)$$

$$d_0 \geq 2(l_5 + l_6 + l_7 + l_8)$$

$$\cdots\cdots\cdots\cdots\cdots\cdots$$

$$d_0 \geq 2(l_{2^{\lceil \lg n \rceil - 2} + 1} + l_{2^{\lceil \lg n \rceil - 2} + 2} + \cdots + l_{2^{\lceil \lg n \rceil - 1}})\ddagger$$

Summing these inequalities, we obtain

$$(\lceil \lg n \rceil - 1)d_0 \geq 2 \left(\sum_{i=2}^{2^{\lceil \lg n \rceil - 1}} l_i \right) \qquad (4.11)$$

Since

$$\left\lceil \frac{n}{2} \right\rceil + 1 \leq 2^{\lceil \lg n \rceil - 1} + 1 \qquad \text{for } n > 1$$

(4.6) yields

$$d_0 \geq 2 \sum_{i=2^{\lceil \lg n \rceil - 1} + 1}^{n} l_i \qquad (4.12)$$

Adding (4.4), (4.11), and (4.12), we obtain

$$(\lceil \lg n \rceil + 1)d_0 \geq 2 \sum_{i=1}^{n} l_i = 2d$$

or

$$\frac{d}{d_0} \leq \tfrac{1}{2} \lceil \lg n \rceil + \tfrac{1}{2} \qquad \square$$

[†] We only need to consider this case when n is odd.
[‡] Note that $2^{\lceil \lg n \rceil - 1}$ is the largest power of 2 that is smaller than n.

*4.9 FACTORS OF A GRAPH

The notion of a hamiltonian circuit of a graph can be extended to that of a 2-factor of a graph. We first present, however, a more general definition: A *k-factor* of a graph is defined to be a spanning subgraph of the graph with the degree of each of its vertices being k. For example, for the graph in Fig. 4.25a, Fig. 4.25b shows a 1-factor and Fig. 4.25c shows a 2-factor of the graph. Note that a graph might have many different k-factors, or it might not have any k-factor at all for some k. For example, Fig. 4.26a and b shows two graphs that have no 1-factor and Fig. 4.26c shows a graph that has no 2-factor.

Although we can show that the graph in Fig. 4.26a has no 1-factor and the graph in Fig. 4.26c has no 2-factor by exhaustion, it probably is not obvious that the graph in Fig. 4.26b does not have a 1-factor. We show now an interesting proof of this result. (For an alternative proof, see Prob. 4.32.) Suppose we are given a chessboard with its two diagonally opposite corners truncated, as shown in Fig. 4.27. Suppose we have some dominoes, each of which covers exactly two of the squares in the chessboard. We want to know whether it is possible to cover the truncated chessboard with 31 dominoes. We note that this is exactly the problem of determining whether there is a 1-factor in the graph in Fig. 4.26b. We recall that the squares of a standard chessboard are alternately colored black and white. Since the two truncated squares are both white, as shown in Fig. 4.27, there are 30 white squares and 32 black squares in the truncated chessboard. Because a domino always covers a black

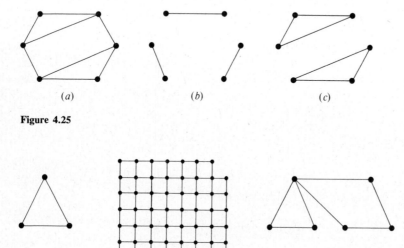

(a)   (b)   (c)

Figure 4.25

(a)   (b)   (c)

Figure 4.26

Figure 4.27

square and a white square no matter how it is placed, the impossibility of covering the chessboard follows immediately.

Consider the problem of scheduling meetings for a number of committees. There are two conference rooms, and it is possible to schedule two meetings concurrently. However, because of overlapping memberships of the committees, certain pairs of meetings cannot be scheduled in the same time slot. Let $G = (V, E)$ be a graph, where V is the set of meetings and an edge $\{a, b\}$ in E means that the two meetings a and b can be scheduled in the same time slot. As the reader will note, a 1-factor of G will be an acceptable pairing.

As was mentioned earlier, the notion of a 2-factor is a generalization of that of a hamiltonian circuit, since, according to Euler's theorem, a 2-factor is a set of vertex-disjoint circuits the union of which contains all the vertices of the graph. As an illustrative example, let us present the solution of a puzzle commonly known as "instant insanity." We are given four cubes, each of their faces being colored with one of the four colors blue, green, red, and yellow, as shown in Fig. 4.28a, where the cubes are numbered 1, 2, 3, 4, and the colors of the faces are abbreviated as b, g, r, y. We are asked to stack the four cubes in a column so that the four colors will show in each of the four sides of the column, as illustrated in Fig. 4.28b. Indeed, the name of this puzzle derives from the possibility of a player becoming insane when trying to sort out the large number of different ways to stack the cubes. We first note that the four cubes can be represented by the graph in Fig. 4.29a, where the four vertices correspond to the four colors. Corresponding to each cube there are three edges in the graph specifying the colors of the three pairs of opposite faces. One should note at this point that, as far as this problem is concerned, the only information we need on each cube is the colors of the three pairs of opposite faces. Indeed, a lesson to learn from this example is the importance of seeing the essentials and not being confused by irrelevant information when trying to solve a problem. If we look at a solution of this problem, we note that the colors of the cubes on two opposite sides of the column can be described by a 2-factor of the graph in Fig. 4.29a such that the 2-factor contains four edges which are labeled 1, 2, 3, 4, respectively. Figure 4.29b shows a 2-factor that satisfies these conditions. Once such an observation is made, the puzzle can be rephrased as that of looking for two edge-disjoint 2-factors of the graph so that the four edges in each 2-factor are labeled distinctly with 1, 2, 3, 4. The 2-factors in Fig. 4.29b

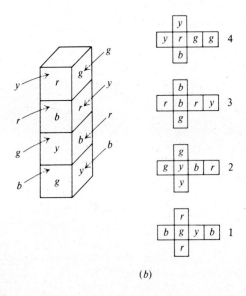

Figure 4.28

and c correspond to the solution in Fig. 4.28b. After extracting the two 2-factors from the graph in Fig. 4.29a, a pleasant surprise is to discover that the edges remaining as shown in Fig. 4.29d also form a 2-factor whose edges are labeled distinctly with 1, 2, 3, 4. What this means is that our solution may contain a bonus feature, namely, the cubes can be oriented with the four "bottoms" colored distinctly and the four "tops" colored distinctly at the same time.

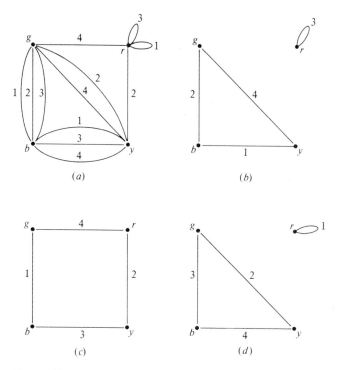

Figure 4.29

*4.10 PLANAR GRAPHS

We study now a class of graphs called *planar graphs*. Not only is this a class of graphs we encounter very frequently, it also has many extremely interesting properties, some of which we shall discuss here.

A graph is said to be *planar* if it can be drawn on a plane in such a way that no edges cross one another, except, of course, at common vertices. Figure 4.30a shows a planar graph. Notice that the graph in Fig. 4.30b is also planar because it can be redrawn as that shown in Fig. 4.30c. Figure 4.30d shows a nonplanar graph. As a matter of fact, the graph in Fig. 4.30d corresponds to the well-known problem of determining whether it is possible to connect three houses *a*, *b*, and *c* to three utilities *d*, *e*, and *f* in such a way that no two connecting pipelines cross one another. Experience shows that this cannot be done. In other words, the graph representing the pipeline connections is indeed a nonplanar graph. (The reader may feel a little bit uncomfortable when we say that a graph is nonplanar simply because, after a certain number of attempts, we find that we cannot draw the graph on a plane so that there will be no crossing edges. This issue will be settled rigorously later.)

Suppose we draw a planar graph on the plane and take a sharp knife to cut along the edges; then the plane will be divided into pieces which are called the

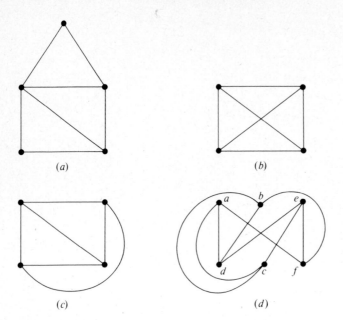

Figure 4.30

regions of the graph. To be more formal, a region of a planar graph is defined to be an area of the plane that is bounded by edges and is not further divided into subareas. For example, the graph in Fig. 4.31*a* has five regions, as shown in Fig. 4.31*b*. Notice that cutting along the edge *a* does not further divide region 1, and cutting along the edges *b*, *c*, and *d* does not further divide region 5. A region is said to be a *finite* region if its area is finite, and is said to be an *infinite* region if its area is infinite. Clearly, a planar graph has exactly one infinite region.

We have the following result:

Theorem 4.7 For any connected planar graph,

$$v - e + r = 2 \tag{4.13}$$

where v, e, and r are the number of vertices, edges, and regions of the graph, respectively.

PROOF The proof proceeds by induction on the number of edges. As the basis of induction, we observe that for the two graphs with a single edge shown in Fig. 4.32, (4.13) is satisfied. As the induction step, we assume that (4.13) is satisfied in all graphs with $n - 1$ edges. Let G be a connected graph with n edges. If G has a vertex of degree 1, the removal of this vertex together with the edge incident with it will yield a connected graph G', as illustrated in Fig. 4.33*a*. Since (4.13) is satisfied in G', it is also satisfied in G,

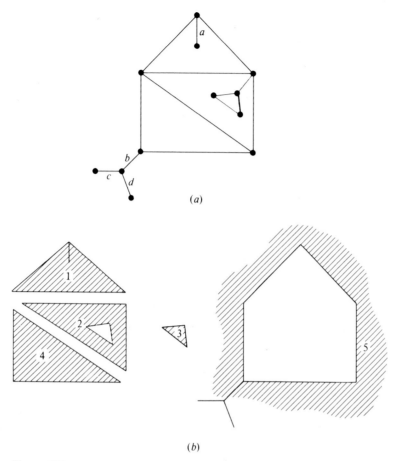

(a)

(b)

Figure 4.31

because putting the removed edge and vertex back into G will increase the count of vertices by 1 and the count of edges by 1, but will not change the count of regions. If G has no vertex of degree 1, the removal of any edge in the boundary of a finite region† will yield a connected graph G', as illustrated in Fig. 4.33b. Since (4.13) is satisfied in G', it is also satisfied in G, because putting the removed edge back into G' will increase the count of edges by

Figure 4.32

---

† It is not difficult to see that, if G has no finite region, G must contain a vertex of degree 1.

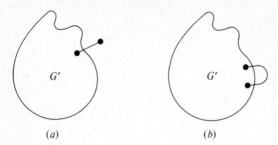

(a)   (b)   **Figure 4.33**

one and the count of regions by one, but will not change the count of vertices.†   □

Equation (4.13) is known as *Euler's formula* for planar graphs. It is truly remarkable that, with no exception, all connected planar graphs must satisfy such a simple formula. We show now some applications of Euler's formula:

We want to show that in any connected linear planar graph that has no loops and has two or more edges,

$$\mathbf{e} \leqq 3\mathbf{v} - 6$$

Let us count the edges in the boundary of a region and then compute the total count for all the regions.‡ Because the graph is linear, each region is bounded by three or more edges. Therefore, the total count is larger than or equal to **3r**. On the other hand, since an edge is in the boundaries of at most two regions, the total count is less than or equal to **2e**. Thus,

$$2\mathbf{e} \geqq 3\mathbf{r}$$

or

$$\frac{2\mathbf{e}}{3} \geqq \mathbf{r}$$

According to Euler's formula, we have

$$\mathbf{v} - \mathbf{e} + \frac{2\mathbf{e}}{3} \geqq 2 \qquad\qquad (4.14)$$

or

$$3\mathbf{v} - 6 \geqq \mathbf{e} \qquad\qquad (4.15)$$

We show now that the graph in Fig. 4.34 is not a planar graph. Since for this graph $\mathbf{v} = 5$ and $\mathbf{e} = 10$, the inequality in (4.15) is not satisfied.

We shall show that the graph in Fig. 4.35 is not a planar graph either. If the graph were planar, every region would be bounded by four or more edges.

---

† (4.13) is not satisfied by disconnected planar graphs. Why is our proof not valid for disconnected planar graphs?

‡ Edges such as *a, b, c, d* in Fig. 4.31*a* are considered to be not in the boundary of any region.

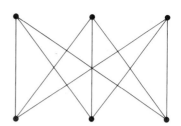

Figure 4.34   Figure 4.35

Thus, the inequality in (4.14) can be tightened as

$$v - e + \frac{e}{2} \geq 2$$

or

$$2v - 4 \geq e \tag{4.16}$$

For the graph in Fig. 4.35, $v = 6$ and $e = 9$. Consequently, (4.16) is contradicted.

Although Euler's formula can sometimes be applied to assert that a given graph is nonplanar, the argument in such applications of the formula can become involved and tricky for graphs containing even a moderate number of vertices and edges. Moreover, except by actually mapping a graph on the plane, we have seen no way of asserting that a given graph is planar. We shall state a theorem due to Kuratowski which enables us to determine the planarity of a graph unequivocally.

The planarity of a graph is clearly not affected if an edge is divided into two edges by the insertion of a new vertex of degree 2, as illustrated in Fig. 4.36a, or if two edges that are incident with a vertex of degree 2 are combined as a single edge by the removal of that vertex, as illustrated in Fig. 4.36b. This suggests the following definition: Two graphs G_1 and G_2 are said to be *isomorphic to within vertices of degree 2* if they are isomorphic or if they can be transformed into isomorphic graphs by repeated insertions and/or removals of vertices of degree 2, as illustrated in Fig. 4.36a and b. For example, the two graphs in Fig. 4.36c are isomorphic to within vertices of degree 2.

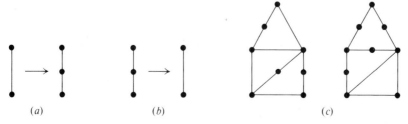

(a)   (b)   (c)

Figure 4.36

Theorem 4.8 (Kuratowski) A graph is planar if and only if it does not contain any subgraph which is isomorphic, to within vertices of degree 2, to either the graph in Fig. 4.34 or the graph in Fig. 4.35.

The graphs in Figs. 4.34 and 4.35 are also called *Kuratowski graphs*. We note that Kuratowski's theorem is indeed a very strong characterization result in the sense that, although there are numerous nonplanar graphs, they all must contain a subgraph that is isomorphic to within vertices of degree 2 to one of the Kuratowski graphs. The proof of this theorem is lengthy although it is elementary. We shall not include the proof here, which can be found in Berge[1] or Liu[8].

4.11 REMARKS AND REFERENCES

Berge[1], Harary[5], and Wilson[11] are useful general references on the theory of graphs. See also Busacker and Saaty[2] and Liu[8] for examples of application of the theory of graphs. See Even[4], Lawler[7], and Reingold, Nievergelt, and Deo[9] for discussion on graph algorithms. Hennie[6] contains a careful and complete treatment of the finite-state model of logical machines. Theorem 4.6 is due to Rosenkrantz, Stearns, and Lewis[10].

[1] Berge, C.: "The Theory of Graphs and Its Applications," John Wiley & Sons, Inc., New York, 1962.
[2] Busacker, R. G., and T. L. Saaty: "Finite Graphs and Networks: An Introduction with Applications," McGraw-Hill Book Company, New York, 1965.
[3] Dijkstra, E. W.: A Note on Two Problems in Connexion with Graphs, Numerische Mathematik, 1: 269–271 (1959).
[4] Even, S.: "Algorithmic Combinatorics," The Macmillan Company, New York, 1973.
[5] Harary, F.: "Graph Theory," Addison-Wesley Publishing Company, Reading, Mass., 1969.
[6] Hennie, F. C.: "Finite-State Models for Logical Machines," John Wiley & Sons, Inc., New York, 1968.
[7] Lawler, E.: "Combinatorial Optimization," Holt, Rinehart and Winston, New York, 1976.
[8] Liu, C. L.: "Introduction to Combinatorial Mathematics," McGraw-Hill Book Company, New York, 1968.
[9] Reingold, E. M., J. Nievergelt, and N. Deo: "Combinatorial Algorithms: Theory and Practice," Prentice-Hall, Inc., Englewood Cliffs, N.J., 1977.
[10] Rosenkrantz, D. J., R. E. Stearns, and P. M. Lewis: Approximate Algorithms for the Travelling Salesman Problem, *Proc. 15th IEEE Symposium on Switching and Automata Theory*: **33–42** (1974).
[11] Wilson, R. J.: "Introduction to Graph Theory," Academic Press, New York, 1972.

PROBLEMS

4.1 A man is supposed to bring a dog, a sheep, and a bag of cabbage across a river on a rowboat. The boat is very small, and he can carry only one of these items on the boat at a time. Furthermore, he cannot leave the dog alone with the sheep nor the sheep alone with the cabbage. In order to determine how he should proceed, we shall describe the possible steps he can take by an undirected graph. Let the vertices of a graph represent all the allowable configurations. For example, the man, the dog, the sheep, and the cabbage are all on one side of the river is the initial configuration; the man, the dog, the sheep, and the cabbage are all on the other side of the river is the final configuration. The man, the dog, and the sheep are on one side of the river while the cabbage is on the other side of the river is one allowable intermediate configuration. Let there be an edge between two vertices if the man can make a trip across the river so that the configuration corresponding to one of the vertices can be transformed into that corresponding to the other vertex, and conversely. Construct the graph and determine all possible ways for the man to transport the items across the river.

4.2 (*a*) Three married couples on a journey come to a river where they find a boat which cannot carry more than two persons at a time. The crossing of the river is complicated by the fact that the husbands are all very jealous and will not permit their wives to be left without them in a company where there are other men present. Construct a graph to show how the transfer can be made.

(*b*) Prove that the puzzle in part (*a*) cannot be solved if there are four couples.

(*c*) Prove that the puzzle in part (*a*) can be solved if there are four couples and the boat holds three persons.

4.3 An electronic circuit is built to recognize sequences of 0s and 1s. In particular it shall accept sequences of the form 010*10, where 0* means any number (including none) of 0s. For example, 0110, 01010, and 010000010 are all acceptable sequences. Construct a directed graph in which every vertex has two outgoing edges labeled 0 and 1, and in which there are two vertices v_i (the initial vertex) and v_f (the final vertex) such that every path from v_i to v_f is a sequence of the form 010*10. (For each sequence the circuit will start at v_i and trace the edges according to the 0s and 1s in the sequence. The circuit will accept the sequence if the path terminates at v_f.)

4.4 Repeat Prob. 4.3 for sequences of the form 01*(10)*10*, where (10)* means any number (including none) of 10 patterns, using each of the following modifications:

(*a*) For each vertex there is no restriction on the number of outgoing edges labeled 0 or 1. The sequence of labels on every path from v_i to v_f is a sequence of the form described above. Furthermore, for every sequence of the form described above there is a corresponding path from v_i to v_f.

(*b*) There may be several final vertices, $v_{f_1}, v_{f_2}, \ldots$. The sequence of labels on every path from v_i to some final v_{f_j} is a sequence of the form described above. Furthermore, for every sequence of the form described above there is a corresponding path from v_i to some final vertex v_{f_j}.

4.5 The head of the Department of Computer Science wanted to call six committee meetings on a certain day. After sending out notices on the starting times of the meetings, as shown in Fig. 4P.1*a*, his secretary, Mrs. Mason, discovered that the agendas of these meetings had been revised. She subsequently made a new estimation on the lengths of these meetings, as shown in Fig. 4P.1*b*. Since these meetings must be held in the order they were originally scheduled, changes in the starting times of some of these meetings became necessary. Hoping to make as few changes as possible, Mrs. Mason constructed the following undirected graph: Let $v_1, v_2, \ldots, v_6$ be six vertices

Meeting	Starting time	Meeting	Estimated length of meeting
1st	8:00 a.m.	1st	90 min.
2nd	9:00 a.m.	2nd	60 min.
3rd	11:00 a.m.	3rd	120 min.
4th	2:00 p.m.	4th	120 min.
5th	3:30 p.m.	5th	90 min.
6th	4:00 p.m.	6th	30 min.

(*a*)   (*b*)

Figure 4P.1

representing the six meetings. Let there be an edge between v_i and v_j for $i < j$ if the sum of the lengths of the ith, the $(i + 1)$st, the $(i + 2)$nd, ..., and the $(j - 1)$st meetings is less than or equal to the difference between the originally chosen starting times of the jth and the ith meeting.

(a) What is the physical significance of an edge between v_i and v_j? of a complete subgraph containing $v_{i_1}, v_{i_2}, ..., v_{i_k}$? of a largest possible complete subgraph?

(b) Suppose in rescheduling the meetings, Mrs. Mason wanted to make sure that the first meeting cannot begin prior to 8:00 a.m., and the last meeting cannot run beyond 5:00 p.m. How can she make use of a computer program that determines a largest possible complete subgraph in a given graph? Determine a new set of starting times for the meetings with as few changes from the original ones as possible.

4.6 n cities are connected by a network of k highways. (A highway is defined to be a road between two cities that does not go through any intermediate cities.) Show that if $k > \frac{1}{2}(n - 1)(n - 2)$, then one can always travel between any two cities through connecting highways.

4.7 An ordered n-tuple $(d_1, d_2, ..., d_n)$ of nonnegative integers is said to be *graphical* if there exists a linear graph with no self-loops that has n vertices with the degrees of the vertices being $d_1, d_2, ..., d_n$.

(a) Show that $(4, 3, 2, 2, 1)$ is graphical.

(b) Show that $(3, 3, 3, 1)$ is not graphical.

(c) Without loss of generality, suppose $d_1 \geq d_2 \geq d_3 \geq \cdots \geq d_n$. Show that $(d_1, d_2, ..., d_n)$ is graphical if and only if $(d_2 - 1, d_3 - 1, ..., d_{d_1} - 1, d_{d_1+1} - 1, d_{d_1+2}, ..., d_n)$ is graphical.

(d) Use the result in part (c) to determine whether $(5, 5, 3, 3, 2, 2, 2)$ is graphical.

4.8 (a) Show that the sum of the in-degrees over all vertices is equal to the sum of the out-degrees over all vertices in any directed graph.

(b) Show that the sum of the squares of the in-degrees over all vertices is equal to the sum of the squares of the out-degrees over all vertices in any directed complete graph.

4.9 A graph is said to be *self-complementary* if it is isomorphic to its complement.

(a) Show a self-complementary graph with four vertices.

(b) Show a self-complementary graph with five vertices.

(c) Is there a self-complementary graph with three vertices? Six vertices?

(d) Show that a self-complementary graph must have either $4k$ or $4k + 1$ vertices.

4.10 A set of vertices in an undirected graph is said to be a *dominating set* if every vertex not in the set is adjacent to one or more vertices in the set. A *minimal dominating set* is a dominating set such that no proper subset of it is also a dominating set.

(a) For the graph in Fig. 4P.2 find two minimal dominating sets of different sizes.

(b) Let the vertices of a graph represent cities and the edges represent communication links between cities. Give a physical interpretation of the notion of a dominating set in this case.

(c) Let the 64 squares of a chessboard be represented by 64 vertices. Let there be an edge between two vertices if their corresponding squares are in the same row, same column, or same (backward or forward) diagonal. It is known that five queens can be placed on the chessboard so that they will dominate all 64 squares. Moreover, five is the minimum number of queens that is needed. Restate this statement in graph theoretic terms.

Figure 4P.2

4.11 A set of vertices in an undirected graph is said to be an *independent set* if no two vertices in it are adjacent. A *maximal independent set* is an independent set which will no longer be one when any vertex is added to the set.

(*a*) For the graph in Fig. 4P.2 find two maximal independent sets of different sizes.

(*b*) How can the problem of placing eight queens on a chessboard so that no one captures another be stated in graph theoretic terms?

4.12 (*a*) The edges of a K_6 are to be painted either red or blue. Show that for any arbitrary way of painting the edges there is either a red K_3 (a K_3 with all its edges painted red) or a blue K_3.

(*b*) Use the result in part (*a*) to show that among a group of six people there are either three who are mutual friends or three who are strangers to each other.

(*c*) The edges of a K_n are painted either red or blue in some arbitrary way. Show that if there are six or more red edges incident with one vertex, then there is either a red K_4 or a blue K_3. Show that if there are four or more blue edges incident with one vertex, then there is either a red K_4 or a blue K_3.

(*d*) Show that for any arbitrary way of painting the edges of K_9 either red or blue there is either a red K_4 or a blue K_3.

4.13 By *properly coloring* a graph we mean to paint the vertices of the graph with one or more distinct colors in such a way that no two adjacent vertices are painted with the same color.

(*a*) What is the minimum number of colors that is needed to properly color the graph in Fig. 4P.3*a*? The graph in Fig. 4P.3*b*? The graph in Fig. 4P.3*c*?

(*b*) Let G be a linear graph with no self-loops. Let the vertices of G represent examinations to be given during the final examination period. Let the edges of G represent constraints such that an edge between two vertices means the corresponding examinations cannot be scheduled in the same time slot. What interpretation can be given to a way of properly coloring G? to the minimum number of colors needed to properly color G?

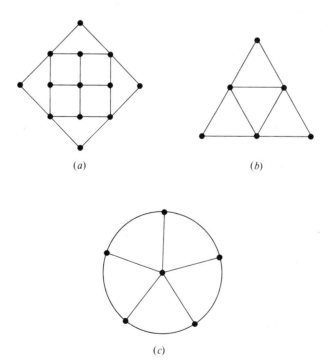

(*a*)

(*b*)

(*c*)

Figure 4P.3

4.14 (*a*) Let $G = (V, E)$ be a linear directed graph where V represents a set of people and E represents a parent-child relationship such that an edge (a, b) in E means a is a parent of b. G is called a *genetic graph*. We note that G has the following properties:

1. The incoming degree of every vertex is at most 2.
2. There is no circuit in G.
3. The vertices of G can be colored with two colors so that for any two edges (a, c) and (b, c) in E, a and b are colored with different colors.

What genetic interpretation can we give to these conditions?

(*b*) Let $\hat{G} = (V, \hat{E})$ be an undirected graph such that there is an edge $\{a, b\}$ in $\hat{E}$ if and only if there are two edges (a, c) and (b, c) in E for some c. Show that condition (3) can be satisfied if and only if $\hat{G}$ can be properly colored with two colors. (See Prob. 4.13 for the definition of proper coloring.)

(*c*) Show that an undirected graph can be properly colored with two colors if and only if it contains no circuit of odd length.

(*d*) We define an alternating circuit in a directed graph as a sequence of edges (v_1, v_2) (v_3, v_2) (v_3, v_4) (v_5, v_4) $\cdots$ (v_{k-1}, v_k) (v_1, v_k) such that reversing the direction of alternating edges in the sequence will yield a (directed) circuit. Show that condition (3) can be satisfied if and only if the length of every alternating circuit in G is divisible by 4.

4.15 The *distance* between two vertices in an undirected graph is defined to be the minimum number of edges in a path between them. The *diameter* of an undirected graph is defined to be the maximum of the distances between all pairs of vertices.

(*a*) What is the diameter of the graph in Fig. 4P.4?

(*b*) Let the vertices of a graph represent computers and the edges represent data communication links between computers. In this case, what is the physical significance of the diameter of the graph?

(*c*) Let d denote the diameter of a graph with n vertices. Let δ denote the maximum of the degrees of the vertices. Show that

$$1 + \delta + \delta(\delta - 1) + \delta(\delta - 1)^2 + \cdots + \delta(\delta - 1)^{d-1} \geqq n$$

[Thus, if we want to design a computer communication network with the constraints: (1) a computer can communicate with any other computer without going through more than $d - 1$ intermediate computers; (2) a computer can communicate directly with at most δ computers, then we cannot have a network containing "too many" computers.]

Figure 4P.4

4.16 Apply the algorithm presented in Sec. 4.5 to determine a shortest path between a and z in the graph in Fig. 4P.5 where the numbers associated with the edges are the distances between vertices.

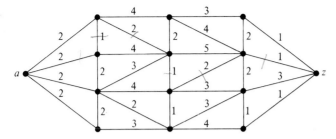

Figure 4P.5

4.17 Apply the algorithm presented in Sec. 4.5 to determine a shortest path between a and z in the graph in Fig. 4P.6 where the numbers associated with the edges are the distances between vertices.

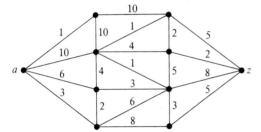

Figure 4P.6

4.18 Let G be a connected graph with no self-loops where the edges represent the streets in a city. A policeman wants to make a round trip to patrol each side of each street exactly once. Furthermore, he wants to patrol the two sides of a street in opposite directions. Show that such a trip can always be designed.

4.19 Among the many rooms in an old mansion, there is a ghost in each room that has an even number of doors. If the mansion has only one entrance, prove that a man entering from outside can always reach a room in which there is no ghost.

4.20 (*a*) The edges in the graph in Fig. 4P.7 can be partitioned into two (edge-disjoint) paths. Show one such partition.

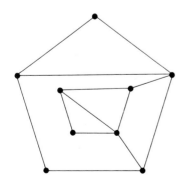

Figure 4P.7

(*b*) Let *G* be a connected graph with *k* vertices of odd degree ($k > 0$). Show that the edges in *G* can be partitioned into $k/2$ (edge-disjoint) paths.

(*c*) Let *G* be a graph with *k* vertices of odd degree ($k > 0$). What is a minimum number of edges that can be added to *G* so that the resultant graph will have an eulerian circuit? Show how this can be done for the graph in Fig. 4P.7. State how this can be done in general.

(*d*) In part (*c*), suppose we are only allowed to add edges that are parallel to existing edges in *G*. What is a minimum number of edges that can be added so that the resultant graph will have an eulerian circuit? Can this always be done? State a necessary and sufficient condition under which this can be done.

4.21 Is it possible to move a knight on an 8×8 chessboard so that it completes every possible move exactly once? A move between two squares of the chessboard is completed when it is made in either direction.

4.22 Find a circular arrangement of nine *a*'s, nine *b*'s, and nine *c*'s such that each of the 27 words of length 3 from the alphabet $\{a, b, c\}$ appears exactly once.

4.23 (*a*) Show a graph that has both an eulerian circuit and a hamiltonian circuit.

(*b*) Show a graph that has an eulerian circuit but has no hamiltonian circuit.

(*c*) Show a graph that has no eulerian circuit but has a hamiltonian circuit.

(*d*) Show a graph that has neither an eulerian circuit nor a hamiltonian circuit.

4.24 Show that the graph in Fig. 4P.8 has no hamiltonian circuit.

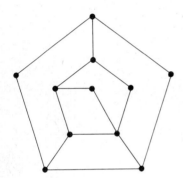

Figure 4P.8

4.25 Let *G* be an undirected linear graph with *n* vertices, $n \geq 3$. Let the vertices of *G* represent *n* people and the edges of *G* represent a friendship relationship among them such that two vertices are connected by an edge if and only if the corresponding persons are friends.

(*a*) What interpretation can be given to the degree of a vertex?

(*b*) What interpretation can be given to the fact that *G* is a connected graph?

(*c*) What interpretation can be given to a subgraph of *G* that is a complete graph with *m* vertices?

(*d*) What interpretation can be given to a 1-factor of *G*?

(*e*) Suppose that between any two people they know all the remaining $n - 2$ people. Show that the *n* people can be stood in line so that everyone stands next to two of his friends, except the two persons at the two ends of the line, each of whom stands next to only one of his friends. (*Hint:* Apply Theorem 4.4.)

(*f*) Extend Theorem 4.4 to show that if the sum of the degrees for each pair of vertices of a linear graph with *n* vertices is *n* or larger, then there exists a hamiltonian circuit in *G*.

(*g*) Use the result in part (*f*) to show that the condition in part (*e*) guarantees that the *n* people can be stood around a circle so that everyone stands next to two of his friends for $n \geq 4$.

4.26 Let *G* be a complete directed graph. A nonempty subset of the vertices of *G* is said to be an *outclassed group* if any edge joining a vertex in the subset and a vertex not in the subset is always directed from the latter to the former. Show that G has a directed circuit containing all the vertices if there is no outclassed group of vertices.

4.27 Eleven students plan to have dinner together for several days. They will be seated at a round table, and the plan calls for each student to have different neighbors at every dinner. For how many days can this be done?

4.28 An *n-cube* is an undirected graph with 2^n vertices which are labeled with the 2^n n-digit binary numbers. There is an edge between two vertices if their binary labels differ exactly at one digit. Show that an n-cube has a hamiltonian circuit for any $n \geq 1$. (A sequential arrangement of the 2^n n-digit binary numbers such that every two adjacent numbers differ in exactly one digit is known as a *Gray code*.)

4.29 An undirected graph is said to be *orientable* if direction can be assigned to each of the edges of the graph so that the resultant directed graph is strongly connected.

(*a*) Show that the graph in Fig. 4P.9 is orientable.

(*b*) Show that any graph with an eulerian circuit is orientable.

(*c*) Show that any graph with a hamiltonian circuit is orientable.

(*d*) Show that a connected undirected graph is orientable if and only if each edge of the graph is contained in at least one circuit.

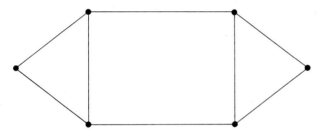

Figure 4P.9

4.30 (*a*) Use the nearest-neighbor method to determine a hamiltonian circuit for the graph in Fig. 4P.10, starting at vertex *a*.

(*b*) Repeat part (*a*), except starting at vertex *d* instead.

(*c*) Determine a minimum hamiltonian circuit for the graph in Fig. 4P.10.

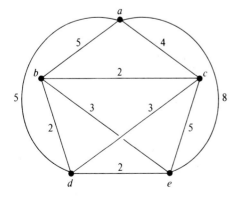

Figure 4P.10

4.31 We present in this problem a procedure that produces a good, but not necessarily best possible, result for the traveling salesperson problem discussed in Sec. 4.8. Again, let $G = (V, E, w)$ be a complete graph of n vertices, where w is a function from E to the set of positive real numbers satisfying the triangle inequality. Let T be a minimum spanning tree of G. (See Sec. 5.7 for the definition of a minimum spanning tree.)

(a) Let V' denote the subset of vertices that have odd degrees in the subgraph T. We claim that there must be an even number of vertices in V'. Why?

(b) Let G' be the complete subgraph of G that contains all the vertices in V'. Let M denote a 1-factor of G' that has minimum weight. (The weight of a 1-factor is defined to be the sum of the weights of the edges in the 1-factor.) Let $T \cup M$ denote the graph obtained by superimposing the graph T and M. (If an edge appears in both T and M, it will appear twice in $T \cup M$.) Show that $T \cup M$ has an eulerian circuit.

(c) Show that the weight of T is less than that of a minimum hamiltonian circuit.

(d) Show that the weight of M is less than or equal to half of that of a minimum hamiltonian circuit.

(e) Show how we can obtain a hamiltonian circuit of G from the eulerian circuit of $T \cup M$ by avoiding to visit vertices that have already been visited such that the weight of the former is less than or equal to that of the latter. Consequently, we conclude that the weight of the hamiltonian circuit obtained is less than 1.5 times that of a minimum hamiltonian circuit.

For example, for the weighted graph in Fig. 4P.11a, the weights of all edges are 1 except the weight of edge $\{f, g\}$ which is 2. Figure 4P.11b shows a minimum spanning tree T. Figure 4P.11c shows a 1-factor of the complete subgraph G' which contains the vertices $\{a, b, c, e, f, g\}$. Figure 4P.11d shows an eulerian circuit of $T \cup M$, and Fig. 4P.11e shows a hamiltonian circuit obtained from the eulerian circuit in Fig. 4P.11d, where the eulerian circuit is traced according to the vertex sequence $(a, b, c, d, e, f, d, a, g, a)$.

4.32 We show in this problem an alternative proof of the result on the impossibility of covering a truncated chessboard with dominoes presented in Sec. 4.9. Let us identify each square of the chessboard by a term $x^i y^j$, $0 \le i \le 7$ and $0 \le j \le 7$, where i is the x coordinate of the square and j is the y coordinate of the square. Thus the sum of the 64 terms corresponding to the 64 squares in a chessboard is

$$(1 + x + x^2 + \cdots + x^7)(1 + y + y^2 + \cdots + y^7)$$

and the sum of the 62 terms corresponding to the 62 squares in the truncated chessboard is

$$(1 + x + x^2 + \cdots + x^7)(1 + y + y^2 + \cdots + y^7) - 1 - x^7 y^7 \qquad (4P.1)$$

Suppose that the truncated chessboard can be covered by 31 dominoes. We note that a domino placed horizontally covers two squares whose corresponding terms are $x^i y^j$ and $x^{i+1} y^j$. Consequently, the sum of the terms corresponding to all squares covered by horizontally placed dominoes is of the form $(1 + x)f(x, y)$. Similarly, the sum of the terms corresponding to all squares covered by vertically placed dominoes is of the form $(1 + y)g(x, y)$. Thus the sum of the 62 terms corresponding to the 62 squares in the truncated chessboard is

$$(1 + x)f(x, y) + (1 + y)g(x, y) \qquad (4P.2)$$

Show that (4P.1) cannot possibly be equal to (4P.2). (*Hint:* Consider the values of (4P.1) and (4P.2) for $x = -1$ and $y = -1$.)

4.33 Show that regardless of where one white and one black square are deleted from a standard 8×8 chessboard, the defective chessboard can always be covered exactly with 31 2×1 dominoes.

4.34 A graph $G = (V, E)$ is called a *bipartite* graph if the set of vertices V can be partitioned into two nonempty subsets X and Y such that there is no edge in E joining two vertices in X or two vertices in Y.

(a) Let $G = (V, E)$ be a bipartite graph. Let X represent a set of workmen and Y represent a set of jobs. Let an edge $\{v_x, v_y\}$ in E denote the fact that workman v_x is qualified to perform job v_y. What interpretation can be given to a 1-factor of G?

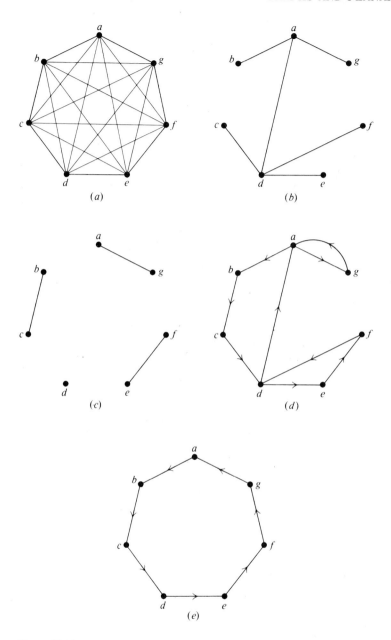

Figure 4P.11

(b) Let X represent a set of committees and Y represent a set of senators. Let an edge $\{v_x, v_y\}$ in E denote the fact that senator v_y is a member of committee v_x. What interpretation can be given to a 1-factor of G?

(c) Let X represent a set of boys and Y represent a set of girls. Let an edge $\{v_x, v_y\}$ in E denote the fact that boy v_x knows girl v_y. What interpretation can be given to a 1-factor of G? a hamiltonian circuit in G?

4.35 (a) Show that a linear planar graph has a vertex of degree 5 or less.

(b) Show that a linear planar graph with less than 30 edges has a vertex of degree 4 or less.

4.36 Show that in a connected planar linear graph with 6 vertices and 12 edges, each of the regions is bounded by 3 edges.

4.37 Let G be a graph with 11 or more vertices. Show that either G or $\bar{G}$ is nonplanar.

4.38 The *thickness* of a graph G is defined to be the minimum number of (edge-disjoint) planar subgraphs into which G can be decomposed. We shall denote the thickness of G by $\theta(G)$.

(a) What is the thickness of a planar graph?

(b) What is $\theta(K_5)$? $\theta(K_8)$?

(c) Show that for any linear graph with no self-loops G,

$$\theta(G) \geq \left\lceil \frac{e}{3v - 6} \right\rceil$$

(d) Show that

$$\theta(K_n) \geq \left\lceil \frac{n + 7}{6} \right\rceil$$

(*Hint:* Show that $\lceil p/q \rceil = \lfloor (p + q - 1)/q \rfloor$ for any positive integers p and q.)

TREES AND CUT-SETS

5.1 TREES

We study in this chapter a class of graphs called *trees*. Consider a group of boxers contending for the title of heavyweight champion of the world. Suppose that each boxer has one chance to challenge the reigning champion, and the loser of any match will be eliminated from contention. Let $G = (V, E)$ be an undirected graph where the vertices in V represent the boxers and the edges in E represent the matches. That is, for x and y in V, $\{x, y\}$ is in E if there was a match between x and y. For example, let $V = \{a, b, c, d, e, f, g, h, i\}$. Suppose a was the reigning champion at the beginning. In the subsequent challenge matches, a beat b, c, and d, and then lost the title to e; e beat f and g, and then lost the title to h; and finally, h lost the title to i. The graph in Fig. 5.1a shows all the matches that took place. As another example, consider four gossipy couples $\{a, A, b, B, c, C, d, D\}$, where a, b, c, d are the husbands and A, B, C, D are, respectively, their wives. Suppose a calls his wife to tell her some gossip, she then calls the other wives to spread the gossip, and each of them, in turn, calls her husband about it. The graph in Fig. 5.1b shows how the gossip was spread where the edges represent the telephone calls. The graphs in these examples share some common properties which we shall identify and study.

We define a *tree* to be a connected (undirected) graph that contains no simple circuit. For example, the graphs in Fig. 5.1a and b are trees. A collection of disjoint trees is called, quite appropriately, a *forest*. A vertex of degree 1 in a tree is called a *leaf* or a *terminal node*, and a vertex of degree larger than 1 is called a *branch node* or an *internal node*. For example, the

(a)   (b)

Figure 5.1

vertices *b, c, d, f, g,* and *i* in Fig. 5.1*a* are leaves and the vertices *a, e,* and *h* are branch nodes. We note first some properties of trees:

1. There is a ~~unique~~ *Simple* path between every two vertices in a tree.
2. The number of vertices is one more than the number of edges in a tree.
3. A tree with two or more vertices has at least two leaves.

Property 1 follows almost directly from the definition of a tree. Because a tree is a connected graph, there is at least one path between every two vertices. However, if there were two or more paths between a pair of vertices, there would be a circuit in the tree. We have thus proved property 1. For the examples in Fig. 5.1*a* and *b* we can check this property immediately.

We shall prove property 2 by induction on the number of vertices in the tree. As the basis of induction, we see that a tree with one vertex contains no edge, and a tree with two vertices contains one edge. Let there be a tree T with v vertices and e edges.† Let $\{a, b\}$ be an edge in T. Suppose we remove the edge $\{a, b\}$ from T. We claim that the remaining edges form a forest of two trees. Let c be a vertex such that the path between a and c in T does not include the edge $\{a, b\}$. Then the path between b and c in T must include the edge $\{a, b\}$ because, otherwise, there is a circuit in T. Thus, after the removal of the edge $\{a, b\}$, there is a path between a and c but no path between b and c. Similarly, let d be a vertex such that the path between a and d in T includes the edge $\{a, b\}$. Then the path between b and d does not include the edge $\{a, b\}$. Thus, after the removal of the edge $\{a, b\}$, there is a path between b and d but no path between a and d. Consequently, the removal of the edge $\{a, b\}$ divides T into two disjoint trees T' and T'', where T' contains a and all the vertices whose paths to a in T do not contain the edge $\{a, b\}$ and T'' contains b and all the vertices whose paths to b in T do not contain the edge $\{a, b\}$. Since both T' and T'' have at most $v - 1$ vertices, it follows from the induction hypothesis that

$$e' = v' - 1$$

$$e'' = v'' - 1$$

where e' and e'' are the numbers of edges and v' and v'' are the numbers of

† We shall use v to denote the number of vertices and e the number of edges in a graph.

vertices in T' and T''. Thus

$$\mathbf{e}' + \mathbf{e}'' = \mathbf{v}' + \mathbf{v}'' - 2$$

Since

$$\mathbf{e} = \mathbf{e}' + \mathbf{e}'' + 1$$

$$\mathbf{v} = \mathbf{v}' + \mathbf{v}''$$

we have

$$\mathbf{e} = \mathbf{v} - 1$$

We note that the tree in Fig. 5.1*a* has nine vertices and eight edges, the tree in Fig. 5.1*b* has eight vertices and seven edges.

Property 3 follows from property 2. We recall that the sum of the degrees of the vertices in any graph is equal to 2e, which is also equal to $2\mathbf{v} - 2$ in a tree. Since a tree with more than one vertex cannot have any isolated vertex, there must be at least two vertices of degree 1 in the tree. Both the examples in Fig. 5.1 have more than two leaves. As an exercise, we ask the reader to characterize all trees that have exactly two leaves. (See Prob. 5.1.)

We show now some results on the characterization of trees. All of these characterizations can be considered as equivalent definitions of trees:

1. A graph in which there is a ~~unique~~ *simple* path between every pair of vertices is a tree.
2. A connected graph with $\mathbf{e} = \mathbf{v} - 1$ is a tree.
3. A graph with $\mathbf{e} = \mathbf{v} - 1$ that has no circuit is a tree.

To show characterization 1, we note first that a graph in which there is a path between every pair of vertices is connected. Moreover, the graph cannot contain a circuit if these paths are unique, since the existence of a circuit implies the existence of two distinct paths between a certain pair of vertices. Thus, we can conclude that a graph in which there is a unique path between every pair of vertices is a tree.

As an example, consider the spies in a spy ring which is organized so that every two spies can communicate with each other either directly or through a unique chain of their colleagues. Let V be the set of spies and E be a set of edges such that the edge $\{a, b\}$ in E means spies a and b can communicate with each other directly. According to characterization 1, $G = (V, E)$ is a tree.

We now show characterization 2. Let G be a connected graph with $\mathbf{e} = \mathbf{v} - 1$. Suppose G contains a simple circuit C.† Let $\mathbf{c}$ denote the number of vertices in C. Clearly, the number of edges in C is equal to $\mathbf{c}$. Since G is connected, every vertex of G that is not in C must be connected to the vertices in C. Now each edge of G that is not in C can connect only one additional vertex to the vertices in C. There are $\mathbf{v} - \mathbf{c}$ vertices that are not in C, so G must contain at least $\mathbf{v} - \mathbf{c}$ edges that are not in C. Thus we must have $\mathbf{e} \geq \mathbf{c} + (\mathbf{v} - \mathbf{c}) = \mathbf{v}$, which is a contradiction. It follows that G does not contain any circuit, and is, therefore, a tree.

† If G contains a circuit, G contains a simple circuit.

Consider an example in which someone sends a batch of chain letters to her friends. Her friends, in turn, send chain letters to their friends who have not received one before, and so on. Let $G = (V, E)$ be a graph where V is the set of people who received a chain letter, together with the originator, and E is a set of edges representing the letters. Since each letter adds one person to the group of people who received chain letters we have $e = v - 1$. Furthermore, G is a connected graph because there is only one originator of the letters. Thus, according to characterization 2, G is a tree.

Let us now show characterization 3. Let G be a graph with $e = v - 1$ that has no circuit. Suppose G is not connected. Let G', G'', ... denote the connected components of G. Since each of G', G'', ... is connected and has no circuits, they are all trees. According to property 2 of trees shown in the foregoing, we have

$$e' = v' - 1$$

$$e'' = v'' - 1$$

$$\cdots\cdots\cdots\cdots$$

where e', e'', ... are the numbers of edges and v', v'', ... are the number of vertices in G', G'', ... We have

$$e' + e'' + \cdots = v' - 1 + v'' - 1 + \cdots \tag{5.1}$$

Since

$$e = e' + e'' + \cdots$$

$$v = v' + v'' + \cdots$$

(5.1) implies

$$e < v - 1$$

which is a contradiction. Thus, G must be connected and is, therefore, a tree.

Consider the example presented earlier of a group of boxers contending for the heavyweight championship. We want to show that the graph describing the challenge matches, G, is always a tree for any group of boxers and for any possible outcomes of the matches. Since each match eliminates exactly one of the boxers from contention, we have $e = v - 1$. Suppose we have a circuit in G. Let $(c_1, c_2, c_3, \ldots, c_k)$ denote the sequence of vertices in the circuit. *Without loss of generality*, we can assume c_2 is the loser in the match between c_1 and c_2. In that case, c_2 must be the winner and c_3 the loser in an earlier match. Similarly, c_3 must be the winner and c_4 must be the loser in a match prior to that between c_1 and c_2. Finally, c_k must be the winner and c_1 must be the loser in a match prior to the match between c_1 and c_2, which clearly is an impossibility. Thus, according to characterization 3, G is a tree.

5.2 ROOTED TREES

A directed graph is said to be a *directed tree* if it becomes a tree when the directions of the edges are ignored. For example, Fig. 5.2*a* shows a directed tree. A directed tree is called a *rooted tree* if there is exactly one vertex whose in-degree is 0 and the in-degrees of all the other vertices are 1. The vertex with in-degree 0 is called the *root* of the rooted tree. For example, Fig. 5.2*b* shows a rooted tree. In a rooted tree, a vertex whose out-degree is 0 is called a *leaf* or a *terminal node*, and a vertex whose out-degree is nonzero is called a *branch node* or an *internal node*. There are many occasions when we encounter structures that can be represented as rooted trees. For example, the organization chart of a corporation in Fig. 5.3 can be represented immediately by a rooted tree.

Let *a* be a branch node in a rooted tree. A vertex *b* is said to be a *son†* of *a* if there is an edge from *a* to *b*. Also, *a* is said to be the *father* of *b*. Two vertices are said to be *brothers* if they are sons of the same vertex. A vertex *c* is said to be a *descendant* of *a* if there is a directed path from *a* to *c*. Also, *a* is said to be an *ancestor* of *c*. These terms indeed remind us that what we commonly call family trees are indeed rooted trees.

Let *a* be a branch node in the tree *T*. By *the subtree with a as the root* we mean the subgraph $T' = (V', E')$ of *T* such that *V'* contains *a* and all of its descendants and *E'* contains the edges in all directed paths emanating from *a*. By a *subtree of a* we mean a subtree that has a son of *a* as root.

When we draw a rooted tree, if we consistently follow the convention of placing the sons of a branch node below it, the arrowheads of the edges may be

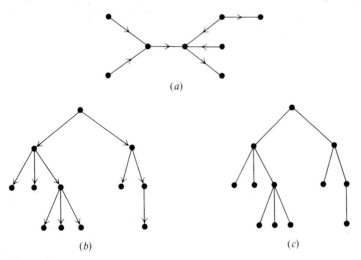

(*a*)

(*b*)   (*c*)

Figure 5.2

† In the literature, the terms "daughter" and "child" were also used. We shall not constantly worry about the question of sex discrimination.

Figure 5.3

omitted, because they will be understood to be pointing downward. Figure 5.2*c* shows an example.

Consider the rooted tree in Fig. 5.4*a* which is the family tree of a man who has two sons, with the older son having no children and the younger son having three. Although the rooted tree in Fig. 5.4*b* is isomorphic to that in Fig. 5.4*a*, it could be the family tree of another man whose older son has three children and whose younger son has none. This example motivates the definition of an *ordered tree*, which allows us to refer to each of the subtrees of a branch node in an unambiguous way. An *ordered tree* is a rooted tree with the edges incident from each branch node labeled with integers 1, 2, ..., *i*, ...† Consequently, the subtrees of a branch node can be referred to as the first, second, ..., and *i*th subtrees of the branch node corresponding to the labels of the edges incident from it. (Note that we do not insist that the labels be consecutive integers. Thus, for example, if the three edges incident from a branch node are labeled 1, 2, 5, we shall say that the node has no third or fourth subtree.) For example, the rooted trees in Fig. 5.4*a* and *b* are labeled as shown in Fig. 5.4*c*. Two ordered trees are said to be isomorphic if there is one-to-one correspondence between their vertices and edges that preserves the incidence relation and if the labels of corresponding edges match. Thus, the two ordered trees in Fig. 5.4*c* are not isomorphic. An ordered tree in which every branch node has at most *m* sons is called an *m-ary tree*.‡ An *m*-ary tree is said to be *regular* if every one of its branch nodes has exactly *m* sons. An important class of *m*-ary trees is *binary trees*. For binary trees, instead of referring to the first subtree or the second subtree of a branch node, we often refer to the *left subtree* or the *right subtree* of the node.

The reader probably has seen how the structure of a sentence can be

---

† Such labels can be implicit when they are obvious from the way the tree was defined or the way the tree was drawn.

‡ Clearly, a corresponding notion can be defined for all rooted trees.

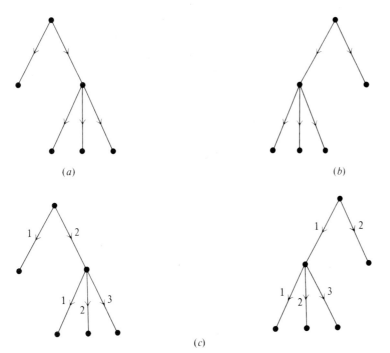

(a)

(b)

(c)

Figure 5.4

Figure 5.5

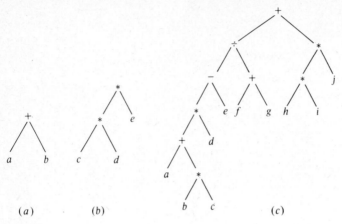

Figure 5.6

diagrammed as an ordered tree. Figure 5.5 shows an example. As another example, we note how an arithmetic expression can be represented as a binary tree. It is clear that simple expressions such as $a + b$ and $c * d * e$ can be represented as shown in Fig. 5.6a and b, where the operands appear in the leaves and the operators appear in the branch nodes of the ordered trees. Note that parentheses are no longer needed when an arithmetic expression is represented as a binary tree. For example, the arithmetic expression $(((a + b * c) * d - e)/(f + g)) + h * i * j$ is represented as shown in Fig. 5.6c.

5.3 PATH LENGTHS IN ROOTED TREES

Consider the problem of determining the number of games played in a single-elimination tennis tournament.† Suppose there are eight players in the tournament. There will be four games to be played in the first round, two games to be played in the second round, and one game to be played in the final round, for a total of seven games. The problem seems to become more complicated when there is an odd number of players, say 11, in the tournament. In this case, there will be five games in the first round (one of the players draws a bye), three games in the second round, one game in the third round (one of the three players remaining after the second round draws a bye), and one game in the final round for a total of 10 games. There is, however, a more direct way to determine the result. If we examine the schedule of a single-elimination tournament such as that shown in Fig. 5.7, we realize immediately that it is exactly a regular binary tree in which the leaves represent the players in the tournament and the branch nodes represent the winners of the matches, or equivalently the matches played in the tournament. Conversely, we see that

---

† A single-elimination tournament is one in which a player will be eliminated after one loss.

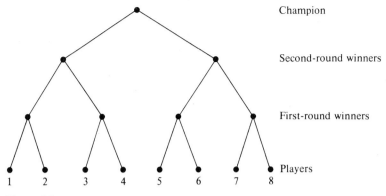

Champion

Second-round winners

First-round winners

Players

Figure 5.7

any regular binary tree can be viewed as the schedule of a single-elimination tournament.† We thus would like to know the relationship between i, the number of branch nodes, and t, the number of leaves, of a regular binary tree. Since in a single-elimination tournament each game played eliminates a player, and at the end of the tournament, all players but the champion are eliminated, the number of games played is one less than the number of players in the tournament. Consequently, we have

$$i = t - 1$$

The result can be extended immediately to the case of regular m-ary trees. Imagine a certain kind of m-player game that has only one winner. In that case, any regular m-ary tree can be viewed as a schedule for a single-elimination tournament. Again, since every game played eliminates $m - 1$ players and only the champion remains at the end of the tournament, we have‡

$$(m - 1)i = t - 1 \tag{5.2}$$

We consider some examples:

Example 5.1 Consider the problem of connecting 19 lamps to a single electric outlet by using extension cords each of which has four outlets. Since any such connection is a quaternary tree with the single outlet connected to the root of the tree, according to (5.2),

$$(4 - 1)i = 19 - 1$$

That is, although there are many ways to connect the lights, six extension cords are always needed. □

---

† By a single-elimination tournament, we *only* mean a tournament in which a player will be eliminated after one loss. We do not insist that every player plays in the first round, for example.

‡ As an alternative proof, we note that mi is the total number of sons of the branch nodes, which is equal to the number of branch nodes and terminal nodes minus one (the root).

Example 5.2 Let us consider a hypothetical computer that has an instruction which computes the sum of three numbers. Suppose we want to find the sum of nine numbers, x_1, x_2, ..., x_9. We realize that any sequence of execution of this instruction to obtain the result is a regular ternary tree with nine leaves. Figure 5.8a and b shows two possible sequences. According to (5.2),

$$(3 - 1)i = 9 - 1$$

It follows that

$$i = 4$$

That is, the addition instruction will always be executed four times. □

There is another aspect of the problem in Example 5.2 that we wish to explore. It is clear that two different ternary trees with nine leaves correspond to different orders in which the nine numbers are added, as illustrated in Fig. 5.8a and b. Although it is always the case that it takes at least four addition operations to compute the sum $x_1 + x_2 + \cdots + x_9$, the possibility of computing some partial sums concurrently (such as in the case of a multiprocessor computing system) makes one sequence superior to another. For example, the sequence of additions shown in Fig. 5.8a can be carried out in two steps, where in each step we compute all partial sums that can be computed concurrently. On the other hand, it takes four steps to carry out the sequence of additions shown in Fig. 5.8b. The *path length* of a vertex in a rooted tree is defined to be the number of edges in the path from the root to the vertex. For example, the path length of vertex x_1 in Fig. 5.8b is 4, and that of vertex x_5 is 3. We define the *height* of a tree to be the maximum of the path lengths in the tree. It is clear that an m-ary tree of height h has at most m^h leaves corresponding to the case shown in Fig. 5.8a. On the other hand, a regular m-ary tree of height h has at least $m + (m - 1)(h - 1)$ leaves corresponding to the case shown in Fig. 5.8b.

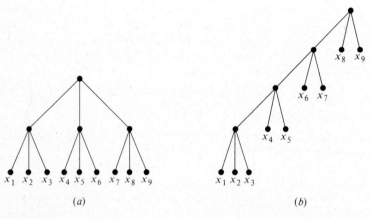

(a)   (b)

Figure 5.8

Let I denote the sum of the path lengths of all the branch nodes and E denote the sum of the path lengths of the leaves in a rooted tree. We shall show that for a regular binary tree $E = I + 2i$, where i is the number of branch nodes. Consider an edge (x, y) in a regular binary tree T. The edge (x, y) is included in the computation of the path lengths of all the branch nodes and leaves of T in the subtree with y as its root. Since there is exactly one more leaf than branch node in the subtree, this edge is counted one more time in the computation of E than in the computation of I. Repeating this argument for all edges in the binary tree, we obtain

$$E = I + \text{number of edges in the binary tree}$$
$$= I + 2i$$

We leave it to the reader to carry out the obvious extension to show that, for a regular m-ary tree,

$$E = (m - 1)I + mi$$

5.4 PREFIX CODES

We present now more examples on the concept of path lengths in rooted trees. Consider a problem in telecommunication in which we want to represent the letters in the English alphabet by sequence of 0s and 1s (or equivalently by sequences of dashes and dots). Since the 26 letters in the alphabet must be represented by distinct sequences of 0s and 1s, they can be represented by sequences of length 5 ($2^4 < 26 < 2^5$). To send a message, we simply transmit a long string of 0s and 1s containing the sequences for the letters in the message. At the receiving end, this string of 0s and 1s will be divided into sequences of length 5 and the corresponding letters can be recognized.

It is well known, however, that the letters in the alphabet are not used with uniform frequencies. For example, the letters e and t are used more frequently than the letters q and z. Consequently, one might wish to represent more frequently used letters with shorter sequences and less frequently used letters with longer sequences so that the overall length of the string will be reduced. An interesting problem arises: When we represent the letters by sequences of various lengths, there is the question of how one at the receiving end can unambiguously divide a long string of 0s and 1s into sequences corresponding to the letters. For example, if we use the sequence 00 to represent the letter e, 01 to represent the letter t, and 0001 to represent the letter w, we will not be able to determine whether the transmitted text was et or w if we receive the string 0001 at the receiving end. A set of sequences is said to be a *prefix code* if no sequence in the set is a prefix of another sequence in the set. For example, the set {000, 001, 01, 10, 11} is a prefix code, whereas the set {1, 00, 01, 000, 0001} is not. We shall show that if we represent the letters in the alphabet by the sequences in a prefix code, it will always be possible to divide a received string into sequences representing the letters in a message unambiguously.

We note first that we can obtain a binary prefix code directly from a binary tree. For a given binary tree, we label the two edges incident from each branch node with 0 and 1. Let us assign to each leaf a sequence of 0s and 1s which is the sequence of labels of the edges in the path from the root to that leaf. For example, Fig. 5.9 shows a binary tree and the sequences assigned to its leaves, which we enclose in boxes for clarity. It is clear that the set of sequences assigned to the leaves in any binary tree is a prefix code.

Conversely, we want to show that corresponding to a prefix code there is a binary tree, with the two edges incident from each of the branch nodes labeled with 0 and 1, such that the sequences of 0s and 1s assigned to the leaves are the sequences in the code.† Let h denote the length of the longest sequence(s) in the prefix code. We construct a *full* regular binary tree of height h ‡ and label the two edges incident from each of the branch nodes with 0 and 1. Let us assign to each vertex a sequence of 0s and 1s which is the sequence of labels of the edges in the path from the root to that vertex. Clearly, each binary sequence of length less than or equal to h is assigned to exactly one vertex. Let us mark all the vertices that are assigned with sequences in the prefix code, and then prune the tree by removing every vertex, together with the edge incident to it, that does not have a marked descendant. Since no marked vertex is an ancestor of another marked vertex, the set of marked vertices will be exactly the set of leaves of the resultant tree. This is, indeed, the tree we are looking for. As an example, for the prefix code {1, 01, 000, 001} we obtain the binary tree in Fig. 5.10*b* by pruning the full regular binary tree of height 3 in Fig. 5.10*a*, where the vertices that are assigned with sequences in the prefix code are boxed.

Finally, we are ready to give a simple argument to show that it is always possible to divide a received string of 0s and 1s into sequences that are in a prefix code. Starting at the root of the binary tree, we shall trace a downward path in the tree according to the digits in the received string. That is, at a branch node, we shall follow the edge labeled with a 0 if we encounter a 0 in the received string, and we shall follow the edge labeled with a 1 if we encounter a 1 in the received string. When the downward path reaches a leaf, we know that

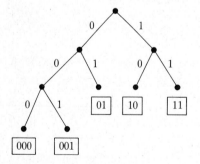

Figure 5.9

---

† For an alternative proof, see Prob. 5.8.

‡ A *full* regular binary tree is a regular binary tree in which all leaves have the same path length (equal to the height of the tree).

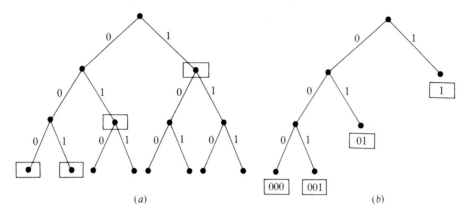

Figure 5.10

a sequence in the prefix code has been detected, and we should return to the root of the tree to start looking for the next sequence. Clearly, this procedure guarantees that there will be no ambiguity in dividing the received string into sequences that are in the prefix code.

Another interesting problem is to be more precise about the idea of using short sequences to represent more frequently used letters. To this end, we need some information on the frequencies of usage of the letters. Indeed, such information is available, and Table 5.1 shows the average number of occurrences of the letters in the English alphabet.† If we use a binary sequence of l_i digits to represent the ith letter in the alphabet, the average length of the binary string that represents an English text of 1000 letters will be $\Sigma_{i=1}^{26} w_i l_i$, where w_i is the average number of occurrences of the ith letter in 1000 letters.

Our discussion leads us to the following problem: Suppose we are given a set of weights $w_1, w_2, \ldots, w_t$. Without loss of generality, let us assume $w_1 \leq w_2 \leq \cdots \leq w_t$. A binary tree that has t leaves with the weights $w_1, w_2, \ldots, w_t$ assigned to the leaves is said to be a *binary tree for the weights* $w_1, w_2, \ldots, w_t$. We define the *weight of a binary tree for the weights* $w_1, w_2, \ldots, w_t$ to be $\Sigma_{i=1}^{t} w_i l(w_i)$, where $l(w_i)$ is the path length of the leaf to which the weight w_i is assigned. The weight of a tree T shall be denoted $W(T)$. A binary tree for the weights $w_1, w_2, \ldots, w_t$ is said to be an optimal tree if its weight is minimum. For example, given the weights 5, 6, 7, 12, Fig. 5.11a shows an optimal tree. (We ask the reader to check that the tree in Fig. 5.11b is inferior.)

There is a very elegant procedure due to D. A. Huffman[5] for constructing an optimal tree for a given set of weights. The key observation that leads to the construction procedure is that we can obtain an optimal tree for the weights $w_1, w_2, \ldots, w_t$, from an optimal tree, T', for the weights $w_1 + w_2, w_3, w_4, \ldots, w_t$.‡ Specifically, we claim that replacing the leaf in T' to which the weight $w_1 + w_2$

---

† See, for example, Zipf[10].

‡ Note that the weights $w_1 + w_2, w_3, w_4, \ldots, w_t$ might no longer be in increasing order.

Table 5.1

Letter	Number of occurrences in 1000 letters
a	82
b	14
c	28
d	38
e	131
f	29
g	20
h	53
i	63
j	1
k	4
l	34
m	25
n	71
o	80
p	20
q	1
r	68
s	61
t	105
u	25
v	9
w	15
x	2
y	20
z	1

is assigned by the subtree shown in Fig. 5.12 will yield an optimal tree for the weights $w_1, w_2, \ldots, w_t$. For example, suppose we want to construct an optimal tree for the weights 3, 4, 5, 6, 12. Since we know the tree in Fig. 5.11a is an optimal tree for the weights 5, 6, 7, 12, we obtain the tree in Fig. 5.13 as an

(a)

(b)

Figure 5.11

w_1 w_2 **Figure 5.12**

optimal tree for the weights 3, 4, 5, 6, 12. To prove our claim, we want to show first that there is an optimal tree for the weights w_1, w_2, ..., w_t in which the leaves to which w_1 and w_2 are assigned are brothers. Suppose we have an optimal tree for the weights w_1, w_2, ..., w_t. Let a be a branch node of largest path length in the tree. Suppose the weights assigned to the sons of a are w_x and w_y. Thus, $l(w_x) \geq l(w_1)$ and $l(w_x) \geq l(w_2)$. On the other hand, since the tree is optimal we must have $l(w_x) \leq l(w_1)$ and $l(w_x) \leq l(w_2)$. [If $l(w_x) > l(w_1)$, interchanging the assignment of w_1 and w_x will yield a tree of smaller weight, which is a contradiction. The same argument applies if $l(w_x) > l(w_2)$.] Consequently, we have $l(w_x) = l(w_y) = l(w_1) = l(w_2)$. Interchanging the assignment of w_x, w_y with that of w_1, w_2, we obtain an optimal tree in which the leaves to which w_1 and w_2 are assigned are brothers.

Let $\hat{T}$ denote an optimal tree for the weights w_1, w_2, ..., w_t in which the leaves to which w_1 and w_2 are assigned are brothers. Replace the subtree in $\hat{T}$ that contains these two leaves and their father by a leaf, and assign to this new leaf the weight $w_1 + w_2$. Let $\hat{T}'$ denote this resultant tree, which is a binary tree for the weights $w_1 + w_2$, w_3, ..., w_t. Clearly,

$$W(\hat{T}) = W(\hat{T}') + w_1 + w_2$$

Let T' be an optimal tree for the weights $w_1 + w_2$, w_3, ..., w_t. Let T be the tree obtained from T' by replacing the leaf in T' to which $w_1 + w_2$ is assigned by the subtree shown in Fig. 5.12. We have

$$W(T) = W(T') + w_1 + w_2$$

If $W(T) > W(\hat{T})$, then $W(T') > W(\hat{T}')$, which is a contradiction because T' is an optimal tree for the weights $w_1 + w_2$, w_3, ..., w_t. Consequently, T is an optimal tree for the weights w_1, w_2, ..., w_t.

According to our observation, the problem of constructing an optimal tree for t weights can be reduced to that of constructing one for $t - 1$ weights,

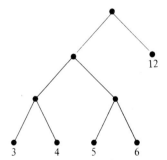

Figure 5.13

which can be reduced to that of constructing one for $t - 2$ weights, and so on. Since the problem of constructing an optimal tree for two weights is a trivial one, the problem of constructing an optimal tree for t weights is solved. For example, for the weights 3, 4, 5, 6, 12, Fig. 5.14*a*, *b*, *c*, *d* shows the step-by-step construction according to Huffman's procedure.

(*a*)

(*b*)

(*c*)

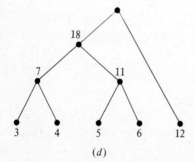

(*d*)   **Figure 5.14**

5.5 BINARY SEARCH TREES

Suppose a friend of ours has picked a number between 1 and 99 and we want to determine what this number is by asking questions of the form, "Is the number 37?" Our friend will respond by telling us the number is indeed 37, or the number is smaller than 37, or the number is larger than 37. Although we can determine the number our friend has picked by asking the sequence of questions, "Is the number 1?", "Is the number 2?", "Is the number 3?", ..., "Is the number 99?", most of us would employ the following strategy. We begin with the question, "Is the number 50?" If we are told that the number is indeed 50, we are done. If we are told that the number is less than 50, we have narrowed down the range containing the number to 1 through 49. If we are told that the number is larger than 50, we have narrowed down the range containing the number to 51 through 99. Accordingly, if the number lies between 1 and 49, our next question is, "Is the number 25?"; and if the number lies between 51 and 99, our next question is, "Is the number 75?" It is abundantly clear that our strategy is to ask a sequence of questions so that our friend's answer to each question will enable us either to determine the number he or she has picked or to reduce the range containing the number to half of what it was before. We trust that the reader would agree that this is a good strategy.

We present two more examples before we go on. Suppose that the records of the employees of a company are arranged according to social security numbers. An important problem is to design a procedure to search for the record of an employee, given her social security number. A similar problem is to find a person's telephone number in a telephone directory or to determine that the number is not listed. Clearly, the strategy in the example in the preceding paragraph can be readily applied to these cases.

Let us now formulate precisely the problem of searching for an item in an ordered list. We assume that we are dealing with items over which there is a linear ordering, $<$. In practical examples, the linear ordering can be numerical, alphabetical, alphanumeric, and so on. Let $K_1, K_2, \ldots, K_n$ be the n items in an ordered list which are known as the *keys*. Assume that $K_1 < K_2 < \cdots < K_n$. Given an item x, our problem is to search the keys and determine whether x is equal to one of the keys or whether x falls between keys K_i and K_{i+1} for some i. We note first that the search has $2n + 1$ possible outcomes, namely, x is less than K_1, x is equal to K_1, x is larger than K_1 but less than K_2, x is equal to K_2, and so on.

A search procedure consists of a sequence of comparisons between x and the keys where each comparison of x with a key tells us whether x is equal to, or less than, or larger than that key.† We show now how we can describe a search procedure using a binary-tree representation. We define a *search tree*

---

† Suppose we compare x with K_i and find out that x is less than K_i. A subsequent step to compare x with K_j for $j > i$ will be a totally wasteful one. We assume the search procedures considered do not contain any such wasteful steps.

for the keys $K_1, K_2, \ldots, K_n$ to be a binary tree with n branch nodes and $n + 1$ leaves. The branch nodes are labeled $K_1, K_2, \ldots, K_n$, and the leaves are labeled $K_0, K_1, K_2, \ldots, K_n$,† such that, for the branch node with the label K_i, its left subtree contains only vertices with labels K_j, $j < i$, and its right subtree contains only vertices with labels K_j, $j \geqq i$. For example, Fig. 5.15 shows a search tree for the keys K_1, K_2, K_3, K_4 where, for clarity, we use circles to denote branch nodes and squares to denote leaves. We note immediately that a search tree corresponds to a search procedure: Starting with the root of the search tree, we compare a given item x with the label of the root K_i. If x is equal to K_i, the search is completed. If x is less than K_i we shall compare x with the left son of the root, and if x is larger than K_i we shall compare x with the right son of the root.‡ Such comparison continues for successive branch nodes until either x matches a key or a leaf is reached. Clearly, if a leaf labeled K_j is reached, it means that x is larger than the key K_j but less than the key K_{j+1}. (If the leaf labeled K_0 is reached, it means that x is less than K_1. If the leaf labeled K_n is reached, it means that x is larger than K_n.) For example, let AB, CF, EG, PP be the keys K_1, K_2, K_3, K_4 in the search tree in Fig. 5.15. Given the item BB, the search steps according to Fig. 5.15 will be:

1. Compare BB with K_3, which is EG.
2. Since BB is less than EG, compare BB with K_1, which is AB.
3. Since BB is larger than AB, compare BB with K_2, which is CF.
4. Since BB is less than CF, the leaf labeled K_1 is reached.

Thus, we conclude that the item BB is larger than AB and less than CF.

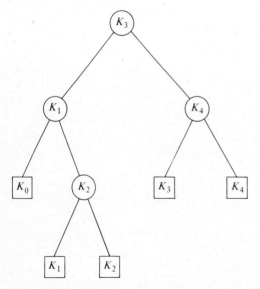

Figure 5.15

---

† The meaning of the label K_0 will become clear later on.

‡ In a binary tree, the left (right) son of a branch node is the root of the left (right) subtree of the node.

One obvious criterion for measuring the effectiveness of a search procedure is the maximum number of comparisons the procedure carries out in the worst case, that is, the height of the corresponding search tree. Consequently, for a given set of n keys a search tree whose height is $\lceil \lg(n+1) \rceil$ will correspond to a best possible search procedure. On the other hand, since there always exists a binary search tree of height $\lceil \lg(n+1) \rceil$ for any n†, the problem of designing a best possible search procedure according to this criterion is not a difficult one.

In general, the outcomes of our searches might occur in different frequencies. For example, in the problem of guessing the number a friend of ours has picked, if we know that it is more likely that he or she picks a number less than 20 than a number larger than or equal to 20, we might want to reconsider our strategy in asking the questions. To be specific, suppose we are given u_1, $u_2, \ldots, u_n$ as the frequencies of occurrences, say out of 1000 searches, of the outcomes that a given item is equal to $K_1, K_2, \ldots, K_n$, and are given $w_0, w_1,$ $w_2, \ldots, w_n$ as the frequencies of occurrences of the outcomes that a given item is less than K_1, is larger than K_1 but less than $K_2, \ldots,$ is larger than K_n. For a given search tree for the keys $K_1, K_2, \ldots, K_n$, the total number of comparisons in 1000 searches will equal

$$\sum_{j=1}^{n} u_j[l(K_j) + 1] + \sum_{j=0}^{n} w_j l'(K_j) \tag{5.3}$$

where $l(K_j)$ is the path length of the branch node that is labeled K_j and $l'(K_j)$ is the path length of the leaf that is labeled K_j in the search tree. Note that $l(K_j) + 1$ is the number of comparisons carried out by the search procedure if the given item is equal to K_j, and $l'(K_j)$ is the number of comparisons carried out by the search procedure if the given item is larger than K_j and less than K_{j+1}.

The problem of constructing a search tree to minimize the quantity in (5.3) for given $u_1, u_2, \ldots, u_n, w_0, w_1, w_2, \ldots, w_n$ was studied in Knuth[8]. A variation of the problem was studied in Hu and Tucker[4] and Hu[3]. To go into the details of their results will be beyond our scope of discussion. The interested reader is referred to these papers as well as chap. 6 of Knuth[7].

5.6 SPANNING TREES AND CUT-SETS

Let G be a connected graph where the vertices represent the buildings in a factory, and the edges represent connecting tunnels between the buildings. One might wish to determine a subset of tunnels which should be kept open all the time so that we can reach one building from another through these tunnels. One might also wish to determine the subsets of tunnels whose blockage will separate some of the buildings from the others. We study in this section the concepts of "subsets of connecting edges" and "subsets of disconnecting edges" in a graph.

† See Problem 5.11.

A *tree of a graph* is a subgraph of the graph which is a tree. A *spanning tree of a connected graph* is a spanning subgraph of the graph which is a tree. For example, Fig. 5.16b shows a tree and Fig. 5.16c shows a spanning tree of the graph in Fig. 5.16a. A *branch* of a tree is an edge of the graph that is in the tree. A *chord*, or a *link*, of a tree is an edge of the graph that is not in the tree. The set of the chords of a tree is referred to as the *complement* of the tree.

We observe that a *connected graph always contains a spanning tree.* Suppose we are given a connected graph. If the graph does not contain any circuit, then it is a tree. If the graph contains one or more circuits, we can remove an edge from one of the circuits and still have a connected subgraph. Such removal of edges from circuits can be repeated until we have a spanning tree. It follows immediately from this argument that a spanning tree is a minimal connecting subgraph† of a connected graph in the sense that from a connecting subgraph which is not a spanning tree, one or more of its edges can be removed so that the resultant graph is still a connecting subgraph, and, on the other hand, no edge can be removed from a spanning tree so that the resultant subgraph is still a connecting subgraph. In the example of buildings connected by tunnels, if we keep the tunnels corresponding to the edges in a spanning tree open, we are assured that we can reach from one building to another through these tunnels. Moreover, this would be a minimal set of tunnels that must be kept open.

For a connected graph with e edges and v vertices, there are $v - 1$ branches in any spanning tree. It follows that, relative to any spanning tree, there are $e - v + 1$ chords.

A *cut-set* is a (minimal) set of edges in a graph such that the removal of the set will increase the number of connected components in the remaining subgraph, whereas the removal of any proper subset of it will not. It follows that in a connected graph, the removal of a cut-set will separate the graph into two parts. This suggests an alternative way of defining a cut-set. Let the vertices in a connected component of a graph be divided into two subsets such that every two vertices in each subset are connected by a path that meets only vertices in that subset. Then, the set of edges joining the vertices in the two subsets is a cut-set. As an example, for the graph in Fig. 5.17a, the set of edges $\{e_1, e_5, e_6, e_7, e_4\}$ is a cut-set, since its removal will leave an unconnected

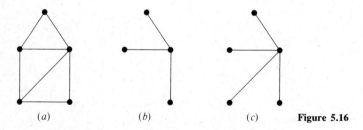

(a)   (b)   (c)   **Figure 5.16**

† A *connecting subgraph* of a graph is a spanning subgraph that is connected.

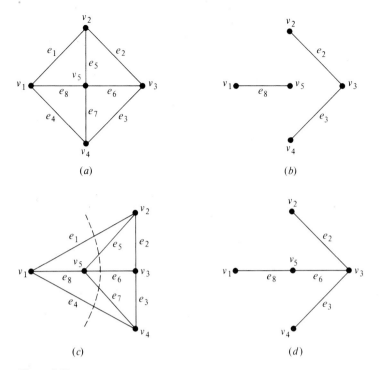

Figure 5.17

subgraph as shown in Fig. 5.17*b*, while the removal of any of its proper subsets will not. Also, this is the set of edges that join the vertices in the two subsets $\{v_1, v_5\}$ and $\{v_2, v_3, v_4\}$. Figure 5.17*a* is redrawn as Fig. 5.17*c* to emphasize such a division of vertices. In the example of the buildings connected by tunnels, if the tunnels corresponding to the edges in a cut-set are blocked, then the buildings will be separated into two clusters with no passage from a building in one cluster to a building in another.

The concepts of spanning trees, circuits, and cut-sets are closely related. Because a spanning tree contains a unique path between any two vertices in the graph, the addition of a chord to the spanning tree yields a subgraph that contains exactly one circuit. Suppose that the chord $\{v_1, v_2\}$ is added to a spanning tree. Because the spanning tree contains a path between v_1 and v_2, this path together with the edge $\{v_1, v_2\}$ form a circuit in the graph. On the other hand, if the addition of the chord $\{v_1, v_2\}$ yields two or more circuits, there must be two or more paths between v_1 and v_2 in the spanning tree, which is clearly impossible. For a given spanning tree, a unique circuit can be obtained by adding to the spanning tree each of the chords. The set of $e - v + 1$ circuits obtained in this way is called the *fundamental system of circuits* relative to the spanning tree. A circuit in the fundamental system is called a *fundamental circuit*. Since a fundamental circuit contains exactly one chord of the spanning tree, it is referred to as the fundamental circuit corresponding to the chord. For example, for the graph in Fig. 5.17a and the spanning tree in Fig. 5.17d the fundamental circuit corresponding to the chord e_1 is the circuit $\{e_1, e_2, e_6, e_8\}$, and the other fundamental circuits are $\{e_5, e_2, e_6\}$, $\{e_4, e_8, e_6, e_3\}$, and $\{e_7, e_6, e_3\}$.

Since the removal of any branch from a spanning tree breaks the spanning tree up into two trees (either or both of which may consist of a single vertex), we say that corresponding to a branch in a spanning tree there is a division of the vertices in the graph into two subsets corresponding to the vertices in the two trees. It follows that for every branch in a spanning tree there is a corresponding cut-set. For example, for the graph in Fig. 5.17a, the removal of the branch e_3 from the spanning tree in Fig. 5.17d divides the vertices into the two subsets $\{v_1, v_2, v_3, v_5\}$ and $\{v_4\}$. The corresponding cut-set is $\{e_3, e_7, e_4\}$. For a given spanning tree, the set of the $v - 1$ cut-sets corresponding to the $v - 1$ branches of the spanning tree is called the *fundamental system of cut-sets* relative to the spanning tree. A cut-set in the fundamental system of cut-sets is called a *fundamental cut-set*. Since a fundamental cut-set contains exactly one tree branch, it is referred to as the fundamental cut-set corresponding to the branch. For the graph in Fig. 5.17a and the spanning tree in Fig. 5.17d the fundamental cut-sets are $\{e_1, e_5, e_2\}$, $\{e_1, e_8, e_4\}$, $\{e_1, e_5, e_6, e_7, e_4\}$, and $\{e_4, e_7, e_3\}$.

We now present some of the properties of circuits and cut-sets. Unless otherwise stated, our discussion will be limited to connected graphs, since its extension to unconnected graphs is straightforward.

Theorem 5.1 A circuit and the complement of any spanning tree must have at least one edge in common.

PROOF If there is a circuit that has no common edge with the complement of a spanning tree, the circuit is contained in the spanning tree. However, this is impossible as a tree cannot contain a circuit. ☐

Theorem 5.2 A cut-set and any spanning tree must have at least one edge in common.

PROOF If there is a cut-set that has no common edge with a spanning tree, the removal of the cut-set will leave the spanning tree intact. However, this means that the removal of the cut-set will not separate the graph into two components, which is in contradiction to the definition of a cut-set. ☐

Theorem 5.3 Every circuit has an even number of edges in common with every cut-set.

PROOF Corresponding to a cut-set, there is a division of the vertices of the graph into two subsets which are the two sets of vertices in the two components of the graph when the edges in the cut-set are removed. Therefore, a path connecting two vertices in one subset must traverse the

Subset of
vertices

Subset of
vertices

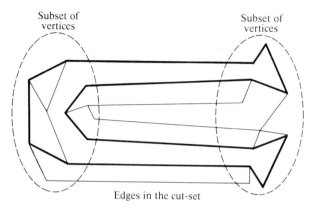

Edges in the cut-set

Figure 5.18

edges in the cut-set an even number of times, as illustrated in Fig. 5.18.
(The edges in the circuit are drawn in heavy lines.) Since a circuit is a path
from some vertex to itself, the theorem follows. □

The following results point out a close relationship between the fundamental system of circuits and the fundamental system of cut-sets relative to a
spanning tree:

Theorem 5.4 For a given spanning tree, let $D = \{e_1, e_2, e_3, \ldots, e_k\}$ be a
fundamental cut-set in which e_1 is a branch and $e_2, e_3, \ldots, e_k$ are chords of
the spanning tree. Then, e_1 is contained in the fundamental circuits
corresponding to e_i for $i = 2, 3, \ldots, k$. Moreover, e_1 is not contained in any
other fundamental circuits.

PROOF Let C be the fundamental circuit corresponding to the chord e_2.
Note that e_2 is in both C and D. Since C and D have an even number of
edges in common and e_1 is the only other edge that can possibly be in both
C and D,† e_1 must be contained in C. A similar argument can be applied to
the fundamental circuits corresponding to the chords $e_3, e_4, \ldots, e_k$. On the
other hand, let C' be the fundamental circuit corresponding to any chord
not in D. C' cannot contain e_1, because otherwise, C' and D will have e_1 as
the only common edge. □

Similarly we have:

Theorem 5.5 For a given spanning tree, let $C = \{e_1, e_2, e_3, \ldots, e_k\}$ be a
fundamental circuit in which e_1 is a chord and $e_2, e_3, \ldots, e_k$ are branches of
the spanning tree. Then, e_1 is contained in the fundamental cut-sets

† All the edges in C with the exception of e_2 are branches, and all the edges in D with the
exception of e_1 are chords.

corresponding to e_i for $i = 2, 3, \ldots, k$. Moreover, e_1 is not contained in any other fundamental cut-set.

We leave the proof of Theorem 5.5 as an exercise (Prob. 5.17), since it is quite similar to that of Theorem 5.4.

5.7 MINIMUM SPANNING TREES

For a given graph, one might want to determine a spanning tree of the graph. We consider here the more general problem of determining a minimum spanning tree of a weighted graph where real numbers are assigned to the edges as their weights. The weight of a spanning tree is defined to be the sum of the weights of the branches of the tree. A *minimum spanning tree* is one with minimum weight. An obvious physical interpretation of this problem is to consider the vertices of a graph as cities and the weights of the edges as the costs of setting up and maintaining communication links between the cities. Suppose we want to set up a communication network connecting all cities at minimum cost. The problem is then that of determining a minimum spanning tree. We present two simple algorithms that do so.

Our first procedure is based on the observation that among all edges in a circuit, the edge with the largest weight is not in a minimum spanning tree.† Let C be a circuit in a weighted graph, and e be the edge of the largest weight in C. Suppose that e is a branch of a spanning tree T. Let D denote the fundamental cut-set corresponding to the branch e. Since the circuit C and the cut-set D must have an even number of edges in common, besides the edge e there must be at least one more edge that is in both C and D. Let f be one such edge. Note that f is a chord of the spanning tree T because D is a fundamental cut-set. Let us add the edge f to the spanning tree T and denote the resultant subgraph U. Clearly, U is a spanning subgraph containing exactly one circuit, the fundamental circuit corresponding to f. According to Theorem 5.4, e is contained in the fundamental circuit corresponding to f. Removing e from U, we obtain a spanning tree whose weight is smaller than that of T.

Our observation suggests a procedure to determine a minimum spanning tree of a connected weighted graph. We shall construct a subgraph of the weighted graph in a step-by-step manner, examining the edges one at a time in increasing ordering of weights. An edge will be added to the partially constructed subgraph if its inclusion does not yield a circuit, and will be discarded otherwise. The construction terminates when all the edges have been examined. It is clear that our construction yields a subgraph that contains no circuit. We note that the subgraph is also connected since, for an edge $\{a, b\}$ in

---

† To simplify the presentation, we assume that the weights of the edges are distinct. In the case that the weights of the edges are not all distinct, this result should be stated in a more general form: Among all edges in a circuit, an edge with largest weight is not contained in *some* minimum spanning tree.

the original graph, either the edge $\{a, b\}$ is included in the subgraph or there is a path between a and b in the subgraph. Thus, the subgraph we have constructed is a tree. Moreover, it is a spanning tree because the original graph is connected. Finally, the spanning tree is minimum because in the construction procedure an edge was excluded in favor of edges of larger weights only if the excluded edge is known to be one that cannot be in a minimum spanning tree. In other words, the $v - 1$ edges in the subgraph are indeed $v - 1$ edges of the smallest weights that can be included in a minimum spanning tree.

As an example, a minimum spanning tree for the weighted graph in Fig. 5.19a is shown in Fig. 5.19b. Note how the edges with weight 4 and weight 7 are excluded in the step-by-step construction.

A second procedure for constructing a minimum spanning tree is based on the observation that among all edges incident with a vertex, the edge with the smallest weight must be in a minimum spanning tree.† Let v_1 be a vertex and $\{v_1, v_2\}$ be the edge with the smallest weight among all edges incident with v_1. Let T be a spanning tree not containing the edge $\{v_1, v_2\}$. Let us add the edge $\{v_1, v_2\}$ to T and denote the resultant subgraph U. Note that U contains exactly one circuit, the fundamental circuit corresponding to the chord $\{v_1, v_2\}$. This circuit is made up of the edge $\{v_1, v_2\}$ and the path from v_1 to v_2 in T. Let $(v_1, v_{i_1}, v_{i_2}, \ldots, v_{i_k}, v_2)$ denote the sequence of vertices in that path. We observe that removing the edge $\{v_1, v_{i_1}\}$ from U yields a spanning tree whose weight is smaller than that of T.

Let $G = (V, E)$ be a graph, and v_1 and v_2 be two vertices in V. We introduce the notion of obtaining a graph G' from G by *coalescing* the vertices v_1 and v_2. Intuitively, G' is a graph obtained from G by combining the vertices v_1 and v_2 into one "supervertex" and retaining all the edges in G. For example, Fig. 5.20b shows a graph obtained from the graph in Fig. 5.20a by coalescing the vertices v_1 and v_2. Because G and G' are, in general, multigraphs, we shall identify the edges in E by edge names such as $e_1, e_2, \ldots$, instead of the vertices with which they are incident. Let $G' = (V', E')$ such that V' contains all the vertices in V except that the vertices v_1 and v_2 are removed and a new vertex v^* is introduced, and E' contains all the edges in E except that, if an edge was

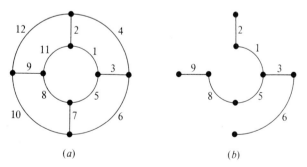

(a)   (b)   **Figure 5.19**

† Again, we assume that the weights of the edges are distinct. In the case that the weights of the edges are not all distinct, this result should be stated in a more general form: Among all edges incident with a vertex, an edge with smallest weight is contained in *some* minimum spanning tree.

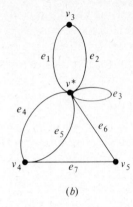

(a) (b) **Figure 5.20**

incident with v_1 or v_2 in G, it is incident with v^* in G'.† We have the following observation: Let e be an edge with smallest weight that is not a self-loop in G. Let G' be the graph obtained from G by coalescing the vertices v_1 and v_2 with which the edge e is incident in G, and T' be a minimum spanning tree of G'. Let T denote a subgraph of G consisting of all the edges in T' together with the edge e. We want to show that T is a minimum spanning tree of G. First of all, we note that T is indeed a spanning tree of G. Secondly, we have

$$W(T) = W(T') + w(e)$$

where $w(e)$ denotes the weight of edge e, and $W(T)$ and $W(T')$ denote the sum of the weights of the edges in T and T', respectively. Finally, if T were not a minimum spanning tree of G, there exists a minimum spanning tree of G, $\hat{T}$, which contains the edge e such that

$$W(\hat{T}) < W(T)$$

Let $\hat{T}'$ denote the tree obtained from $\hat{T}$ by coalescing the vertices v_1 and v_2 and removing the edge e. Clearly, $\hat{T}'$ is a spanning tree of G'. Since

$$W(\hat{T}) = W(\hat{T}') + w(e)$$

we obtain

$$W(\hat{T}') < W(T')$$

which is a contradiction of the assumption that T' is a minimum spanning tree of the graph G'.

Our observations suggest immediately another procedure for determining a minimum spanning tree in a connected weighted graph. Let $\{v_1, v_2\}$ be the edge of the smallest weight in a graph G. Since $\{v_1, v_2\}$ must be included in a minimum spanning tree of G, we can coalesce the two vertices v_1 and v_2 to obtain the graph G', and then try to determine a minimum spanning tree of G'.

---

† An edge between v_1 and v_2 becomes a self-lóop at v^*. (For the purpose of constructing a minimum spanning tree, we could choose to remove all self-loops in G'. However, we do not do so because we do not wish to deviate from the standard definition of coalescing two vertices in a graph.)

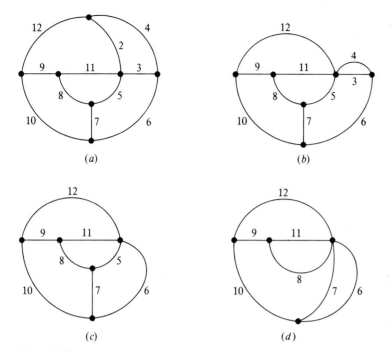

Figure 5.21

The step can then be repeated until we terminate with a graph that has a single vertex.

As an example, for the weighted graph in Fig. 5.19a, several steps of the construction are shown in Fig. 5.21a through d. Note how the edges with weights 4 and 7 are excluded in the step-by-step construction.

*5.8 TRANSPORT NETWORKS

A weighted directed graph is said to be a *transport network* if the following conditions are satisfied:

1. It is connected and contains no loops.
2. There is one and only one vertex in the graph that has no incoming edge.
3. There is one and only one vertex in the graph that has no outgoing edge.
4. The weight of each edge is a nonnegative real number.

In a transport network, the vertex that has no incoming edge is called the *source* and is denoted by a; the vertex that has no outgoing edge is called the *sink* and is denoted by z. The weight of an edge is called the *capacity* of the edge. The capacity of the edge (i, j) is denoted by $w(i, j)$.

Clearly, a transport network represents a general model for the transportation of material from the origin of supply to the destination through shipping routes, where there are upper limits on the amount of material that can be shipped through the routes. Figure 5.22a shows an example of a transport network.

A *flow* in a transport network, ϕ, is an assignment of a nonnegative number $\phi(i, j)$ to each edge (i, j) such that the following conditions are satisfied:

1. $\phi(i, j) \leq w(i, j)$ for each edge (i, j).
2. $\sum_{\text{all } i} \phi(i, j) = \sum_{\text{all } k} \phi(j, k)$ for each vertex j except the source a and the sink z.†

In terms of the transportation of material, $\phi(i, j)$ is the amount of material to be shipped through the route (i, j). Condition 1 means that the amount of material to be shipped through a route cannot exceed the capacity of the route. Condition 2 means that, except at the source and at the sink, the amount of material flowing into a vertex must equal the amount of material flowing out of the vertex. For example, Fig. 5.22b shows a flow in the transport network in Fig. 5.22a. The first number associated with an edge is the capacity of the edge, and the second number associated with an edge is the flow in the edge. The quantity $\sum_{\text{all } i} \phi(a, i)$ is said to be the *value of the flow* ϕ and is denoted by ϕ_v. Intuitively, it is clear that

$$\phi_v = \sum_{\text{all } i} \phi(a, i) = \sum_{\text{all } k} \phi(k, z)$$

that is, the total outgoing flow at the source is equal to the total incoming flow at the sink. This result is proved rigorously in the proof of Theorem 5.6. For a given flow, an edge (i, j) is said to be *saturated* if $\phi(i, j) = w(i, j)$, and is said to be *unsaturated* if $\phi(i, j) < w(i, j)$. A *maximum flow* in a transport network is a flow that achieves the largest possible value. It is conceivable that there might be more than one maximum flow in a transport network. In other words, there might be a number of different flows, all of which attain the largest possible value.

One frequently wants to determine the largest amount of material that can be shipped from the source to the sink of a given transport network. Moreover, it is desirable to have an algorithm for constructing a flow in the network that

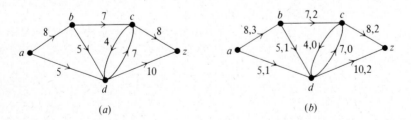

(a)

(b)

Figure 5.22

† We define $\phi(x, y)$ to be zero if there is no edge from x to y.

achieves the largest possible value. We shall present an algorithm for doing so. To this end, let us introduce first some useful concepts and results:

A *cut* in a transport network is a cut-set of the undirected graph, obtained from the transport network by ignoring the direction of the edges, that separates the source from the sink. The notation $(P, \bar{P})$ is used to denote a cut that divides the vertices into two subsets P and $\bar{P}$, where the subset P contains the source and the subset $\bar{P}$ contains the sink. The *capacity of a cut*, denoted by $w(P, \bar{P})$, is defined to be the sum of the capacities of those edges incident from the vertices in P to the vertices in $\bar{P}$; that is,

$$w(P, \bar{P}) = \sum_{i \in P, j \in \bar{P}} w(i, j)$$

For example, the dashed line in Fig. 5.23 identifies a cut that separates the subset of vertices $P = \{a, d\}$ from the subset of vertices $\bar{P} = \{b, c, z\}$. The capacity of this cut is equal to $8 + 7 + 10 = 25$.

The following result gives an upper bound on the values of flows in a transport network.

Theorem 5.6 The value of any flow in a given transport network is less than or equal to the capacity of any cut in the network.

PROOF Let ϕ be a flow and $(P, \bar{P})$ be a cut in a transport network. For the source a,

$$\sum_{\text{all } i} \phi(a, i) - \sum_{\text{all } j} \phi(j, a) = \sum_{\text{all } i} \phi(a, i) = \phi_v \qquad (5.3)$$

since $\phi(j, a) = 0$ for any j. For a vertex p other than a in P,

$$\sum_{\text{all } i} \phi(p, i) - \sum_{\text{all } j} \phi(j, p) = 0 \qquad (5.4)$$

Combining (5.3) and (5.4), we have

$$\phi_v = \sum_{p \in P} \left[\sum_{\text{all } i} \phi(p, i) - \sum_{\text{all } j} \phi(j, p) \right]$$

$$= \sum_{p \in P; \text{all } i} \phi(p, i) - \sum_{p \in P; \text{all } j} \phi(j, p)$$

$$= \sum_{p \in P; i \in P} \phi(p, i) + \sum_{p \in P; i \in \bar{P}} \phi(p, i)$$

$$\quad - \left[\sum_{p \in P; j \in P} \phi(j, p) + \sum_{p \in P; j \in \bar{P}} \phi(j, p) \right] \qquad (5.5)$$

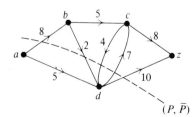

$(P, \bar{P})$ **Figure 5.23**

Note that

$$\sum_{p \in P; i \in P} \phi(p, i) = \sum_{p \in P; j \in P} \phi(j, p)$$

because both sums run through all the vertices in P. Thus, (5.5) becomes

$$\phi_v = \sum_{p \in P; i \in \bar{P}} \phi(p, i) - \sum_{p \in P; j \in \bar{P}} \phi(j, p) \tag{5.6}$$

But, since $\sum_{p \in P; j \in \bar{P}} \phi(j, p)$ is always a nonnegative quantity, we have

$$\phi_v \leq \sum_{p \in P; i \in \bar{P}} \phi(p, i) \leq \sum_{p \in P; i \in \bar{P}} w(p, i) = w(P, \bar{P}) \qquad \square$$

Equation (5.6) is a useful result which can be stated as: For any cut $(P, \bar{P})$ the values of a flow in a transport network are equal to the sum of the flows in the edges from the vertices in P to the vertices in $\bar{P}$ minus the sum of the flows in the edges from the vertices in $\bar{P}$ to the vertices in P.

We are now ready to present an algorithm for constructing a maximum flow in a transport network. In view of the result that the value of any flow in a transport network is less than or equal to the capacity of any cut, whenever we can construct a flow ϕ the value of which is equal to the capacity of some cut $(P, \bar{P})$, we can be certain that ϕ is a maximum flow. Because if there were a larger flow, its value would exceed the capacity of the cut $(P, \bar{P})$. As it turns out, we can always construct a flow the value of which is equal to the capacity of a cut using the following procedure, which is known as the *labeling procedure*:

To start the procedure, we must construct an initial flow ϕ in the network. However, such construction poses no problem as we can always start, trivially, with zero flow in every edge.

At first, the source a is labeled $(-, \infty)$. (The significance of such a label will become clear later.) Next, all the vertices that are adjacent from a are scanned. A vertex b that is adjacent from a is labeled $(a^+, \Delta b)$, where Δb is equal to $w(a, b) - \phi(a, b)$, if $w(a, b) > \phi(a, b)$; it is not labeled if $w(a, b) = \phi(a, b)$. After all the vertices that are adjacent from the source a are scanned and labeled (if possible), those vertices that are adjacent to or from the labeled vertices are scanned. Let b be a labeled vertex, and let q be a vertex that is adjacent *from* b. The vertex q is labeled $(b^+, \Delta q)$, where Δq is equal to the smaller of the two quantities Δb and $[w(b, q) - \phi(b, q)]$ if $w(b, q) > \phi(b, q)$. The vertex q is not labeled if $w(b, q) = \phi(b, q)$. Let b be a labeled vertex, and let q be a vertex that is adjacent to b. The vertex q is labeled $(b^-, \Delta q)$, where Δq is equal to the smaller of the two quantities Δb and $\phi(q, b)$ if $\phi(q, b) > 0$. The vertex q is not labeled if $\phi(q, b) = 0$. Such a labeling procedure is not necessarily unique. The vertex q might be adjacent to or from more than one labeled vertex. Also, there might even be an edge incident from b to q as well as an edge incident from q to b. In any case, when a vertex can be labeled in more than one way, an arbitrary choice of these ways is made.

Let us examine the meanings of these labels before proceeding with the presentation of the remaining steps in the procedure. For a vertex that is adjacent from the source (like the vertex b), the label $(a^+, \Delta b)$ means that the flow into b can be increased by an amount equal to Δb. Moreover, such an increment can be drawn from the source a. Similarly, for a vertex q that is adjacent from a labeled vertex b, the label $(b^+, \Delta q)$ means that, by drawing the increment from the vertex b, the total incoming flow into q from the labeled vertices can be increased by Δq. For a vertex q that is adjacent to a labeled vertex b, the label $(b^-, \Delta q)$ means that, by decreasing the flow from q to b, the total outgoing flow from q to the labeled vertices can be decreased by Δq. In either of these cases, an increase in the flow equal to Δq from the vertex q to the unlabeled vertices is assured. The meaning of the label of the source, $(-, \infty)$ should also become clear now. It means that (out from nowhere) the source can supply an infinite amount of material to the other vertices.

If we repeat the procedure of labeling the vertices that are adjacent to or from the labeled vertices, one of the following two cases shall arise:

Case 1 The sink z is labeled, say, with a label $(y^+, \Delta z)$. [Of course, z will never have a label like $(y^-, \Delta z)$.] We can increase the flow in the edge (y, z) from $\phi(y, z)$ to $\phi(y, z) + \Delta z$, as the increment is guaranteed by the vertex y. Note that the vertex y must be labeled either $(q^+, \Delta y)$ or $(q^-, \Delta y)$, with $\Delta y \geqq \Delta z$, for some vertex q. If y is labeled $(q^+, \Delta y)$, we shall, in turn, draw the increment from the vertex q by increasing the flow in the edge (q, y) from $\phi(q, y)$ to $\phi(q, y) + \Delta z$. On the other hand, if y is labeled $(q^-, \Delta y)$, we shall decrease the flow in the edge (y, q) from $\phi(y, q)$ to $\phi(y, q) - \Delta z$ so that the increment Δz from y to z is compensated. The process is continued back to the source a, and the value of the flow in the transport network will be increased by the amount Δz. The labeling procedure can now be started all over again to further increase the value of the flow in the network.

Case 2 The sink z is not labeled. Let us denote all the labeled vertices by P and all the unlabeled vertices by $\bar{P}$. The fact that the sink z is not labeled means that the flow in each of the edges incident from the vertices in P to the vertices in $\bar{P}$ is equal to the capacity of that edge, and that the flow in each of the edges incident from the vertices in $\bar{P}$ to the vertices in P is equal to zero. We have thus obtained a flow, the value of which is equal to the capacity of the cut $(P, \bar{P})$. The flow, therefore, is a maximum flow.

Consider the following illustrative example. For the transport network in Fig. 5.24a, we start with zero flow in every edge. (The first number associated with an edge is its capacity, and the second number is the flow in the edge.) Figure 5.24b shows the first pass of the labeling procedure. Notice that the sink z can be labeled either with $(d^+, 3)$ or with $(b^+, 2)$. We choose arbitrarily the label $(d^+, 3)$. Figure 5.24c shows the second pass, and Fig. 5.24d shows the third pass. Notice that in Fig. 5.24d the vertex b is labeled $(c^+, 6)$ and the

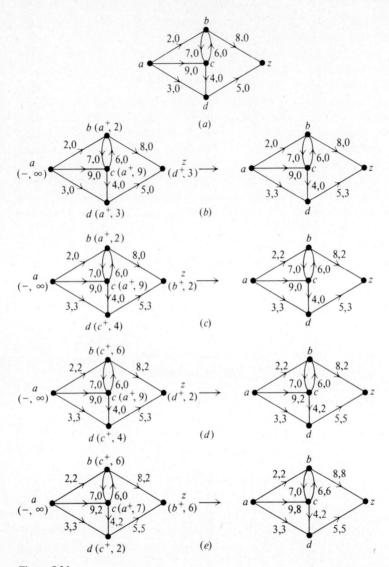

Figure 5.24

vertex d is labeled $(c^+, 4)$; that is, the vertex c has guaranteed a total flow of 10 to the vertices b and d although Δc is only equal to 9. However, since the vertex c would have to supply either the increment of flow at b or the increment of flow at d, but not at both, in the augmentation step no difficulty will arise. Figure 5.24e shows the last pass of the labeling procedure, which yields a maximum flow of 13.

As another example, consider the transport network in Fig. 5.25a where an initial flow has been found. Figure 5.25b shows how we obtain a maximum flow by the labeling procedure.

(a)

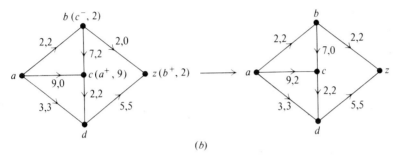

(b)

Figure 5.25

5.9 REMARKS AND REFERENCES

See chap. 2 of Knuth[7] and chap. 6 of Knuth[6] for a very complete treatment of trees and search trees. Ford and Fulkerson[1] and Hu[2] are two excellent references on network flow problems. See chap. 7 of Liu[9] for a further discussion on circuits and cut-sets.

[1] Ford, L. R., Jr., and D. R. Fulkerson: "Flows in Networks," Princeton University Press, Princeton, N.J., 1962.

[2] Hu, T. C.: "Integer Programming and Network Flows," Addison-Wesley Publishing Company, Reading, Mass., 1969.

[3] Hu, T. C.: A New Proof of the T–C Algorithm, *SIAM Journal on Applied Mathematics*, **25**: 83–94 (1971).

[4] Hu, T. C., and A. C. Tucker: Optimum Computer Search Trees, *SIAM Journal on Applied Mathematics*, **21**: 514–532 (1971).

[5] Huffman, D. A.: A Method for the Construction of Minimum Redundancy Codes, *Proc. IRE*, **40**: 1098–1101 (1952).

[6] Knuth, D. E.: "The Art of Computer Programming, Vol. 1, Fundamental Algorithms," 2d ed., Addison-Wesley Publishing Company, Reading, Mass., 1973.

[7] Knuth, D. E.: "The Art of Computer Programming, Vol. 3, Sorting and Searching," Addison-Wesley Publishing Company, Reading, Mass., 1973.

[8] Knuth, D. E.: Optimum Binary Search Trees, *Acta Informatica*, **1**: 14–25 (1971).

[9] Liu, C. L.: "Introduction to Combinatorial Mathematics," McGraw-Hill Book Company, New York, 1968.

[10] Zipf, G. K.: "Human Behavior and the Principle of Least Effort, an Introduction to Human Ecology," Addison-Wesley Publishing Company, Reading, Mass., 1949.

PROBLEMS

5.1 Characterize all trees that have exactly two leaves.

5.2 (*a*) A tree has two vertices of degree 2, one vertex of degree 3, and three vertices of degree 4. How many vertices of degree 1 does it have?

    (*b*) A tree has n_2 vertices of degree 2, n_3 vertices of degree 3, ..., and n_k vertices of degree k. How many vertices of degree 1 does it have?

5.3 (*a*) Show that the sum of the degrees of the vertices of a tree with n vertices is $2n - 2$.

    (*b*) For $n \geq 2$, let $d_1, d_2, \ldots, d_n$ be n positive integers such that $\Sigma_{i=1}^{n} d_i = 2n - 2$. Show that there exists a tree whose vertices have degrees $d_1, d_2, \ldots, d_n$.

5.4 The center of a (connected) graph is defined to be a vertex v with the property that the maximum distance between v and any other vertex is as small as possible. (See Prob. 4.15 for the definition of the distance between two vertices in a graph.)

    (*a*) Show an example of a graph that has one center.

    (*b*) Show an example of a graph that has two or more centers.

    (*c*) Show that a tree has either one or two centers. Moreover, if there are two centers they must be adjacent. Show an example of a tree that has two centers.

5.5 Let v be a vertex in a tree T. Clearly, direction can be assigned to the edges in T to obtain a rooted tree with v being the root. Let $v_1, v_2, \ldots, v_k$ denote the sons of v, and $s_1, s_2, \ldots, s_k$ denote the numbers of vertices in the subtrees with $v_1, v_2, \ldots, v_k$ as the roots. The weight of v is defined to be the maximum of $s_1, s_2, \ldots, s_k$.

    (*a*) For the tree in Fig. 5P.1, compute the weights of all the vertices.

    (*b*) A vertex that has minimum weight is said to be a *centroid* of the tree. Show an example of a tree that has one centroid. Show an example of a tree that has two centroids.

    (*c*) Show that a tree has either one or two centroids. Moreover, if there are two centroids, they must be adjacent.

    (*d*) Show that v is the only centroid of a tree if

$$s_j \leq s_1 + s_2 + \cdots + s_k - s_j \qquad \text{for all } j, \quad 1 \leq j \leq k$$

    (*e*) Show that a tree with two centroids has an even number of vertices. Moreover, the weight of each centroid is equal to half the number of vertices in the tree.

Figure 5P.1

5.6 Show that a regular binary tree has an odd number of vertices.

5.7 By traversing a tree, we mean to visit each of the vertices of the tree exactly once in some sequential order. We describe here three principal ways that may be used to traverse a binary tree.

1. *Preorder traversal*: Visit the root, traverse the left subtree, then traverse the right subtree.
2. *Inorder (symmetric order) traversal*: Traverse the left subtree, visit the root, then traverse the right subtree.
3. *Postorder traversal*: Traverse the left subtree, traverse the right subtree, then visit the root.

Show the sequential orders in which the vertices of the tree in Fig. 5P.2 are visited in a preorder traversal, an inorder traversal, and a postorder traversal.

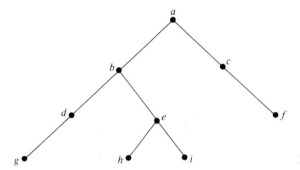

Figure 5P.2

5.8 Let A be a set of binary sequences. Let us partition A into two subsets A_0 and A_1, where A_0 is the set of sequences in A whose first digit is a 0 and A_1 is the set of sequences in A whose first digit is a 1. Let us then partition A_0 into two subsets according to the second digit in the sequences, and also partition A_1 in the same way. Employ this idea of repeatedly partitioning a set of sequences into subsets to show that, if A is a prefix code, then there is a binary tree, with the two edges incident from each of the branch nodes labeled with 0 and 1, such that the sequences of 0s and 1s assigned to the leaves are the sequences in A.

5.9 By a *sorted list* of numbers we mean a list of numbers arranged in ascending order. By *merging* two sorted lists we mean to combine them into one sorted list. We describe one way to merge two sorted lists: Since the smaller one of the smallest numbers of the two lists must be the smallest of all numbers, we can remove this number from the list it is in and place it somewhere else as the first number of the merged list. We can now compare the smallest numbers of the two lists of remaining numbers and place the smaller one of the two as the second number of the merged list. This step can be repeated until the merged list is completely built up. Clearly, it takes $n_1 + n_2 - 1$ comparisons to merge two sorted lists with n_1 and n_2 numbers, respectively. Given m sorted lists, we can select two of them and merge these two lists into one. We can then select two lists from the $m - 1$ sorted lists and merge them into one. By repeating this step, we shall eventually end up with one merged list.

Let A_1, A_2, A_3, A_4 be four sorted lists with 73, 44, 100, 55 numbers, respectively.

(*a*) Determine the total number of comparisons it takes to merge the four lists by merging A_1 and A_2, merging the resultant list with A_3, and then merging the resultant list with A_4.

(*b*) Determine the total number of comparisons it takes to merge the four lists by merging A_1 and A_2, merging A_3 and A_4, and then merging the two resultant lists.

(*c*) Determine an order to merge the four lists so that the total number of comparisons is minimum.

(*d*) Describe a general procedure for determining the order in which m sorted lists $A_1, A_2, \ldots, A_m$ are to be merged so that the total number of comparisons is minimum. (*Hint:* What is a convenient way to describe a certain order in which the lists are merged?)

5.10 (*a*) How can the procedure in Sec. 5.4 for constructing an optimal binary tree be extended to that for constructing an optimal m-ary tree?

(*b*) Construct an optimal ternary tree for the weights 1, 2, 3, 4, 5, 6, 7, 8, 9.

(*c*) Construct an optimal ternary tree for the weights 1, 2, 3, 4, 5, 6, 7, 8. In general, when $t - 1$, where t is the number of weights, is not a multiple of $m - 1$ [see Eq. (5.2)], how should we proceed to construct an optimal m-ary tree?

5.11 Show how a binary search tree of height $\lceil \lg(n+1) \rceil$ can be constructed for n keys, $K_1, K_2, \ldots, K_n$.

5.12 We define a *3-2 tree* to be a rooted tree in which the out-degree of a branch node is either 3 or 2. Furthermore, the path lengths of all leaves must be the same. For example, Fig. 5P.3*a* shows a 3-2 tree. A 3-2 tree can be used to describe a class of search procedures in which we compare an item either with one key or with two keys in each step of search. Using the 3-2 tree in Fig. 5P.3*b* as an example, explain in detail how such a search procedure works. How can we be sure that the number of leaves in a 3-2 tree is always equal to the number of keys plus one?

(*a*)

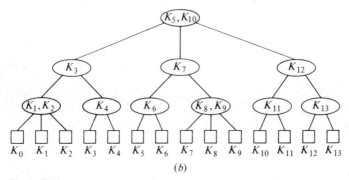

(*b*)

Figure 5P.3

5.13 Prove that the complement of a spanning tree does not contain a cut-set and that the complement of a cut-set does not contain a spanning tree.

5.14 Let L be a circuit in a graph G. Let a and b be any two edges in L. Prove that there exists a cut-set C such that $L \cap C = \{a, b\}$.

5.15 Let T_1 and T_2 be two spanning trees of a connected graph G. Let a be an edge that is in T_1 but not T_2. Prove that there is an edge b in T_2 but not T_1 such that both $(T_1 - \{a\}) \cup \{b\}$ and $(T_2 - \{b\}) \cup \{a\}$ are spanning trees of G.

5.16 (*a*) Let L_1 and L_2 be two circuits in a graph G. Let a be an edge that is in both L_1 and L_2, and let b be an edge that is in L_1 but not L_2. Prove that there exists a circuit L_3 which is such that $L_3 \subseteq (L_1 \cup L_2) - \{a\}$ and $b \in L_3$.

(*b*) Repeat part (*a*) when the term *circuit* is replaced by the term *cut-set*.

5.17 Prove Theorem 5.5.

5.18 We show in this problem that there are n^{n-2} spanning trees in a complete graph with n distinctly labeled vertices. Without loss of generality, let the labels of the vertices be $1, 2, \ldots, n$. We shall demonstrate a one-to-one correspondence between the spanning trees and the n^{n-2} $n-2$ digit sequences over the alphabet $\{1, 2, \ldots, n\}$. Let T denote a spanning tree whose vertices are labeled $1, 2, \ldots, n$. We construct a sequence $a_1 a_2 \cdots a_{n-2}$ as follows:

1. Let $i = 1$. Let T be the tree currently under examination.
2. Among all vertices of degree 1 in the tree currently under examination, select the one with the smallest label. Remove the edge that is incident with this vertex and let a_i equal the label of the *other* vertex with which this edge is incident.
3. The resultant tree in (2) becomes now the tree currently under examination. Increase i by 1, repeat (2) until a sequence of $n-2$ digits is formed.

(*a*) Determine the 4-digit sequence corresponding to the tree in Fig. 5P.4.

(*b*) Show that the degree of the vertex with label i in T is equal to the number of times the letter i appears in $a_1a_2 \cdots a_{n-2}$ plus one.

(*c*) Determine an algorithm for reconstructing a tree from a sequence $a_1a_2 \cdots a_{n-2}$.

(*d*) Show that the tree reconstructed by the algorithm in part (*c*) is the only tree that will yield the sequence $a_1a_2 \cdots a_{n-2}$ according to the foregoing procedure.

Figure 5P.4

5.19 Determine a minimum spanning tree for the graph shown in Fig. 5P.5.

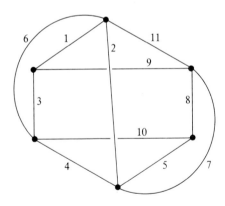

Figure 5P.5

5.20 Use the labeling procedure to find a maximum flow in the transport network in Fig. 5P.6.

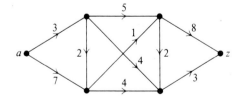

Figure 5P.6

5.21 Use the labeling procedure to find a maximum flow in the transport network in Fig. 5P.7.

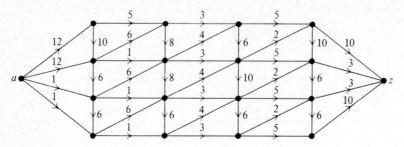

Figure 5P.7

5.22 Equipment is manufactured at three factories x_1, x_2, and x_3, and is to be shipped to three depots y_1, y_2, and y_3 through the transport network shown in Fig. 5P.8.

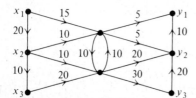

Figure 5P.8

Factory x_1 can make 40 units, factory x_2 can make 20 units, and factory x_3 can make 10 units. Depot y_1 needs 15 units, depot y_2 needs 25 units, and depot y_3 needs 10 units. How many units should each factory make so that they can be transported to the depots?

5.23 Find a maximum flow in the transport network in Fig. 5P.9 in which flows in the unoriented edges can be in either direction.

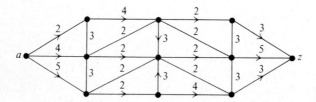

Figure 5P.9

5.24 (*a*) Seven kinds of military equipment are to be flown to a destination by five cargo planes. There are four units of each kind, and the five planes can carry eight, eight, five, four, and four units, respectively. Can the equipment be loaded in such a way that no two units of the same kind are on one plane?

(*b*) Give a solution to part (*a*) when the capacities of the planes are seven, seven, six, four, and four units, respectively.

5.25 Construct a directed graph with four vertices and with no more than two edges from one vertex to another such that the outgoing and incoming degrees of the vertices are (5, 4), (3, 3), (1, 2), and (2, 2), respectively, by solving the corresponding network flow problem.

5.26 Given a transport network, we wish to find a flow such that the flow in each edge is larger than or equal to the capacity of the edge and the value of the flow is minimum.

(*a*) Define the "capacity of a cut," and state the minimum-flow maximum-cut condition which is analogous to the maximum-flow minimum-cut condition presented in Sec. 5.8.

(*b*) Prove the minimum-flow maximum-cut condition by designing an algorithm to find a minimum-flow.

5.27 Engineers and technicians are to be hired by a company to participate in three projects. The personnel requirements of these three projects are listed in the following table:

	Minimum number of people needed in each project	Minimum number in each category			
		Mechanical engineers	Mechanical technicians	Electrical engineers	Electrical technicians
Project I	40	5	10	10	5
Project II	40	10	5	15	5
Project III	20	5	0	10	5

Moreover, to prepare for later expansion, the company wants to hire at least 30 mechanical engineers, 20 mechanical technicians, 20 electrical engineers, and 20 electrical technicians. What is a minimum number of persons in each category that the company should hire, and how should they be allocated to the three projects?

5.28 In the graph in Fig. 5P.10 a minimum set of edges is to be selected such that every vertex is incident with at least one of the edges in the set. Solve this problem as a minimum-flow problem associated with a transport network.

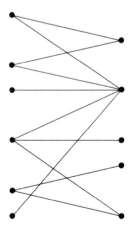

Figure 5P.10

5.29 Three candidates x_1, x_2, and x_3 have been promised minimum amounts of campaign money of $40,000, $23,000, and $50,000, respectively. Each candidate, in turn, has promised three campaign areas at least the amounts of money shown in the table below. In addition, each candidate will need at least $5,000 for his own expenses.

Candidate	Campaign area		
	C_1	C_2	C_3
x_1	$20,000	$10,000	$10,000
x_2	10,000	5,000	2,000
x_3	5,000	10,000	20,000

If the three campaign areas C_1, C_2, and C_3 require a minimum of $30,000, $25,000, and $50,000, respectively, to conduct a thorough campaign, what is a minimum amount of campaign money each candidate must obtain and how should it be distributed?

DISCRETE NUMERIC FUNCTIONS AND GENERATING FUNCTIONS

6.1 INTRODUCTION

We recall that a function is a binary relation that assigns to each element in the domain a unique value which is an element in the range. In this and the next chapter we shall study the class of functions whose domain is the set of natural numbers and whose range is the set of real numbers. We shall refer to these functions as *discrete numeric functions* or, briefly, *numeric functions*. Numeric functions are of particular interest to us because we encounter them very often in digital computation.

We shall use bold face lower case letters to denote numeric functions. For a numeric function **a** we use a_0, a_1, a_2, ..., a_r, ... to denote the values of the function at 0, 1, 2, ..., r, ...† A numeric function can be specified by an exhaustive listing of its values $(a_0, a_1, a_2, ...)$ or, if possible, by closed form expressions for a_r.

Let us consider some examples:

Example 6.1 Suppose we deposit $100 in a savings account at an interest rate of 7 percent per year, compounded annually. At the end of the first year, the total amount in the account is $107, at the end of the second year the total amount in the account is $114.49, at the end of the third year the

† This is merely a simpler notation than that we introduced in Sec. 3.6, where we used $a(0)$, $a(1)$, $a(2)$, ..., $a(r)$, ...

total amount in the account is $122.50, and so on. The amount in the account at the end of each year can be described by a numeric function **a**, which can either be specified as (100, 107, 114.49, 122.50, ...) or as

$$a_r = 100(1.07)^r \qquad r \geq 0 \qquad \qquad \square$$

Example 6.2 Let a_r denote the altitude of an aircraft, in thousands of feet, at the rth minute. Suppose the aircraft takes off after spending 10 min on the ground, climbs up at a uniform speed to a cruising altitude of 30,000 ft in 10 min, starts to descend uniformly after 110 min of flying time, and lands 10 min later. We have

$$a_r = \begin{cases} 0 & 0 \leq r \leq 10 \\ 3(r-10) & 11 \leq r \leq 20 \\ 30 & 21 \leq r \leq 120 \\ 3(130-r) & 121 \leq r \leq 130 \\ 0 & r \geq 131 \end{cases} \qquad \qquad \square$$

6.2 MANIPULATION OF NUMERIC FUNCTIONS

The *sum* of two numeric functions is a numeric function whose value at r is equal to the sum of the values of the two numeric functions at r. The *product* of two numeric functions is a numeric function whose value at r is equal to the product of the values of the two numeric functions at r. For example, consider the two functions **a** and **b** where

$$a_r = \begin{cases} 0 & 0 \leq r \leq 2 \\ 2^{-r} + 5 & r \geq 3 \end{cases}$$

and

$$b_r = \begin{cases} 3 - 2^r & 0 \leq r \leq 1 \\ r + 2 & r \geq 2 \end{cases}$$

Let **c** be the sum of the two functions **a** and **b**, which we shall denote $a + b$. Then

$$c_r = a_r + b_r = \begin{cases} 3 - 2^r & 0 \leq r \leq 1 \\ 4 & r = 2 \\ 2^{-r} + r + 7 & r \geq 3 \end{cases}$$

Let **d** be the product of the two functions **a** and **b**, which we shall denote **ab**. Then

$$d_r = a_r b_r = \begin{cases} 0 & 0 \leq r \leq 2 \\ r \cdot 2^{-r} + 2^{-r+1} + 5r + 10 & r \geq 3 \end{cases}$$

For example, if **a** denotes the monthly income of a husband and **b** denotes the monthly income of his wife, then $a + b$ will be their joint monthly income. Let **a** denote the balance in a savings account in each month and **b** denote the monthly interest rate which fluctuates from month to month, then **ab** will be the interest earned in each month.

Let **a** be a numeric function and α be a real number. We use α**a** to denote a numeric function whose value at r is equal to α times a_r. The numeric function α**a** is called a *scaled version* of **a** with *scaling factor* α. For example, let **a** be a numeric function whose value at r is $(1.07)^r$. Then 100**a** is a numeric function whose value at r is $100(1.07)^r$. Clearly, if **a** describes the total amount in a savings account through the years for an initial deposit of $1, then 100**a** describes the total amount in the account for an initial deposit of $100.

Let **a** be a numeric function and i a positive integer. We use S^i**a** to denote a numeric function such that its value at r is 0 for $r = 0, 1, 2, \ldots, i - 1$ and is a_{r-i} for $r \geq i$. For example, let

$$a_r = \begin{cases} 1 & 0 \leq r \leq 10 \\ 2 & r \geq 11 \end{cases}$$

Let **b** be S^5**a**. Then

$$b_r = \begin{cases} 0 & 0 \leq r \leq 4 \\ 1 & 5 \leq r \leq 15 \\ 2 & r \geq 16 \end{cases}$$

As another example, let **a** describe the altitude of an aircraft as given in the foregoing. Then S^{17}**a** describes the altitude when takeoff is delayed by 17 min.

Let **a** be a numeric function and i a positive integer. We use S^{-i}**a** to denote a numeric function such that its value at r is a_{r+i} for $r \geq 0$. For example, let

$$a_r = \begin{cases} 1 & 0 \leq r \leq 10 \\ 2 & r \geq 11 \end{cases}$$

Let **b** be S^{-7}**a**. Then

$$b_r = \begin{cases} 1 & 0 \leq r \leq 3 \\ 2 & r \geq 4 \end{cases}$$

As another example, let **a** be a numeric function where a_r is the number of applicants whose height is r inches above the minimum height requirement to join the Army. It follows that the numeric function S^{-2}**a** describes the numbers of applicants in the height groups if two inches have been added to the minimum height requirement.

The *accumulated sum* of a numeric function **a** is a numeric function whose value at r is equal to $\sum_{i=0}^{r} a_i$. For example, let **a** describe the monthly earnings of an employee. Let **b** be the accumulated sum of **a**. Then **b** gives her accumulated earnings by month. As another example, let

$$a_r = 100(1.07)^r \qquad r \geq 0$$

Let **b** be the accumulated sum of **a**. Then

$$b_r = \sum_{i=0}^{r} a_i = \sum_{i=0}^{r} 100(1.07)^i = \frac{10,000}{7}[(1.07)^{r+1} - 1] \qquad r \geq 0$$

Note that if we deposit $100 each year in a savings account at an annual compounded interest rate of 7 percent, then b_r is the total amount we have in the account after r years.

The *forward difference* of a numeric function $\mathbf{a}$ is a numeric function, denoted $\Delta\mathbf{a}$, whose value at r is equal to $a_{r+1} - a_r$. The *backward difference* of a numeric function $\mathbf{a}$ is a numeric function, denoted $\nabla\mathbf{a}$, whose value is equal to a_0 at 0 and is equal to $a_r - a_{r-1}$ at $r \geq 1$. Thus, if $\mathbf{a}$ describes the total business income of a company in each month, $\Delta\mathbf{a}$ will describe the increase of income from the rth month to the $(r+1)$st month and $\nabla\mathbf{a}$ will describe the increase of income for the rth month over the $(r-1)$st month. For example, let $\mathbf{a}$ be a numeric function such that

$$a_r = \begin{cases} 0 & 0 \leq r \leq 2 \\ 2^{-r} + 5 & r \geq 3 \end{cases}$$

Let $\mathbf{b}$ denote the forward difference of $\mathbf{a}$, $\Delta\mathbf{a}$; then

$$b_r = \begin{cases} 0 & 0 \leq r \leq 1 \\ \frac{41}{8} & r = 2 \\ -2^{-(r+1)} & r \geq 3 \end{cases}$$

Let $\mathbf{c}$ denote the backward difference of $\mathbf{a}$, $\nabla\mathbf{a}$; then

$$c_r = \begin{cases} 0 & 0 \leq r \leq 2 \\ \frac{41}{8} & r = 3 \\ -2^{-r} & r \geq 4 \end{cases}$$

We note that $S^{-1}(\nabla\mathbf{a}) = \Delta\mathbf{a}$.

Let $\mathbf{a}$ and $\mathbf{b}$ be two numeric functions. The *convolution* of $\mathbf{a}$ and $\mathbf{b}$, denoted $\mathbf{a} * \mathbf{b}$, is a numeric function $\mathbf{c}$ such that

$$c_r = a_0 b_r + a_1 b_{r-1} + a_2 b_{r-2} + \cdots + a_i b_{r-i} + \cdots + a_{r-1} b_1 + a_r b_0 = \sum_{i=0}^{r} a_i b_{r-i}$$

For example, let $\mathbf{a}$ and $\mathbf{b}$ be the two numeric functions such that

$$a_r = 3^r \qquad r \geq 0$$

and

$$b_r = 2^r \qquad r \geq 0$$

Then, for $\mathbf{c} = \mathbf{a} * \mathbf{b}$

$$c_r = \sum_{i=0}^{r} 3^i 2^{r-i} \qquad r \geq 0$$

Example 6.3 Consider the problem of determining c_r, the number of sequences of length r that are made up of the letters $\{x, y, z, \alpha, \beta\}$, with the first portion of each sequence made up of English letters and the second portion made up of Greek letters. Since

$$a_r = 3^r$$

is the number of sequences of length r made up of the letters $\{x, y, z\}$ and

$$b_r = 2^r$$

is the number of sequences of length r made up of the letters $\{\alpha, \beta\}$,

$$c_r = \sum_{i=0}^{r} 3^i 2^{r-i}$$

is the answer to our question. ☐

Example 6.4 Consider again the example of a savings account that receives annually compounded interest at a rate of 7 percent per year. Suppose we deposit \$100 in the account at the beginning, \$110 at the end of the first year, \$120 at the end of the second year, $\ldots$, \$$100(1 + 0.1r)$ at the end of the rth year, and so on. We want to know the total amount in the account at the end of the rth year. Let the numeric function $\mathbf{a}$ describe the yearly deposit, that is,

$$a_r = 100(1 + 0.1r) \qquad r \geq 0$$

Let $\mathbf{b}$ be a numeric function such that

$$b_r = (1.07)^r \qquad r \geq 0$$

Note that b_r is the amount in the savings account at the end of the rth year if \$1 was deposited into the account at the beginning (or at the end of the 0th year). Thus, if a_i dollars was deposited in the savings account at the end of the ith year, it will become $a_i b_{r-i}$ dollars $r - i$ years later (or at the end of the rth year). It follows that

$$c_r = \sum_{i=0}^{r} a_i b_{r-i}$$

is the total amount in the savings account at the end of the rth year. ☐

Example 6.5 We consider the financial operation of a company. For every \$100,000 of orders it receives, the company must borrow \$70,000 as working capital for materials, machinery, wages, and so on. In particular, the company will need \$30,000 immediately, \$20,000 in the next month, and \$5,000 in each of the subsequent four months. Thus, the needs for new working capital, in thousands of dollars, can be described by a numeric function $\mathbf{a}$ such that

$$a_r = \begin{cases} 30 & r = 0 \\ 20 & r = 1 \\ 5 & 2 \leq r \leq 5 \\ 0 & r \geq 6 \end{cases}$$

For every \$1000 the company borrows, it must repay a total of \$1100 in 11 installments of \$100 each for 11 months starting with the month after the money was borrowed. Let **b** be a numeric function describing the repayment schedule, in thousands of dollars. We have

$$b_r = \begin{cases} 0 & r = 0 \\ 0.1 & r = 1, 2, \ldots, 11 \\ 0 & r \geq 12 \end{cases}$$

Suppose the company has total orders of $200,000 in each of the first ten months and total orders of $300,000 in subsequent months. We want to know the total repayment in the rth month. Let **c** be a numeric function such that

$$c_r = \begin{cases} 2 & 0 \leq r \leq 9 \\ 3 & r \geq 10 \end{cases}$$

represents the orders received in the rth month, in hundreds of thousands of dollars. We note that $\mathbf{a} * \mathbf{b}$ will give the monthly repayment corresponding to a single order of $100,000 at the beginning. We ask the reader to check that $\mathbf{a} * \mathbf{b} = (0, 3, 5, 5.5, 6, 6.5, 7, 7, 7, 7, 7, 7, 4, 2, 1.5, 1, .5, 0, 0, \ldots)$. Consequently, $\mathbf{c} * (\mathbf{a} * \mathbf{b})$ will give the repayment schedule of all debts in thousands of dollars.† Again, we ask the reader to check that $\mathbf{c} * (\mathbf{a} * \mathbf{b}) = (0, 6, 16, 27, \ldots)$. □

6.3 GENERATING FUNCTIONS

We introduce now an alternative way to represent numeric functions. For a numeric function $(a_0, a_1, a_2, \ldots, a_r, \ldots)$, we define an infinite series

$$a_0 + a_1 z + a_2 z^2 + \cdots + a_r z^r + \cdots$$

which is called the *generating function* of the numeric function **a**. Indeed, we can take the viewpoint that we have introduced nothing more than a new notation at this moment. Instead of writing down the values of a numeric function one by one and using commas as "separators" as in $(a_0, a_1, a_2, \ldots, a_r, \ldots)$, we choose a formal variable z and use the powers of z as "indicators" in an infinite series such that the coefficient of z^r is the value of the numeric function at r. For a numeric function **a**, we shall use the corresponding upper case letter and write $A(z)$ to denote the generating function of **a**. Clearly, given a numeric function, we can easily obtain its generating function, and conversely. For example, the generating function of the numeric function $(3^0, 3^1, 3^2, \ldots, 3^r, \ldots)$ is

$$3^0 + 3z + 3^2 z^2 + 3^3 z^3 + \cdots + 3^r z^r + \cdots \tag{6.1}$$

† We ask the reader to ascertain that expressing a_r and b_r in thousands of dollars and c_r in hundreds of thousands of dollars will yield the repayment schedule in thousands of dollars.

We note that the infinite series in (6.1) can be written in closed form as

$$\frac{1}{1-3z}$$

which is a rather compact way to represent the numeric function $(1, 3, 3^2, \ldots, 3^r, \ldots)$. Indeed, as we shall see, the reason we introduce the generating-function representation of numeric functions is the possibility of expressing infinite series in closed form so that they can be manipulated conveniently.

The reader can readily check that, corresponding to $\mathbf{b} = \alpha \mathbf{a}$, we have $B(z) = \alpha A(z)$. Thus, for example, the generating function of the numeric function

$$a_r = 7 \cdot 3^r \qquad r \geq 0$$

is

$$A(z) = \frac{7}{1-3z}$$

and the generating function of the numeric function

$$a_r = 3^{r+2} \qquad r \geq 0$$

is

$$A(z) = \frac{9}{1-3z}$$

It is also obvious that corresponding to $\mathbf{c} = \mathbf{a} + \mathbf{b}$, we have $C(z) = A(z) + B(z)$. Thus, the generating function of the numeric function

$$a_r = 2^r + 3^r \qquad r \geq 0$$

is

$$A(z) = \frac{1}{1-2z} + \frac{1}{1-3z}$$

Also, corresponding to the generating function

$$A(z) = \frac{2 + 3z - 6z^2}{1-2z}$$

which can be written as

$$A(z) = 3z + \frac{2}{1-2z}$$

we have

$$a_r = \begin{cases} 2 & r = 0 \\ 7 & r = 1 \\ 2^{r+1} & r \geq 2 \end{cases}$$

Let $\mathbf{a}$ be a numeric function and $A(z)$ its generating function. Let $\mathbf{b}$ be a numeric function such that

$$b_r = \alpha^r a_r$$

for some constant α. Since the generating function of $\mathbf{b}$ is

$$a_0 + \alpha a_1 z + \alpha^2 a_2 z^2 + \cdots + \alpha^r a_r z^r + \cdots$$
$$= a_0 + a_1(\alpha z) + a_2(\alpha z)^2 + \cdots + a_r(\alpha z)^r + \cdots$$

we have $B(z) = A(\alpha z)$. For example, since the generating function of the numeric function

$$a_r = 1 \qquad r \geq 0$$

is

$$A(z) = \frac{1}{1-z}$$

the generating function of the numeric function

$$a_r = \alpha^r \qquad r \geq 0$$

is

$$A(z) = \frac{1}{1 - \alpha z}$$

As another example, corresponding to the generating function

$$A(z) = \frac{2}{1 - 4z^2}$$

which can be written as

$$A(z) = \frac{1}{1-2z} + \frac{1}{1+2z}$$

we have

$$a_r = 2^r + (-2)^r \qquad r \geq 0$$

or

$$a_r = \begin{cases} 0 & r \text{ odd} \\ 2^{r+1} & r \text{ even} \end{cases}$$

Unfortunately, for the case $\mathbf{c} = \mathbf{ab}$, there is no simple form in which we can express $C(z)$ in terms of $A(z)$ and $B(z)$.

Let $A(z)$ be the generating function of $\mathbf{a}$. It follows that $z^i A(z)$ is the generating function of $S^i \mathbf{a}$ for any positive integer i. For example, corresponding to

$$A(z) = \frac{z^4}{1 - 2z}$$

we have

$$a_r = \begin{cases} 0 & 0 \leq r \leq 3 \\ 2^{r-4} & r \geq 4 \end{cases}$$

Also, $z^{-i}[A(z) - a_0 - a_1 z - a_2 z^2 - \cdots - a_{i-1} z^{i-1}]$ is the generating function for $S^{-i}\mathbf{a}$. Thus, the generating function for

$$a_r = 3^{r+2} \qquad r \geq 0$$

can be computed as

$$A(z) = z^{-2}\left(\frac{1}{1-3z} - 1 - 3z\right)$$

$$= z^{-2}\frac{9z^2}{1-3z}$$

$$= \frac{9}{1-3z}$$

For another example, let

$$a_r = \begin{cases} 0 & 0 \le r \le 8 \\ 2 & r = 9 \\ 3 & r = 10 \\ 3 & r = 11 \\ 2 & r = 12 \\ 0 & r = 13 \\ 1 & r = 14 \\ 3 & r = 15 \\ 2 & r = 16 \\ 2 & r = 17 \\ 0 & r \ge 18 \end{cases}$$

represent the hourly productivity of a machinist who comes to work at 8 a.m., where r indicates the hour of the day. Then

$$A(z) = z^9(2 + 3z + 3z^2 + 2z^3 + z^5 + 3z^6 + 2z^7 + 2z^8)$$

Suppose there are three machinists. One comes to work at 6 a.m., one comes to work at 8 a.m., and one comes to work at 9 a.m., and they simply shift their productivity according to their starting time. If we let **b** be a numeric function describing the total hourly productivity, we have

$$B(z) = (z^{-2} + 1 + z)A(z)$$

For $\mathbf{b} = \Delta\mathbf{a}$, we have

$$B(z) = \frac{1}{z}[A(z) - a_0] - A(z)$$

and for $\mathbf{b} = \nabla\mathbf{a}$, we have

$$B(z) = A(z) - zA(z)$$

The generating function representation of numeric functions is particularly useful in dealing with the convolution of numeric functions. Let $\mathbf{c} = \mathbf{a} * \mathbf{b}$. Since

$$c_r = a_0b_r + a_1b_{r-1} + a_2b_{r-2} + \cdots + a_{r-1}b_1 + a_rb_0$$

is the coefficient of z^r in the product

$$(a_0 + a_1z + a_2z^2 + \cdots + a_rz^r + \cdots)(b_0 + b_1z + b_2z^2 + \cdots + b_rz^r + \cdots)$$

we conclude that the generating function $C(z)$ is equal to $A(z)B(z)$.

Example 6.6 Let $c = a * b$ where

$$a_r = 3^r \qquad r \geq 0$$

and

$$b_r = 2^r \qquad r \geq 0$$

Since

$$A(z) = \frac{1}{1 - 3z}$$

and

$$B(z) = \frac{1}{1 - 2z}$$

we have

$$C(z) = A(z)B(z) = \frac{1}{1 - 3z} \frac{1}{1 - 2z} = \frac{3}{1 - 3z} - \frac{2}{1 - 2z}$$

or

$$c_r = 3(3)^r - 2(2)^r = 3^{r+1} - 2^{r+1} \qquad \square$$

Example 6.7 Let a be an arbitrary numeric function and b be the numeric function $(1, 1, 1, \ldots, 1, \ldots)$. Let

$$c = a * b$$

so that

$$c_r = \sum_{i=0}^{r} a_i b_{r-i} = \sum_{i=0}^{r} a_i$$

Thus c is the accumulated sum of a, and the generating function of c is

$$C(z) = \frac{1}{1 - z} A(z)$$

For example, by letting $A(z)$ be $1/(1 - z)$ we obtain the result that $1/[(1 - z)^2]$ is the generating function of the numeric function $(1, 2, 3, \ldots, r, \ldots)$. $\qquad \square$

Example 6.8 We wish to evaluate the sum

$$1^2 + 2^2 + 3^2 + \cdots + r^2$$

Let us first determine the generating function of the numeric function $(0^2, 1^2, 2^2, 3^2, \ldots, r^2, \ldots)$. Differentiating both sides of the identity

$$\frac{1}{1 - z} = 1 + z + z^2 + z^3 + z^4 + \cdots + z^r + \cdots$$

we obtain

$$\frac{1}{(1 - z)^2} = 1 + 2z + 3z^2 + 4z^3 + \cdots + rz^{r-1} + \cdots$$

It follows that

$$\frac{d}{dz} \frac{z}{(1 - z)^2} = 1^2 + 2^2 \cdot z + 3^2 \cdot z^2 + 4^2 \cdot z^3 + \cdots + r^2 \cdot z^{r-1} + \cdots$$

and that

$$z \frac{d}{dz} \frac{z}{(1-z)^2} = 0^2 + 1^2 \cdot z + 2^2 \cdot z^2 + 3^2 \cdot z^3 + 4^2 \cdot z^4 + \cdots + r^2 \cdot z^r + \cdots$$

Thus, $z(d/dz)[z/(1-z)^2]$, which is equal to $[z(1+z)]/(1-z)^3$, is the generating function of the numeric function $(0^2, 1^2, 2^2, 3^2, \ldots, r^2, \ldots)$, and it follows that $[z(1+z)]/(1-z)^4$ is the generating function of the numeric function $(0^2, 0^2 + 1^2, 0^2 + 1^2 + 2^2, 0^2 + 1^2 + 2^2 + 3^2, \ldots, 0^2 + 1^2 + 2^2 + 3^2 + \cdots + r^2, \ldots)$. According to the binomial theorem,† the coefficient of z^r in $1/(1-z)^4$ is

$$\frac{(-4)(-4-1)(-4-2) \cdots (-4-r+1)}{r!}(-1)^r = \frac{4 \times 5 \times 6 \times \cdots \times (r+3)}{r!}$$

$$= \frac{(r+1)(r+2)(r+3)}{1 \cdot 2 \cdot 3}$$

Therefore, the coefficient of z^r in the expansion of $[z(1+z)]/(1-z)^4$ is

$$\frac{r(r+1)(r+2)}{1 \cdot 2 \cdot 3} + \frac{(r-1)r(r+1)}{1 \cdot 2 \cdot 3} = \frac{r(r+1)(2r+1)}{6}$$

that is,

$$1^2 + 2^2 + 3^2 + \cdots + r^2 = \frac{r(r+1)(2r+1)}{6} \qquad \square$$

*6.4 COMBINATORIAL PROBLEMS

The generating function representation of numeric functions is most useful in solving combinatorial problems. We shall present in this section some illustrative examples. Consider the numeric function **a** such that

$$a_r = C(n, r)$$

for a fixed n.‡ The generating function of **a** is

$$A(z) = C(n, 0) + C(n, 1)z + C(n, 2)z^2 + \cdots + C(n, r)z^r + \cdots + C(n, n)z^n$$

Let us consider a roundabout way to select r objects from n objects. We can first divide the n objects into two piles, one with k objects and the other with $n - k$ objects, for some fixed k less than n. We can then select i objects from the first pile, where there are $C(k, i)$ ways, and $r - i$ objects from the second pile, where there are $C(n - k, r - i)$ ways, for $i = 0, 1, 2, \ldots, r$. Thus we have

† The binomial theorem is

$$(1+z)^n = 1 + \sum_{r=1}^{} \frac{n(n-1)(n-2) \cdots (n-r+1)}{r!} z^r$$

where the upper limit of the summation is n if n is a positive integer, and is ∞ otherwise.
‡ We recall that $C(n, r) = 1$ for $r = 0$ and $C(n, r) = 0$ for $r > n$.

the equation

$$C(n, r) = \sum_{i=0}^{r} C(k, i)C(n - k, r - i)$$

It follows that we can write $\mathbf{a} = \mathbf{d} * \mathbf{e}$ where

$$D(z) = C(k, 0) + C(k, 1)z + C(k, 2)z^2 + \cdots + C(k, k)z^k$$

and

$$E(z) = C(n - k, 0) + C(n - k, 1)z + C(n - k, 2)z^2 + \cdots + C(n - k, n - k)z^{n-k}$$

Repeating the argument a sufficient number of times, we obtain

$$A(z) = (1 + z)^n \tag{6.2}$$

because the generating function of the numeric function $(C(1, 0), C(1, 1), 0, 0, \ldots)$ is $1 + z$. Let us also point out a very simple combinatorial argument that can also be used to derive (6.2). Consider the coefficient of the term z^r in the expansion of $(1 + z)^n$. In computing the product of the n factors $1 + z$, each factor will "contribute" either a 1 or a z. In particular, to make up the term z^r, r of the factors contribute a z each and $n - r$ of the factors contribute a 1 each. Consequently, the coefficient of z^r is the number of ways of selecting r of the n $1 + z$ factors to make up the term z^r, which, of course, is equal to $C(n, r)$. We show now some results that can be derived directly from (6.2).

Example 6.9 Setting z to 1 in (6.2), we obtain

$$C(n, 0) + C(n, 1) + \cdots + C(n, r) + \cdots + C(n, n) = 2^n$$

That is, the number of ways to select none, or one, or two, $\ldots$, or n objects from n objects is 2^n.

Setting z to -1 in (6.2) we obtain

$$C(n, 0) - C(n, 1) + C(n, 2) + \cdots + (-1)^r C(n, r) + \cdots$$
$$\cdots + (-1)^n C(n, n) = 0$$

or

$$C(n, 0) + C(n, 2) + C(n, 4) + \cdots = C(n, 1) + C(n, 3) + C(n, 5) + \cdots$$

That is, the number of ways of selecting an even number of objects from n objects is equal to the number of ways of selecting an odd number of objects. $\qquad\square$

Example 6.10 The relation

$$C(n, r) = C(n - 1, r) + C(n - 1, r - 1) \tag{6.3}$$

can be proved in several ways. By straightforward algebraic manipulation, we obtain

$$C(n, r) = \frac{n!}{r!(n-r)!}$$

$$= \frac{n!}{r!(n-r)!} \left(\frac{n-r}{n} + \frac{r}{n} \right)$$

$$= \frac{(n-1)!}{r!(n-r-1)!} + \frac{(n-1)!}{(r-1)!(n-r)!}$$

$$= C(n-1, r) + C(n-1, r-1)$$

We can also prove (6.3) using a combinatorial argument: If we are to select r objects from n objects, there are $C(n-1, r)$ ways to select r objects so that a particular object is always excluded, and $C(n-1, r-1)$ ways so that this particular object is always included. Thus, (6.3) follows. The use of generating functions provides a third proof of (6.3): Note that $C(n, r)$ is the coefficient of z^r in $(1+z)^n$, which can be written as $(1+z)^{n-1} + z(1+z)^{n-1}$. Since the coefficient of z^r in $(1+z)^{n-1}$ is $C(n-1, r)$, and the coefficient of z^r in $z(1+z)^{n-1}$ is $C(n-1, r-1)$, (6.3) follows immediately. □

We present now some further examples:

Example 6.11 Let $(a_0, a_1, a_2, \ldots)$ be a numeric function such that a_r is equal to the number of ways to select r objects from 10 objects among which one, which will be denoted X, can be selected at most twice, one, which will be denoted Y, can be selected at most thrice, and the others can be selected only once. We claim that the generating function for **a** is

$$A(z) = (1 + z + z^2)(1 + z + z^2 + z^3)(1 + z)^8$$

Note that the coefficient of z^r in $A(z)$ is the number of ways to make up the term z^r from the factors $1 + z + z^2$, $1 + z + z^2 + z^3$, and the eight factors $1 + z$. The contribution from the factor $1 + z + z^2$ can be 1, z, or z^2, corresponding to selecting the object X zero times, one time, or twice. The contribution from the factor $1 + z + z^2 + z^3$ can be 1 or z or z^2 or z^3, corresponding to selecting the object Y zero times, one time, twice, or thrice. The contribution of each of the eight factors $1 + z$ can be 1 or z, corresponding to selecting each of the eight remaining objects zero times or one time. □

Example 6.12 Let a_r denote the number of ways of selecting r objects from n objects with unlimited repetitions. Since each object can be selected as many times as we wish, a_r is equal to the coefficient of z^r in

$$(1 + z + z^2 + \cdots)^n$$

Note that the contribution from each of the factors $1 + z + z^2 + \cdots$ will correspond to the number of times one of the objects is selected. Since

$$(1 + z + z^2 + \cdots)^n = \left(\frac{1}{1-z}\right)^n = (1-z)^{-n}$$

we obtain

$$a_r = (-1)^r \frac{(-n)(-n-1) \cdots (-n-r+1)}{r!}$$

$$= \frac{(n+r-1)!}{r!(n-r)!}$$

$$= C(n+r-1, r)$$

This result is a result we derived earlier in Chap. 2. $\qquad\square$

Example 6.13 Suppose we want to determine the number of ways in which $2t + 1$ marbles can be distributed among three distinct boxes so that no box will contain more than t marbles. We claim that the coefficient of z^{2t+1} in

$$A(z) = (1 + z + z^2 + \cdots + z^t)^3$$

will be our answer. Note that the coefficient of z^{2t+1} in $A(z)$ is the number of ways to make up the term z^{2t+1} from the three factors $1 + z + z^2 + \cdots + z^t$. The contribution from each factor $1 + z + z^2 + \cdots + z^t$ can be 1, z, z^2, ..., or z^t corresponding to having none, one, two, ..., or t marbles in a box. Since

$$(1 + z + z^2 + \cdots + z^t)^3 = \left(\frac{1-z^{t+1}}{1-z}\right)^3$$

$$= (1 - 3z^{t+1} + 3z^{2t+2} - z^{3t+3})(1-z)^{-3} \qquad (6.4)$$

the coefficient of z^{2t+1} in (6.4) is the coefficient of z^{2t+1} in $(1-z)^{-3}$ minus three times the coefficient of z^t in $(1-z)^{-3}$. Thus, it is

$$C(3 + 2t + 1 - 1, 2t + 1) - 3C(3 + t - 1, t)$$

which is simplified as

$$C(2t + 3, 2t + 1) - 3C(t + 2, t) \qquad\qquad\square$$

6.5 REMARKS AND REFERENCES

See chap. 3 of Beckenbach[1], chap. 2 of Liu[4], chaps. 1 and 5 of Liu and Liu[5], chaps. 1 and 2 of Riordan[6] on numeric functions and their generating functions. See chap. 6 of Chung[2] and chap. 11 of Feller[3] on application of the technique of generating functions in probability theory.

[1] Beckenbach, E. F. (Ed.): "Applied Combinatorial Mathematics," John Wiley & Sons, Inc., New York, 1964.

[2] Chung, K. L.: "Elementary Probability Theory with Stochastic Processes," Springer-Verlag, New York, 1974.

[3] Feller, W.: "An Introduction to Probability Theory and Its Applications," Vol. 1, 2d ed., John Wiley & Sons, Inc., New York, 1957.

[4] Liu, C. L.: "Introduction to Combinatorial Mathematics," McGraw-Hill Book Company, New York, 1968.

[5] Liu, C. L., and J. W. S. Liu: "Linear Systems Analysis," McGraw-Hill Book Company, New York, 1975.

[6] Riordan, J.: "An Introduction to Combinatorial Analysis," John Wiley & Sons, Inc., New York, 1958.

PROBLEMS

6.1 A ping-pong ball is dropped to the floor from a height of 20 m. Suppose that the ball always rebounds to reach half of the height from which it falls.

(a) Let a_r denote the height it reaches in the rth rebound. Sketch the numeric function **a**.

(b) Let b_r denote the loss in height during the rth rebound. Express b_r in terms of a_r. Sketch the numeric function **b**.

(c) A second ping-pong ball is dropped from a height of 6 m. to the same floor at the same time as the first ball reaches the highest point of its third rebound. Let c_r denote the height the second ball reaches in its rth rebound. Express c_r in terms of a_r.

6.2 In a process control system, a monitoring device measures the temperature inside a chemical reaction chamber once every 30 s. Let a_r denote the rth reading in degrees in centigrade. Determine an expression for a_r if it is known that the temperature rises from 100 to $120°$ at a constant rate in the first 300 s and stays at $120°$ from then on.

6.3 Let a be a numeric function such that a_r is equal to the remainder when the integer r is divided by 17. Let **b** be a numeric function such that b_r is equal to 0 if the integer r is divisible by 3, and is equal to 1 otherwise.

(a) Let $c_r = a_r + b_r$. For what values of r is $c_r = 0$? For what values of r is $c_r = 1$?

(b) Let $d_r = a_r b_r$. For what values of r is $d_r = 0$? For what values of r is $d_r = 1$?

6.4 Let a be a numeric function such that

$$a_r = \begin{cases} 2 & 0 \leq r \leq 3 \\ 2^{-r} + 5 & r \geq 4 \end{cases}$$

(a) Determine $S^2 \mathbf{a}$ and $S^{-2}\mathbf{a}$.

(b) Determine $\Delta \mathbf{a}$ and $\nabla \mathbf{a}$.

6.5 We introduce the notation

$$\Delta^2 \mathbf{a} = \Delta(\Delta \mathbf{a})$$

$$\Delta^3 \mathbf{a} = \Delta(\Delta^2 \mathbf{a})$$

$$\cdots\cdots\cdots\cdots$$

$$\Delta^i \mathbf{a} = \Delta(\Delta^{i-1}\mathbf{a})$$

(a) Let a be a numeric function such that

$$a_r = r^3 - 2r^2 + 3r + 2$$

Determine $\Delta \mathbf{a}$, $\Delta^2 \mathbf{a}$, $\Delta^3 \mathbf{a}$, $\Delta^4 \mathbf{a}$.

(b) Let a be a numeric function such that a_r is a polynomial of the form

$$\alpha_0 + \alpha_1 r + \alpha_2 r^2 + \cdots + \alpha_k r^k$$

Show that $\Delta^{k+1}\mathbf{a}$ is equal to 0.

6.6 (*a*) Let $c = ab$. Show that

$$\Delta c_r = a_{r+1}(\Delta b_r) + b_r(\Delta a_r)$$

(*b*) Let **a** and **b** be two numeric functions such that $a_r = r + 1$ and $b_r = \alpha^r$ for all $r \geq 0$. Determine $\Delta(\mathbf{ab})$.

(*c*) Let **a** and **b** be two numeric functions. The quotient of **a** and **b** denoted **a/b** is a numeric function whose value at r is equal to a_r/b_r. Let $\mathbf{d} = \mathbf{a/b}$. Show that

$$\Delta d_r = \frac{b_r(\Delta a_r) - a_r(\Delta b_r)}{b_r b_{r+1}}$$

(*d*) Determine $\Delta(\mathbf{a/b})$ for the numeric functions **a** and **b** in part (*b*).

6.7 Let **a** be a numeric function. Let $\Delta^{-1}\mathbf{a}$ denote the accumulated sum of **a**. Are the two numeric functions $\Delta(\Delta^{-1}\mathbf{a})$ and $\Delta^{-1}(\Delta\mathbf{a})$ the same?

6.8 Interest for money deposited in a savings account is paid at a rate of 0.5 percent per month, with interest compounded monthly. Suppose \$50 is deposited into a savings account each month for a period of five years. What is the total amount in the account four years after the first deposit? Twenty years after the first deposit?

6.9 Determine $\mathbf{a} * \mathbf{b}$ in the following. Give either a sketch or a closed-form expression of the resulting numeric function, whichever is more convenient.

(*a*)
$$a_r = \begin{cases} 1 & 0 \leq r \leq 2 \\ 0 & r \geq 3 \end{cases}$$

$$b_r = \begin{cases} 1 & 0 \leq r \leq 2 \\ 0 & r \geq 3 \end{cases}$$

(*b*)
$$a_r = 1 \quad \text{for all } r$$

$$b_r = \begin{cases} 1 & r = 1 \\ 2 & r = 3 \\ 3 & r = 5 \\ -6 & r = 7 \\ 0 & \text{otherwise} \end{cases}$$

(*c*)
$$a_r = 2^r \quad \text{for all } r$$

$$b_r = \begin{cases} 0 & 0 \leq r \leq 2 \\ 2^r & r \geq 3 \end{cases}$$

6.10 Let **a**, **b**, and **c** be numeric functions such that $\mathbf{a} * \mathbf{b} = \mathbf{c}$. Given that

$$a_r = \begin{cases} 1 & r = 0 \\ 2 & r = 1 \\ 0 & r \geq 2 \end{cases}$$

$$c_r = \begin{cases} 1 & r = 0 \\ 0 & r \geq 1 \end{cases}$$

determine **b**.

6.11 Let

$$a_r = \begin{cases} 1 & r = 0 \\ 3 & r = 1 \\ 2 & r = 2 \\ 0 & r \geq 3 \end{cases}$$

$$c_r = 5^r \quad \text{for all } r$$

Determine **b** so that $\mathbf{a} * \mathbf{b} = \mathbf{c}$.

6.12 (*a*) Every particle inside a nuclear reactor splits into two particles in each second. Suppose one particle is injected into the reactor every second beginning at $t = 0$. How many particles are there in the reactor at the nth second?

(*b*) When a particle splits into two, one of them is actually the original particle and the other is a newly created particle. Suppose a particle has a lifetime of 10 s. That is, a particle created in the 0th second will vanish after the 10th second. Repeat part (*a*) and assume that all injected particles are newly created.

6.13 For a fee of $1000, a radio station will provide a store with 10 min of commercial time each day for five consecutive days and then 2 min of commercial time each day for the next five consecutive days. Suppose that every minute of commercial time on a particular day will produce a total sale of $500 on that day and a total sale of 500×2^{-n} on the nth day from that day.

(*a*) Suppose a fee of $1000 is paid to the radio station on the 0th day. Determine the numeric function **a**, where a_r is the total sale on the rth day.

(*b*) Suppose a fee of $2000 is paid to the radio station daily for 10 consecutive days starting on the 0th day. Determine the numeric function **b** where b_r is the total sale on the rth day.

6.14 Determine the numeric functions corresponding to the following generating functions:

(*a*)
$$A(z) = \frac{z^5}{5 - 6z + z^2}$$

(*b*)
$$A(z) = \frac{1 + z^2}{4 - 4z + z^2}$$

(*c*)
$$A(z) = \frac{1}{(1 - z)(1 - z^2)(1 - z^3)}$$

6.15 (*a*) Let a_r denote the number of ways a sum of r can be obtained when two indistinguishable dice are rolled. Determine $A(z)$.

(*b*) Let b_r denote the number of ways a sum of r can be obtained when a single die is rolled once and then two indistinguishable dice are rolled once. Determine $B(z)$.

6.16 The life span of a rabbit is exactly 10 years. Suppose that there are two newborn rabbits at the beginning, and the number of newborn rabbits in each year is two times that in the previous year. Determine the number of rabbits there are in the rth year.

6.17 Let a_r denote the number of ways to select r balls from an infinite supply of red, blue, and white balls with the constraint that the number of red balls selected is always an even number. Determine the generating function for the numeric function **a**.

6.18 In how many ways can $3r$ balls be selected from $2r$ red balls, $2r$ blue balls, and $2r$ white balls?

6.19 Let a_r denote the number of ways to select r balls from 100 red balls, 50 blue balls, and 50 white balls with the constraints that the number of red balls selected is not equal to two, the number of blue balls selected is not equal to three, and the number of white balls selected is not equal to four. Determine the generating function for the numeric function **a**.

6.20 A set of r balls is selected from an infinite supply of red, blue, white, and yellow balls. A selection must satisfy the condition that either the number of red balls is even and the number of blue balls is odd, or the number of white balls is even and the number of yellow balls is odd. Let a_r denote the total number of such selections.

(*a*) Determine the generating function $A(z)$.

(*b*) Find a closed form expression for a_r. Evaluate the expression for $r = 23$.

6.21 Let a_r denote the number of ways to divide r identical marbles into four distinct piles so that each pile has an odd number of marbles that is larger than or equal to three.

(*a*) Determine $A(z)$.

(*b*) Determine a closed-form expression for a_r.

6.22 (*a*) Determine the sum

$$\binom{n}{1} + 2\binom{n}{2} + \cdots + i\binom{n}{i} + \cdots + n\binom{n}{n}$$

(b) If we write down all the combinations of n letters $\{1, 2, \ldots, n\}$, i.e., all 1-combinations, all 2-combinations, $\ldots$, how many times does each letter occur?

6.23 Determine the sum

$$\binom{n}{0}\binom{m}{k} + \binom{n}{1}\binom{m}{k-1} + \binom{n}{2}\binom{m}{k-2} + \cdots + \binom{n}{k}\binom{m}{0}$$

where $k \leq n$ and $k \leq m$.

6.24 (a) Let

$$A(z) = (1+z)^{2n} + z(1+z)^{2n-1} + \cdots + z^i(1+z)^{2n-i} + \cdots + z^n(1+z)^n$$

Show that

$$a_r = \begin{cases} \binom{2n+1}{r} & 0 \leq r \leq n \\ \binom{2n+1}{r} - \binom{n}{r-n-1} & n+1 \leq r \leq 2n+1 \\ 0 & \text{otherwise} \end{cases}$$

(b) Evaluate the sum

$$\binom{2n}{n} + \binom{2n-1}{n-1} + \cdots + \binom{2n-i}{n-i} + \cdots + \binom{n}{0}$$

6.25 (a) Show that

$$\binom{r}{0}^2 + \binom{r}{1}^2 + \binom{r}{2}^2 + \cdots + \binom{r}{i}^2 + \cdots + \binom{r}{r}^2 = \binom{2r}{r}$$

(b) Show that the generating function for $\mathbf{a}$, where

$$a_r = \binom{2r}{r}$$

is $(1 - 4z)^{-1/2}$.

RECURRENCE RELATIONS

7.1 INTRODUCTION

Consider the numeric function $\mathbf{a} = (3^0, 3^1, 3^2, \ldots, 3^r, \ldots)$. Clearly, the function can be specified by a general expression for a_r, namely,

$$a_r = 3^r \qquad r \geq 0$$

As was pointed out in Chap. 6, the function can also be specified by its generating function, namely,

$$A(z) = \frac{1}{1 - 3z}$$

We note that there is another way to specify the numeric function: Since the value of a_r is equal to three times the value of a_{r-1} for all r, once we know the value of a_0, which is equal to 1, we can determine the value of a_1, and then the value of a_2, and eventually the value of a_r in a step-by-step manner for any r. Thus, the relation

$$a_r = 3a_{r-1} \tag{7.1}$$

together with the information that $a_0 = 1$ also completely specifies the numeric function $\mathbf{a}$.

As another example, consider the sequence of numbers† known as the *Fibonacci sequence of numbers*. The sequence starts with the two numbers 1, 1 and contains numbers which are equal to the sum of their two immediate predecessors. A portion of the sequence is

$$1, 1, 2, 3, 5, 8, 13, 21, 34, \ldots$$

---

† Clearly, a numeric function can be viewed simply as a sequence of real numbers, and conversely.

It is quite difficult in this case to obtain a general expression for the rth number in the sequence by observation. On the other hand, the sequence can be described by the relation

$$a_r = a_{r-1} + a_{r-2}$$

together with the conditions that $a_0 = 1$ and $a_1 = 1$.

For a numeric function $(a_0, a_1, a_2, \ldots, a_r, \ldots)$, an equation relating a_r, for any r, to one or more of the a_i's, $i < r$, is called a *recurrence relation*. A recurrence relation is also called a *difference equation*, and these two terms will be used interchangeably. In many discrete computation problems, it is sometimes easier to obtain a specification of a numeric function in terms of a recurrence relation than to obtain a general expression for the value of the numeric function at r or a closed form expression for its generating function. It is clear that according to the recurrence relation, we can carry out a step-by-step computation to determine a_r from $a_{r-1}, a_{r-2}, \ldots$, to determine a_{r+1} from a_r, $a_{r-1}, \ldots$, and so on, provided that the value of the function at one or more points is given so that the computation can be initiated. These given values of the function are called *boundary conditions*. In the first example above, the boundary condition is $a_0 = 1$, and in the second example above, the boundary conditions are $a_0 = 1$ and $a_1 = 1$. We thus conclude that a numeric function can be described by a recurrence relation together with an appropriate set of boundary conditions.

One step beyond determining the values of a numeric function in a step-by-step computation according to a given recurrence relation is to obtain from the recurrence relation either a general expression for the value of the function at r or a closed form expression for its generating function. This is known as *solving* a recurrence relation. Unfortunately, no general method is known for solving all recurrence relations. In the following, we shall study the solution of a class of recurrence relations known as *linear recurrence relations with constant coefficients*.

7.2 LINEAR RECURRENCE RELATIONS WITH CONSTANT COEFFICIENTS

A recurrence relation of the form

$$C_0 a_r + C_1 a_{r-1} + C_2 a_{r-2} + \cdots + C_k a_{r-k} = f(r) \tag{7.2}$$

where the C_i's are constants, is called a *linear recurrence relation with constant coefficients*. The recurrence relation in (7.2) is known as a *kth-order recurrence relation*, provided that both C_0 and C_k are nonzero. For example,

$$3a_r - 5a_{r-1} + 2a_{r-2} = r^2 + 5 \tag{7.3}$$

is a second-order linear recurrence relation with constant coefficients.

If k consecutive values of the numeric function **a**, a_{m-k}, $a_{m-k+1}, \ldots$, a_{m-1}, are known for some m, the value of a_m can be calculated according to (7.2), namely,

$$a_m = -\frac{1}{C_0}[C_1 a_{m-1} + C_2 a_{m-2} + \cdots + C_k a_{m-k} - f(m)]$$

Furthermore, the value of a_{m+1} can be computed as

$$a_{m+1} = -\frac{1}{C_0}[C_1 a_m + C_2 a_{m-1} + \cdots + C_k a_{m-k+1} - f(m+1)]$$

and the values of a_{m+2}, a_{m+3}, ... can be computed in a similar manner. Also, the value of a_{m-k-1} can be computed as

$$a_{m-k-1} = -\frac{1}{C_k}[C_0 a_{m-1} + C_1 a_{m-2} + \cdots + C_{k-1} a_{m-k} - f(m-1)]$$

and the values of a_{m-k-2}, a_{m-k-3}, ... can be computed in a similar manner. For example, for the recurrence relation in (7.3), given that $a_3 = 3$ and $a_4 = 6$, we can compute a_5 as

$$a_5 = -\tfrac{1}{3}[-5 \times 6 + 2 \times 3 - (5^2 + 5)] = 18$$

we can then compute a_6 as

$$a_6 = -\tfrac{1}{3}[-5 \times 18 + 2 \times 6 - (6^2 + 5)] = \tfrac{119}{3}$$

and so on. Also, we can compute a_2 as

$$a_2 = -\tfrac{1}{2}[3 \times 6 - 5 \times 3 - (4^2 + 5)] = 9$$

and so on. We thus conclude that the solution to (7.2) is determined uniquely by the values of k consecutive a_i's. Indeed, for a kth-order linear recurrence relation, the values of k consecutive a_i's constitute an appropriate set of boundary conditions. In other words, the values of k consecutive a_i's will always be sufficient to determine the numeric function **a** uniquely. However, the values of k nonconsecutive a_i's might or might not constitute an appropriate set of boundary conditions, depending on the specific recurrence relation we have. We shall not study the problem of what constitutes an appropriate set of boundary conditions, since it is not a significantly important one. See, however, Prob. 7.5.

The (total) solution of a linear difference equation with constant coefficients is the sum of two parts, the *homogeneous solution*, which satisfies the difference equation when the right-hand side of the equation is set to 0, and the *particular solution*, which satisfies the difference equation with $f(r)$ on the right-hand side.† Let $\mathbf{a}^{(h)} = (a_0^{(h)}, a_1^{(h)}, \ldots, a_r^{(h)}, \ldots)$ denote the homogeneous solution and $\mathbf{a}^{(p)} = (a_0^{(p)}, a_1^{(p)}, \ldots, a_r^{(p)}, \ldots)$ denote the particular solution to the difference equation. Since

$$C_0 a_r^{(h)} + C_1 a_{r-1}^{(h)} + \cdots + C_k a_{r-k}^{(h)} = 0$$

---

† For a reader who has prior exposure to the topic of differential equations, the analogy between the solution of linear differential equations with constant coefficients and that of linear difference equations with constant coefficients should be quite obvious.

and

$$C_0 a_r^{(p)} + C_1 a_{r-1}^{(p)} + \cdots + C_k a_{r-k}^{(p)} = f(r)$$

we have

$$C_0(a_r^{(h)} + a_r^{(p)}) + C_1(a_{r-1}^{(h)} + a_{r-1}^{(p)}) + \cdots + C_k(a_{r-k}^{(h)} + a_{r-k}^{(p)}) = f(r)$$

Clearly, the total solution, $\mathbf{a} = \mathbf{a}^{(h)} + \mathbf{a}^{(p)}$, satisfies the difference equation.

A homogeneous solution of a linear difference equation with constant coefficients is of the form $A\alpha_1^r$, where α_1 is called a *characteristic root* and A is a constant determined by the boundary conditions. Substituting $A\alpha^r$ for a_r in the difference equation with the right-hand side of the equation set to 0, we obtain

$$C_0 A\alpha^r + C_1 A\alpha^{r-1} + C_2 A\alpha^{r-2} + \cdots + C_k A\alpha^{r-k} = 0$$

This equation can be simplified to

$$C_0\alpha^k + C_1\alpha^{k-1} + C_2\alpha^{k-2} + \cdots + C_k = 0$$

which is called the *characteristic equation* of the difference equation. Therefore, if α_1 is one of the roots of the characteristic equation (it is for this reason that α_1 is called a characteristic root), $A\alpha_1^r$ is a homogeneous solution to the difference equation.

A characteristic equation of kth degree has k characteristic roots. Suppose the roots of the characteristic equation are distinct. In this case it is easy to verify that

$$a_r^{(h)} = A_1\alpha_1^r + A_2\alpha_2^r + \cdots + A_k\alpha_k^r$$

is also a homogeneous solution to the difference equation, where $\alpha_1, \alpha_2, \ldots, \alpha_k$ are the distinct characteristic roots and $A_1, A_2, \ldots, A_k$ are constants which are to be determined by the boundary conditions.†

Example 7.1 Let us revisit the example of the Fibonacci sequence of numbers discussed in Sec. 7.1. The recurrence relation for the Fibonacci sequence of numbers is

$$a_r = a_{r-1} + a_{r-2}$$

The corresponding characteristic equation is

$$\alpha^2 - \alpha - 1 = 0$$

which has the two distinct roots

$$\alpha_1 = \frac{1 + \sqrt{5}}{2} \qquad \alpha_2 = \frac{1 - \sqrt{5}}{2}$$

It follows that

$$a_r^{(h)} = A_1\left(\frac{1 + \sqrt{5}}{2}\right)^r + A_2\left(\frac{1 - \sqrt{5}}{2}\right)^r$$

is a homogeneous solution where the two constants A_1 and A_2 are to be determined from the boundary conditions $a_0 = 1$ and $a_1 = 1$. □

† The question of the uniqueness of the solution will be discussed later.

Now suppose that some of the roots of the characteristic equation are multiple roots. Let α_1 be a root of multiplicity m. We shall show that the corresponding homogeneous solution is

$$(A_1 r^{m-1} + A_2 r^{m-2} + \cdots + A_{m-2} r^2 + A_{m-1} r + A_m) \alpha_1^r$$

where the A_i's are constants to be determined by the boundary conditions. It is clear that $A_m \alpha_1^r$ is a homogeneous solution of the difference equation in (7.2). To show that $A_{m-1} r \alpha_1^r$ is also a homogeneous solution, we recall that α_1 not only is a root of the equation

$$C_0 \alpha^r + C_1 \alpha^{r-1} + C_2 \alpha^{r-2} + \cdots + C_k \alpha^{r-k} = 0 \tag{7.4}$$

but also is a root of the derivative equation of (7.4),

$$C_0 r \alpha^{r-1} + C_1 (r-1) \alpha^{r-2} + C_2 (r-2) \alpha^{r-3} + \cdots + C_k (r-k) \alpha^{r-k-1} = 0 \quad (7.5)$$

because α_1 is a multiple root of (7.4). Multiplying (7.5) by $A_{m-1} \alpha$ and replacing α by α_1, we obtain

$$C_0 A_{m-1} r \alpha_1^r + C_1 A_{m-1} (r-1) \alpha_1^{r-1}$$
$$+ C_2 A_{m-1} (r-2) \alpha_1^{r-2} + \cdots + C_k A_{m-1} (r-k) \alpha_1^{r-k} = 0$$

which shows that $A_{m-1} r \alpha_1^r$ is indeed a homogeneous solution.

The fact that α_1 satisfies the second, third, ..., $(m-1)$st derivative equations of (7.4) enables us to prove that $A_{m-2} r^2 \alpha_1^r$, $A_{m-3} r^3 \alpha_1^r$, ..., $A_1 r^{m-1} \alpha_1^r$ are also homogeneous solutions in a similar manner.

Example 7.2 Consider the difference equation:

$$a_r + 6a_{r-1} + 12a_{r-2} + 8a_{r-3} = 0$$

The characteristic equation is

$$\alpha^3 + 6\alpha^2 + 12\alpha + 8 = 0$$

Thus,

$$a_r = (A_1 r^2 + A_2 r + A_3)(-2)^r$$

is a homogeneous solution since -2 is a triple characteristic root. □

There is no general way of finding the particular solution. However, in simple cases, it can be obtained by the method of inspection which we shall discuss briefly here. In the next section, we shall also present a method for determining both the homogeneous solution and the particular solution at the same time. Consider the difference equation

$$a_r + 5a_{r-1} + 6a_{r-2} = 3r^2 \tag{7.6}$$

We assume that the general form of the particular solution is[†]

$$P_1r^2 + P_2r + P_3 \qquad (7.7)$$

where P_1, P_2, and P_3 are constants to be determined. Substituting the expression in (7.7) into the left-hand side of (7.6), we obtain

$$P_1r^2 + P_2r + P_3 + 5P_1(r-1)^2 + 5P_2(r-1) + 5P_3 + 6P_1(r-2)^2 + 6P_2(r-2) + 6P_3$$

which simplifies to

$$12P_1r^2 - (34P_1 - 12P_2)r + (29P_1 - 17P_2 + 12P_3) \qquad (7.8)$$

Comparing (7.8) with the right-hand side of (7.6), we obtain the equations

$$12P_1 = 3$$

$$34P_1 - 12P_2 = 0$$

$$29P_1 - 17P_2 + 12P_3 = 0$$

which yield

$$P_1 = \tfrac{1}{4} \qquad P_2 = \tfrac{17}{24} \qquad P_3 = \tfrac{115}{288}$$

Therefore, the particular solution is

$$a_r^{(p)} = \tfrac{1}{4}r^2 + \tfrac{17}{24}r + \tfrac{115}{288}$$

In general, when the right-hand side of the difference equation is of the form r^t, the corresponding particular solution will be of the form

$$P_1r^t + P_2r^{t-1} + \cdots + P_tr + P_{t+1}$$

As another example, consider the difference equation

$$a_r + 5a_{r-1} + 6a_{r-2} = 42 \cdot 4^r \qquad (7.9)$$

We assume that the general form of the particular solution is

$$a_r = P4^r \qquad (7.10)$$

Substituting the expression in (7.10) into the left-hand side of (7.9), we obtain

$$P4^r + 5P4^{r-1} + 6P4^{r-2}$$

which simplifies to

$$\tfrac{21}{8}P4^r \qquad (7.11)$$

Comparing (7.11) with the right-hand side of (7.9), we obtain

$$P = 16$$

Therefore, the particular solution is

$$a_r = 16 \cdot 4^r$$

[†] There is no systematic way to determine a general form of the particular solution of a difference equation. In the present case, it is not difficult to see that if the right-hand side of the difference equation is a polynomial in r, the particular solution will also be a polynomial in r.

In general, when the right-hand side of the difference equation is of the form β^r, the corresponding particular solution is of the form $P\beta^r$ if β is not a characteristic root of the difference equation; it is of the form $P_r\beta^r$ if β is a distinct characteristic root of the difference equation; and it is of the form

$$P_r^{m-1}\beta^r$$

if β is a characteristic root of multiplicity $m-1$ of the difference equation. For further details, see Prob. 7.6.

Finally, we must determine the undetermined coefficients in the homogeneous solution. For a kth-order difference equation, the k undetermined coefficients $A_1, A_2, \ldots, A_k$ in the homogeneous solution can be determined by the boundary conditions, $a_{r_0}, a_{r_0+1}, \ldots, a_{r_0+k-1}$, for any r_0. Suppose the characteristic roots of the difference equation are all distinct. The total solution is of the form

$$a_r = A_1\alpha_1^r + A_2\alpha_2^r + \cdots + A_k\alpha_k^r + p(r)$$

where $p(r)$ is the particular solution. Thus, for $r = r_0, r_0+1, \ldots, r_0+k-1$, we have the system of linear equations:

$$a_{r_0} = A_1\alpha_1^{r_0} + A_2\alpha_2^{r_0} + \cdots + A_k\alpha_k^{r_0} + p(r_0)$$

$$a_{r_0+1} = A_1\alpha_1^{r_0+1} + A_2\alpha_2^{r_0+1} + \cdots + A_k\alpha_k^{r_0+1} + p(r_0+1) \qquad (7.12)$$

$$\cdots\cdots\cdots\cdots\cdots\cdots\cdots\cdots\cdots\cdots\cdots\cdots\cdots$$

$$a_{r_0+k-1} = A_1\alpha_1^{r_0+k-1} + A_2\alpha_2^{r_0+k-1} + \cdots + A_k\alpha_k^{r_0+k-1} + p(r_0+k-1)$$

These k linear equations can be solved for $A_1, A_2, \ldots, A_k$. For example, for the difference equation in (7.9), the total solution is

$$a_r = A_1(-2)^r + A_2(-3)^r + 16 \cdot 4^r$$

Suppose we are given the boundary conditions $a_2 = 278$ and $a_3 = 962$. Solving the equations

$$278 = 4A_1 + 9A_2 + 256$$

$$962 = -8A_1 - 27A_2 + 1024$$

we obtain

$$A_1 = 1 \qquad A_2 = 2$$

Thus,

$$a_r = (-2)^r + 2(-3)^r + 16 \cdot 4^r$$

is the total solution of the difference equation.

One might question how we can be sure that solutions of the k equations in (7.12) are always unique. It can be shown† that this is indeed the case. On the other hand, if we are given the value of the numeric function at k not necessarily successive points, although we can set up k equations for the undetermined coefficients $A_1, A_2, \ldots, A_k$ similar to that in (7.12), it is not always the case that these equations can be solved uniquely.

---

† See, for example, chap. 3 of Liu[5].

When the characteristic roots of the difference equation are not all distinct, a derivation similar to the foregoing can be carried out. Again, the undetermined coefficients in the homogeneous solution can be determined uniquely by the value of the numeric function at k successive points.

Although we have shown that the total solution obtained according to the procedure presented above satisfies the difference equation and the given boundary conditions, one might wonder whether there are other solutions that will also satisfy the difference equation and the boundary conditions not covered by our procedure. The answer to this question is a negative one. It can be shown that the solution obtained according to the procedure just presented is indeed unique.

7.3 SOLUTION BY THE METHOD OF GENERATING FUNCTIONS

Instead of solving a difference equation for an expression for the value of a numeric function as we did in Sec. 7.2, we can also determine the generating function of the numeric function from the difference equation. In many cases, once the generating function is determined, an expression for the value of the numeric function can easily be obtained.

For a given kth-order difference equation that specifies a numeric function, we should know for what values of r the equation is valid. First of all, we note that the equation is valid only if $r \geq k$ because for $r < k$ the equation will involve a_{-i}'s which are not defined. Furthermore, in many cases a difference equation arises from a physical problem in which the value of a_i is meaningful only for $i \geq t$ for some t larger than or equal to 0. In that case, the difference equation is valid only for $r - k \geq t$.

Suppose we toss a coin r times. There are 2^r possible sequences of outcomes. We want to know the number of sequences of outcomes in which heads never appear on successive tosses. Let a_r denote the number of such sequences. To each sequence of $r - 1$ heads and tails in which there are no consecutive heads, we can append a tail to obtain a sequence of r heads and tails in which there are no consecutive heads. To each sequence of $r - 2$ heads and tails in which there are no consecutive heads, we can append a tail and then a head to obtain a sequence of r heads and tails in which there are no consecutive heads. Moreover, these exhaust all sequences of r heads and tails in which there are no consecutive heads. We thus have the recurrence relation

$$a_r = a_{r-1} + a_{r-2} \tag{7.13}$$

Note that $a_1 = 2$ and $a_2 = 3$. The value of a_0 has no physical significance, and a reasonable choice of its value is 0. In this case, the difference equation is *not* valid for $r = 2$, and is only valid for $r \geq 3$.

We now proceed to show how we can obtain $A(z)$ from (7.13). Multiplying both sides of (7.13) by z^r and summing from $r = 3$ to $r = \infty$, we obtain

$$\sum_{r=3}^{\infty} a_r z^r = \sum_{r=3}^{\infty} a_{r-1} z^r + \sum_{r=3}^{\infty} a_{r-2} z^r$$

That is,

$$A(z) - a_2 z^2 - a_1 z - a_0 = z[A(z) - a_1 z - a_0] + z^2[A(z) - a_0]$$

which simplifies to

$$A(z) = \frac{z^2 + 2z}{1 - z - z^2} = -1 + \frac{1 + z}{1 - z - z^2}$$

Writing $A(z)$ as

$$A(z) = -1 + \frac{(5 + 3\sqrt{5})/10}{1 - [(1 + \sqrt{5})/2]z} - \frac{(-5 + 3\sqrt{5})/10}{1 - [(1 - \sqrt{5})/2]z}$$

we see that

$$a_r = \begin{cases} 0 & r = 0 \\ \dfrac{5 + 3\sqrt{5}}{10}\left(\dfrac{1 + \sqrt{5}}{2}\right)^r - \dfrac{-5 + 3\sqrt{5}}{10}\left(\dfrac{1 - \sqrt{5}}{2}\right)^r & r \geq 1 \end{cases}$$

Since the value of a_0 has no physical significance, we can choose to set a_0 to 1 instead of 0, so that the difference equation in (7.13) is valid for $r \geq 2$. In that case, we can multiply both sides of (7.13) by z^r and sum from $r = 2$ to $r = \infty$. We obtain

$$\sum_{r=2}^{\infty} a_r z^r = \sum_{r=2}^{\infty} a_{r-1} z^r + \sum_{r=2}^{\infty} a_{r-2} z^r$$

That is,

$$A(z) - a_1 z - a_0 = z[A(z) - a_0] + z^2 A(z)$$

or

$$A(z) = \frac{1 + z}{1 - z - z^2}$$

which yields

$$a_r = \frac{5 + 3\sqrt{5}}{10}\left(\frac{1 + \sqrt{5}}{2}\right)^r - \frac{-5 + 3\sqrt{5}}{10}\left(\frac{1 - \sqrt{5}}{2}\right)^r \qquad r \geq 0$$

We state now a general procedure for determining the generating function of the numeric function **a** from the difference equation

$$C_0 a_r + C_1 a_{r-1} + C_2 a_{r-2} + \cdots + C_k a_{r-k} = f(r)$$

which is valid for $r \geq s$, where $s \geq k$. Multiplying both sides of this equation by z^r and summing from $r = s$ to $r = \infty$, we obtain

$$\sum_{r=s}^{\infty} (C_0 a_r + C_1 a_{r-1} + C_2 a_{r-2} + \cdots + C_k a_{r-k})z^r = \sum_{r=s}^{\infty} f(r)z^r$$

Since

$$\sum_{r=s}^{\infty} C_0 a_r z^r = C_0[A(z) - a_0 - a_1 z - a_2 z^2 - \cdots - a_{s-1} z^{s-1}]$$

$$\sum_{r=s}^{\infty} C_1 a_{r-1} z^r = C_1 z[A(z) - a_0 - a_1 z - a_2 z^2 - \cdots - a_{s-2} z^{s-2}]$$

$$\cdots\cdots\cdots\cdots\cdots\cdots\cdots\cdots\cdots\cdots\cdots\cdots\cdots\cdots\cdots\cdots\cdots\cdots\cdots\cdots$$

$$\sum_{r=s}^{\infty} C_k a_{r-k} z^r = C_k z^k[A(z) - a_0 - a_1 z - a_2 z^2 - \cdots - a_{s-k-1} z^{s-k-1}]$$

we have

$$A(z) = \frac{1}{C_0 + C_1 z + \cdots + C_k z^k} \left[\sum_{r=s}^{\infty} f(r) z^r \right.$$

$$+ C_0(a_0 + a_1 z + a_2 z^2 + \cdots + a_{s-1} z^{s-1})$$

$$+ C_1 z(a_0 + a_1 z + a_2 z^2 + \cdots + a_{s-2} z^{s-2})$$

$$+ \cdots$$

$$\left. + C_k z^k(a_0 + a_1 z + a_2 z^2 + \cdots + a_{s-k-1} z^{s-k-1}) \right]$$

As another example, consider a certain nuclear reaction inside of a reactor containing nuclei and high- and low-energy free particles. There are two kinds of events: (1) A high-energy particle strikes a nucleus and is absorbed, causing it to emit three high-energy particles and one low-energy particle. (2) A low-energy particle strikes a nucleus and is absorbed, causing it to emit two high-energy particles and one low-energy particle. We assume that every free particle causes an event 1 μs after it is emitted. Suppose a single high-energy particle is injected at time 0 into a system containing only nuclei. We want to determine the numbers of high-energy and low-energy particles in the system at time equal to $r \mu$s. Let a_r and b_r denote the number of high-energy and low-energy particles at time r, respectively. We have

$$a_r = 3a_{r-1} + 2b_{r-1}$$

$$b_r = a_{r-1} + b_{r-1} \tag{7.14}$$

Note that these are two simultaneous difference equations for the numeric functions **a** and **b**. The equations are valid for $r \geq 1$, and we have the boundary conditions

$$a_0 = 1 \qquad b_0 = 0$$

From (7.14) we obtain

$$\sum_{r=1}^{\infty} a_r z^r = \sum_{r=1}^{\infty} 3a_{r-1} z^r + \sum_{r=1}^{\infty} 2b_{r-1} z^r$$

$$\sum_{r=1}^{\infty} b_r z^r = \sum_{r=1}^{\infty} a_{r-1} z^r + \sum_{r=1}^{\infty} b_{r-1} z^r$$

or

$$A(z) - 1 = 3zA(z) + 2zB(z)$$
$$B(z) = zA(z) + zB(z)$$

(7.15)

Solving (7.15) for $A(z)$ and $B(z)$, we obtain

$$A(z) = \frac{1-z}{1-4z+z^2} = \frac{(3+\sqrt{3})/6}{1-(2+\sqrt{3})z} + \frac{(3-\sqrt{3})/6}{1-(2-\sqrt{3})z}$$

$$B(z) = \frac{z}{1-4z+z^2} = \frac{\sqrt{3}/6}{1-(2+\sqrt{3})z} - \frac{\sqrt{3}/6}{1-(2-\sqrt{3})z}$$

It follows that

$$a_r = \frac{3+\sqrt{3}}{6}(2+\sqrt{3})^r + \frac{3-\sqrt{3}}{6}(2-\sqrt{3})^r$$

$$b_r = \frac{\sqrt{3}}{6}(2+\sqrt{3})^r - \frac{\sqrt{3}}{6}(2-\sqrt{3})^r$$

We demonstrated in this example how we can solve a set of *simultaneous difference equations* by the technique of generating functions. From a set of simultaneous difference equations relating several numeric functions we can obtain a set of simultaneous (algebraic) equations relating their generating functions. These equations can then be solved for closed-form expressions for the generating functions.

7.4 SORTING ALGORITHMS

Let us return to the simple example in Sec. 7.1 in which we showed that the numeric function $(3^0, 3^1, 3^2, \ldots, 3^r, \ldots)$ can be specified by the recurrence relation

$$a_r = 3a_{r-1}$$

(7.16)

together with the boundary condition $a_0 = 1$. Equation (7.16) is a *recursive* specification of the value of **a** at r. It says that if a_{r-1} is known, then a_r can be computed as three times a_{r-1}. However, if a_{r-2} is known, then a_{r-1} can be computed as three times a_{r-2}; if a_{r-3} is known, then a_{r-2} can be computed as three times a_{r-3}; and so on. Consequently, knowing a_0 is sufficient to determine a_r for any r. Such a point of view is not restricted to the description of numeric functions. It also suggests a very powerful way to describe and design computation algorithms. We present in this section some examples from the problem of sorting.

Assume that there are n registers $x_1, x_2, \ldots, x_n$ in which n numbers are stored. A number stored in a register will be referred to as the content of the register. Without loss of generality, we assume these numbers are distinct. By sorting the n numbers stored in the n registers we mean to rearrange them so that the rearranged contents of the registers are in ascending order. A sorting procedure consists of a sequence of comparison steps, each of which compares the contents of registers x_i and x_j and places the smaller number in x_i and the larger number in x_j. We shall denote a comparison step $A(x_i, x_j)$.

The first procedure we shall present is known as the *bubble sort*. Suppose we are able to find the smallest number among the n numbers in $x_1, x_2, \ldots, x_n$ and place it in register x_1. The sorting problem is then reduced to that of sorting the $n - 1$ numbers in the registers $x_2, x_3, \ldots, x_n$. Let $S(x_1, x_2, \ldots, x_n)$ denote a procedure that sorts the numbers in registers $x_1, x_2, \ldots, x_n$ in ascending order. Let $M(x_1, x_2, \ldots, x_n)$ denote a procedure that finds the smallest number among the n numbers in $x_1, x_2, \ldots, x_n$ and places it in x_1. We have

$$S(x_1, x_2, \ldots, x_n) \triangleq M(x_1, x_2, \ldots, x_n)S(x_2, x_3, \ldots, x_n) \qquad (7.17)$$

The symbol $\triangleq$ means "is defined to be" and concatenation of the symbols of procedures means sequential execution of the corresponding procedures from left to right. Note that $M(x_1, x_2, \ldots, x_n)$ can be defined as:

$$M(x_1, x_2, \ldots, x_n) \triangleq A(x_{n-1}, x_n)A(x_{n-2}, x_{n-1}) \cdots A(x_2, x_3)A(x_1, x_2)$$

since after the step $A(x_{n-1}, x_n)$, the number in x_{n-1} will be the smaller of the two numbers that were in x_{n-1} and x_n; after the steps $A(x_{n-1}, x_n)A(x_{n-2}, x_{n-1})$, the number in x_{n-2} will be the smallest of the three numbers that were in x_{n-2}, x_{n-1}, and x_n; and so on. Equation (7.17) can be written as

$$S(x_1, x_2, \ldots, x_n) \triangleq A(x_{n-1}, x_n)A(x_{n-2}, x_{n-1}) \cdots$$
$$A(x_2, x_3)A(x_1, x_2)S(x_2, x_3, \ldots, x_n) \qquad (7.18)$$

which is a recursive definition of the bubble-sort procedure. With the "boundary condition"

$$S(x_{n-1}, x_n) \triangleq A(x_{n-1}, x_n)$$

our specification of the sorting procedure is completed. (This procedure can be described quite vividly as letting the smallest number "bubble up" to register x_1, then letting the next smallest integer "bubble up" to register x_2, and so on. Hence, the name.)

As an explicit example, we note that

$$S(x_1, x_2, x_3, x_4) \triangleq A(x_3, x_4)A(x_2, x_3)A(x_1, x_2)S(x_2, x_3, x_4)$$

$$\triangleq A(x_3, x_4)A(x_2, x_3)A(x_1, x_2)A(x_3, x_4)A(x_2, x_3)S(x_3, x_4)$$

$$\triangleq A(x_3, x_4)A(x_2, x_3)A(x_1, x_2)A(x_3, x_4)A(x_2, x_3)A(x_3, x_4)$$

Let a_n denote the number of comparison steps the bubble-sort procedure takes to sort n numbers. According to (7.18), we have

$$a_n = (n - 1) + a_{n-1} \qquad (7.19)$$

Solving the recurrence relation in (7.19) and using the boundary condition $a_1 = 0$, we obtain†

$$a_n = \frac{n(n-1)}{2}$$

We present now another sorting procedure known as the *Bose-Nelson procedure*, due to R. C. Bose and R. J. Nelson (see Bose and Nelson[2]), which we shall denote $T(x_1, x_2, \ldots, x_n)$. To simplify our discussion, we assume that the number of numbers to be sorted is a power of 2. Let $P[(x_1, x_2, \ldots, x_n), (x_{n+1}, x_{n+2}, \ldots, x_{2n})]$ denote a procedure that will arrange the numbers in registers $x_1, x_2, \ldots, x_{2n}$ in ascending order given that the numbers in registers $x_1, x_2, \ldots, x_n$ have already been arranged in ascending order, as have the numbers in registers $x_{n+1}, x_{n+2}, \ldots, x_{2n}$. Clearly,

$$T(x_1, x_2, \ldots, x_{2n}) \triangleq T(x_1, x_2, \ldots, x_n) T(x_{n+1}, x_{n+2}, \ldots, x_{2n})$$
$$P[(x_1, x_2, \ldots, x_n), (x_{n+1}, x_{n+2}, \ldots, x_{2n})] \qquad (7.20)$$

In other words, to sort the $2n$ numbers in registers $x_1, x_2, \ldots, x_{2n}$, we can first sort the n numbers in $x_1, x_2, \ldots, x_n$, and the n numbers in $x_{n+1}, x_{n+2}, \ldots, x_{2n}$, and then combine these two sorted sequences into one. The procedure $P[(x_1, x_2, \ldots, x_n), (x_{n+1}, x_{n+2}, \ldots, x_{2n})]$ can be defined recursively as:

$$P[(x_1, x_2, \ldots, x_n), (x_{n+1}, x_{n+2}, \ldots, x_{2n})]$$
$$\triangleq P[(x_1, x_2, \ldots, x_{n/2}), (x_{n+1}, x_{n+2}, \ldots, x_{3n/2})]$$
$$P[(x_{n/2+1}, x_{n/2+2}, \ldots, x_n)(x_{3n/2+1}, x_{3n/2+2}, \ldots, x_{2n})]$$
$$P[(x_{n/2+1}, x_{n/2+2}, \ldots, x_n), (x_{n+1}, x_{n+2}, \ldots, x_{3n/2})] \qquad (7.21)$$

Note that $P[(x_1, x_2, \ldots, x_{n/2}), (x_{n+1}, x_{n+2}, \ldots, x_{3n/2})]$ places the smallest $n/2$ numbers, among the $2n$ numbers in the registers, in registers $x_1, x_2, \ldots, x_{n/2}$ in ascending order. Similarly, $P[(x_{n/2+1}, x_{n/2+2}, \ldots, x_n), (x_{3n/2+1}, x_{3n/2+2}, \ldots, x_{2n})]$ places the largest $n/2$ numbers, among the $2n$ numbers in the registers, in registers $x_{3n/2+1}, x_{3n/2+2}, \ldots, x_{2n}$ in ascending order. Finally, $P[(x_{n/2+1}, x_{n/2+2}, \ldots, x_n), (x_{n+1}, x_{n+2}, \ldots, x_{3n/2})]$ places the remaining n of the $2n$ numbers in the registers in registers $x_{n/2+1}, x_{n/2+2}, \ldots, x_{3n/2}$ in ascending order. Applying the relations in (7.20) and (7.21) repeatedly, we can express $T(x_1, x_2, \ldots, x_n)$, $n = 2^r$, as a sequence of procedures of the forms $T(x_i, x_j)$ and $P[(x_i), (x_j)]$. However, both of $T(x_i, x_j)$ and $P[(x_i), (x_j)]$ are equal to $A(x_i, x_j)$.

For example, we note that

$$T(x_1, x_2, x_3, x_4) \triangleq T(x_1, x_2) T(x_3, x_4) P[(x_1, x_2), (x_3, x_4)]$$

Since

$$T(x_1, x_2) \triangleq A(x_1, x_2)$$

$$T(x_3, x_4) \triangleq A(x_3, x_4)$$

and

$$P[(x_1, x_2), (x_3, x_4)] \triangleq P[(x_1), (x_3)] P[(x_2), (x_4)] P[(x_2), (x_3)]$$
$$\triangleq A(x_1, x_3) A(x_2, x_4) A(x_2, x_3)$$

† See Prob. 7.7.

we have

$$T(x_1, x_2, x_3, x_4) \triangleq A(x_1, x_2)A(x_3, x_4)A(x_1, x_3)A(x_2, x_4)A(x_2, x_3)$$

as a complete specification of a procedure that sorts four numbers stored in the registers x_1, x_2, x_3, x_4.

Let c_r denote the number of comparison steps that $P[(x_1, x_2, \ldots, x_{2^r}),$ $(x_{2^r+1}, x_{2^r+2}, \ldots, x_{2^{r+1}})]$ takes. According to the relation in (7.21), we have

$$c_r = 3c_{r-1} \qquad (7.22)$$

Solving (7.22) with the boundary condition $c_0 = 1$, we obtain

$$c_r = 3^r$$

Let d_r denote the number of comparison steps the procedure $T(x_1, x_2, \ldots, x_{2^r})$ takes. According to the relation in (7.20), we have

$$d_r = 2d_{r-1} + c_{r-1}$$

or

$$d_r = 2d_{r-1} + 3^{r-1} \qquad (7.23)$$

Solving (7.23) by using the methods of Sec. 7.2, we obtain

$$d_r = B2^r + 3^r$$

From the boundary condition $d_0 = 0$, we determine that $B = -1$. Thus,

$$d_r = 3^r - 2^r$$

Another procedure we shall present is known as the *odd-even merge*, due to K. E. Batcher (see Batcher[1]), which we shall denote $U(x_1, x_2, \ldots, x_n)$. Again, we assume that n is a power of 2. Let

$$U(x_1, x_2, \ldots, x_{2n}) \triangleq U(x_1, x_2, \ldots, x_n) U(x_{n+1}, x_{n+2}, \ldots, x_{2n})$$
$$B[(x_1, x_2, \ldots, x_n), (x_{n+1}, x_{n+2}, \ldots, x_{2n})] \qquad (7.24)$$

where the procedure B is defined to be

$$B[(x_1, x_2, x_3, x_4, \ldots, x_{2n}), (y_1, y_2, y_3, y_4, \ldots, y_{2n})]$$
$$\triangleq B[(x_1, x_3, x_5, \ldots, x_{2n-1}), (y_1, y_3, y_5, \ldots, y_{2n-1})]$$
$$B[(x_2, x_4, x_6, \ldots, x_{2n}), (y_2, y_4, y_6, \ldots, y_{2n})]$$
$$A(x_2, x_3)A(x_4, x_5)A(x_6, x_7) \cdots A(x_{2n}, y_1)$$
$$A(y_2, y_3) A(y_4, y_5) \cdots A(y_{2n-2}, y_{2n-1}) \qquad (7.25)$$

It is not immediately obvious that the procedure defined in (7.25) indeed merges two sorted sequences of numbers into one. We present first an illustrative example before we proceed to verify the validity of the procedure. Suppose eight numbers, 2, 3, 6, 7, 1, 4, 5, 8, are stored in registers x_1, x_2, x_3, x_4,

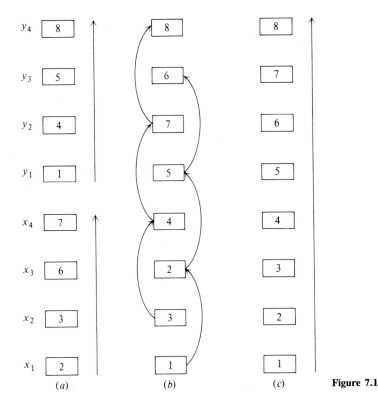

y_4 8   8   8

y_3 5   6   7

y_2 4   7   6

y_1 1   5   5

x_4 7   4   4

x_3 6   2   3

x_2 3   3   2

x_1 2   1   1

(a)   (b)   (c)   **Figure 7.1**

y_1, y_2, y_3, y_4, as shown in Fig. 7.1a, such that the numbers in registers x_1, x_2, x_3, x_4 are in ascending order, and the numbers in registers y_1, y_2, y_3, y_4 are in ascending order. According to (7.25),

$$B[(x_1, x_2, x_3, x_4), (y_1, y_2, y_3, y_4)]$$
$$\triangleq B[(x_1, x_3), (y_1, y_3)]B[(x_2, x_4), (y_2, y_4)] A(x_2, x_3)A(x_4, y_1)A(y_2, y_3)$$

$B[(x_1, x_3), (y_1, y_3)]$ arranges the numbers in x_1, x_3, y_1, y_3 in ascending order, and $B[(x_2, x_4), (y_2, y_4)]$ arranges the numbers in x_2, x_4, y_2, y_4 in ascending order as shown in Fig. 7.1b. Finally, $A(x_2, x_3)$ arranges the numbers in x_2, x_3; $A(x_4, y_1)$ arranges the numbers in x_4, y_1; and $A(y_2, y_3)$ arranges the numbers in y_2, y_3 in ascending order as shown in Fig. 7.1c.

To verify the validity of (7.25), we want first to show that after the execution of $B[(x_1, x_3, x_5, \ldots, x_{2n-1}), (y_1, y_3, y_5, \ldots, y_{2n-1})]$ and $B[(x_2, x_4, x_6, \ldots, x_{2n}), (y_2, y_4, y_6, \ldots, y_{2n})]$:

1. The number in x_1 is the smallest of the $4n$ numbers.
2. The number in y_{2n} is the largest of the $4n$ numbers.
3. The two numbers in x_{2i} and x_{2i+1}, $1 \leqq i \leqq n - 1$, are the $2i$th and the $(2i + 1)$st smallest of the $4n$ numbers; the two numbers in x_{2n} and y_1 are the $2n$th and the $(2n + 1)$st smallest of the $4n$ numbers; and the two numbers in y_{2i} and y_{2i+1},

$1 \leq i \leq n-1$, are the $(2n+2i)$th and $(2n+2i+1)$st smallest of the $4n$ numbers.

If this is indeed the case, the comparisons $A(x_2, x_3)A(x_4, x_5) \cdots A(y_{2n-2}, y_{2n-1})$ will yield a sorted sequence of $4n$ numbers.

(1) and (2) are obvious as can be ascertained by the reader.

To show that (3) is true, we shall only show that the two numbers in x_{2i} and x_{2i+1}, $1 \leq i \leq n-1$, are the $2i$th and the $(2i+1)$st smallest of the $4n$ numbers. The other two cases can be shown in a similar manner.† Consider the number that is in x_{2i} after the execution. This number is larger than exactly $i-1$ of the numbers that were in $x_2, x_4, x_6, \ldots, x_{2n}, y_2, y_4, y_6, \ldots, y_{2n}$. Assume that this number is larger than k of the numbers that were in $x_2, x_4, x_6, \ldots, x_{2n}$, and is larger than $i-k-1$ of the numbers that were in $y_2, y_4, y_6, \ldots, y_{2n}$. That is, this number is larger than the numbers that were in $x_2, x_4, x_6, \ldots, x_{2k}$ and those that were in $y_2, y_4, y_6, \ldots, y_{2(i-k-1)}$. In other words, this number is larger than the numbers that were in $x_1, x_2, x_3, x_4, \ldots, x_{2k}$ and those that were in $y_1, y_2, y_3, y_4, \ldots, y_{2(i-k-1)}$. Now, if this number was in $x_{2(k+1)}$, it is also larger than x_{2k+1}, and it might or might not be larger than the number that was in $y_{2(i-k-1)+1}$. On the other hand, if this number was in $y_{2(i-k)}$, it is also larger than $y_{2(i-k)-1}$, and it might or might not be larger than the number that was in x_{2k+1}. Consequently, we can conclude that this number is larger than $2i-1$ or $2i$ of the $4n$ numbers in the registers.

In a similar way, let us consider the number that is in x_{2i+1} after the execution. Assume that this number is larger than k of the numbers that were in $x_1, x_3, \ldots, x_{2n-1}$, and is larger than $i-k$ of the numbers that were in $y_1, y_3, \ldots, y_{2n-1}$. That is, this number is larger than the numbers that were in $x_1, x_3, \ldots, x_{2k-1}$, and is larger than the numbers that were in $y_1, y_3, \ldots, y_{2(i-k)-1}$. Thus, this number is larger than the numbers that were in $x_1, x_2, x_3, \ldots, x_{2k-2}, x_{2k-1}$, and those that were in $y_1, y_2, y_3, \ldots, y_{2(i-k)-2}, y_{2(i-k)-1}$. Now, if this number was in x_{2k+1}, it is also larger than the number that was in x_{2k}, and it might or might not be larger than the number that was in $y_{2(i-k)}$. If this number was in $y_{2(i-k)+1}$, it is larger than the number that was in $y_{2(i-k)}$, and it might or might not be larger than the number that was in x_{2k}. Consequently, we can conclude that this number is larger than $2i-1$ or $2i$ of the $4n$ numbers in the registers.

We now complete the proof of (3) by noting that if the number in x_{2i} is the $2i$th smallest number, then the number in x_{2i+1} must be the $(2i+1)$st smallest number, and vice versa.

Let c_r denote the number of comparison steps the procedure $B[(x_1, x_2, \ldots, x_{2^r}), (y_1, y_2, \ldots, y_{2^r})]$ takes. According to the relation in (7.25), we have

$$c_r = 2c_{r-1} + (2^r - 1) \qquad (7.26)$$

---

† We do not really have three cases in (3). It is only because of the notation we have chosen that we have to make three parallel statements.

Solving (7.26), with the boundary condition $c_0 = 1$, we obtain

$$c_r = r2^r + 1$$

Let d_r denote the number of comparison steps the procedure $U(x_1, x_2, \ldots, x_{2^r})$ takes. According to the relation in (7.24), we have

$$d_r = 2d_{r-1} + c_{r-1}$$

or

$$d_r = 2d_{r-1} + (r - 1)2^{r-1} + 1 \tag{7.27}$$

Solving the recurrence relation in (7.27), with the boundary condition $d_0 = 0$, we obtain

$$d_r = (r^2 - r + 4)2^{r-2} - 1$$

7.5 REMARKS AND REFERENCES

See Levy and Lessman[4] on the subject of finite-difference equations. See also chaps. 2 and 3 of Liu and Liu[6]. The sorting algorithms discussed in Sec. 7.4 have two distinct features: (1) They are *nonadaptive* in the sense that the sequence of comparison steps in an algorithm is predetermined and does not vary in accordance with the outcome of any particular comparison step. (2) No additional registers for storing intermediate results is needed. We note that in a nonadaptive algorithm, some of the comparison steps can be carried out simultaneously. [For example, $A(x_1, x_2)$ and $A(x_3, x_4)$ can be carried out simultaneously while $A(x_1, x_2)$ and $A(x_1, x_3)$ cannot be.] Consequently, there is the possibility of speeding up the computation by parallel processing. The advantage of algorithms that do not use additional intermediate storage registers is obvious when the number of items to be sorted is large. For a most complete discussion on sorting algorithms and related topics, see chap. 5 of Knuth[3].

[1] Batcher, K. E.: Sorting Networks and Their Applications, *AFIPS Proc. of the 1968 SJCS*, **32**: 307–314 (1968).
[2] Bose, R. C., and R. J. Nelson: A Sorting Problem, *Journal of ACM*, **9**: 282–296 (1962).
[3] Knuth, D. E.: "The Art of Computer Programming, Vol. 3, Sorting and Searching," Addison-Wesley Publishing Company, Reading, Mass., 1973.
[4] Levy, H., and F. Lessman: "Finite Difference Equations," The Macmillan Company, New York, 1961.
[5] Liu, C. L.: "Introduction to Combinatorial Mathematics," McGraw-Hill Book Company, New York, 1968.
[6] Liu, C. L., and J. W. S. Liu: "Linear Systems Analysis," McGraw-Hill Book Company, New York, 1975.

PROBLEMS

7.1 Solve the following recurrence relations:

(a) $a_r - 7a_{r-1} + 10a_{r-2} = 3^r$ given that $a_0 = 0$, $a_1 = 1$.

(b) $a_r + 6a_{r-1} + 9a_{r-2} = 3$ given that $a_0 = 0$, $a_1 = 1$.

(c) $a_r + a_{r-2} = 0$ given that $a_0 = 0$, $a_1 = 2$.

7.2 Solve the following recurrence relations:

(a) $a_r - a_{r-1} - a_{r-2} = 0$ given that $a_0 = 1$, $a_1 = 1$.

(b) $a_r - 2a_{r-1} + 2a_{r-2} - a_{r-3} = 0$ given that $a_0 = 2$, $a_1 = 1$, $a_2 = 1$.

7.3 Given that $a_0 = 0$, $a_1 = 1$, $a_2 = 4$, $a_3 = 12$ satisfy the recurrence relation

$$a_r + C_1 a_{r-1} + C_2 a_{r-2} = 0$$

determine a_r.

7.4 The solution of the recurrence relation

$$C_0 a_r + C_1 a_{r-1} + C_2 a_{r-2} = f(r)$$

is

$$3^r + 4^r + 2$$

Given that $f(r) = 6$ for all r, determine C_0, C_1, and C_2.

7.5 Consider the recurrence relation

$$a_r = a_{r-1} - a_{r-2}$$

(a) Solve the recurrence relation, given that $a_1 = 1$, $a_2 = 0$.

(b) Can you solve the recurrence relation if it is given that $a_0 = 0$, $a_3 = 0$?

(c) Repeat part (b) if it is given that $a_0 = 1$, $a_3 = 2$.

7.6 (a) Determine the particular solution for the difference equation

$$a_r - 3a_{r-1} + 2a_{r-2} = 2^r$$

(b) Determine the particular solution for the difference equation

$$a_r - 4a_{r-1} + 4a_{r-2} = 2^r$$

7.7 (a) Determine the particular solution for the difference equation

$$a_r - 2a_{r-1} = f(r)$$

where $f(r) = 7r$.

(b) Repeat part (a) for $f(r) = 7r^2$.

(c) Determine the particular solution for the difference equation

$$a_r - a_{r-1} = 7r$$

(d) Repeat part (b) if $f(r) = 7r^2$.

(e) Let

$$C_0 a_r + C_1 a_{r-1} + \cdots + C_k a_{r-k} = f(r)$$

be a difference equation with a characteristic root 1. Let $f(r) = r^t$. What can be said about the general form of the particular solution $a_r^{(p)}$?

7.8 (a) Solve the recurrence relation

$$a_r + 3a_{r-1} + 2a_{r-2} = f(r)$$

where

$$f(r) = \begin{cases} 1 & r = 2 \\ 0 & \text{otherwise} \end{cases}$$

with the boundary condition $a_0 = a_1 = 0$.

(b) Repeat part (a) for

$$f(r) = \begin{cases} 1 & r = 5 \\ 0 & \text{otherwise} \end{cases}$$

(c) Consider the recurrence relation

$$C_0 a_r + C_1 a_{r-1} + C_2 a_{r-2} + \cdots + C_k a_{r-k} = f(r)$$

Let $\hat{a}_r$ denote the solution of the recurrence relation for $f(r) = \hat{f}(r)$ with the boundary conditions

$\hat{a}_0 = \hat{a}_1 = \hat{a}_2 = \cdots = \hat{a}_{k-1} = 0$. Let $\tilde{a}_r$ denote the solution of the recurrence relation for $f(r) = \tilde{f}(r)$ with the boundary conditions $\tilde{a}_0 = \tilde{a}_1 = \tilde{a}_2 = \cdots = \tilde{a}_{k-1} = 0$. Given that $\hat{f}(r) = 0$ for $r < k$, and

$$\tilde{f}(r) = \begin{cases} 0 & 0 \leq r \leq l \\ \hat{f}(r-l) & r \geq l \end{cases}$$

for some fixed l, what can we conclude about $\hat{a}_r$ and $\tilde{a}_r$?

7.9 (a) Consider the recurrence relation

$$C_0 a_r + C_1 a_{r-1} + C_2 a_{r-2} + \cdots + C_k a_{r-k} = f(r)$$

Let $\hat{a}_r$ denote the solution of the recurrence relation for $f(r) = \hat{f}(r)$ with the boundary conditions $\hat{a}_0 = \hat{a}_1 = \hat{a}_2 = \cdots = \hat{a}_{k-1} = 0$. Let $\tilde{a}_r$ denote the solution of the recurrence relation for $f(r) = \tilde{f}(r)$ with the boundary conditions $\tilde{a}_0 = \tilde{a}_1 = \tilde{a}_2 = \cdots = \tilde{a}_{k-1} = 0$. Show that $\bar{a}_r = \hat{a}_r + \tilde{a}_r$ is the solution of the recurrence relation for $f(r) = \hat{f}(r) + \tilde{f}(r)$ with the boundary condition $\bar{a}_0 = \bar{a}_1 = \bar{a}_2 = \cdots = \bar{a}_{k-1} = 0$, provided that $\hat{f}(r) = \tilde{f}(r) = 0$ for $r < k$.

(b) Solve the recurrence equation

$$a_r + 5a_{r-1} + 6a_{r-2} = f(r)$$

where

$$f(r) = \begin{cases} 0 & r = 0, 1, 5 \\ 6 & \text{otherwise} \end{cases}$$

given that $a_0 = a_1 = 0$.

7.10 Gossip is spread among r people via telephone. Specifically, in a telephone conversation between A and B, A tells B all the gossip he has heard, and B reciprocates. Let a_r denote the minimum number of telephone calls among r people so that all gossip will be known to everyone.

(a) Show that $a_2 = 1$, $a_3 = 3$, and $a_4 = 4$.

(b) Show that

$$a_r \leq a_{r-1} + 2$$

(c) Show that

$$a_r \leq 2r - 4 \qquad \text{for } r \geq 4$$

[Indeed, it can be shown that $a_r = 2r - 4$. See B. Baker and R. Shostak, Gossips and Telephones, *Discrete Mathematics*, **2**: 191–193, (1972).]

7.11 Let a_r denote the number of partitions of a set of r elements. Show that

$$a_{r+1} = \sum_{i=0}^{r} \binom{r}{i} a_i$$

where $a_0 = 1$.

7.12 Consider an air-traffic-control system in which the desired altitude of an aircraft, a_r, is computed by a computer every second and is compared with the actual altitude of the aircraft, b_{r-1}, determined by a tracking radar 1 s earlier. Depending on whether a_r is larger or smaller than b_{r-1}, the altitude of the aircraft will be changed accordingly. Specifically, the change in altitude at the rth second, $b_r - b_{r-1}$, is proportional to the difference $a_r - b_{r-1}$. That is,

$$b_r - b_{r-1} = K(a_r - b_{r-1})$$

where K is a proportional constant.

(a) Determine b_r, given that $a_r = 1000 \left(\frac{3}{2}\right)^2$, $K = 3$, and $b_0 = 0$.

(b) Determine b_r, given that

$$a_r = \begin{cases} 1000 \left(\frac{3}{2}\right)^r & 0 \leq r \leq 10 \\ 1000 \left(\frac{3}{2}\right)^{10} & r \geq 10 \end{cases}$$

$k = 3$ and $b_0 = 0$.

7.13 *The Tower of Hanoi problem. r* circular rings of tapering sizes are slipped onto a peg with the largest ring at the bottom, as shown in Fig. 7P.1. These rings are to be transferred one at a time onto another peg, and there is a third peg available on which rings can be left temporarily. If, during the course of transferring the rings, no ring may ever be placed on top of a smaller one, in how many moves can these rings be transferred with their relative positions unchanged?

Figure 7P.1

7.14 Consider the multiplication of bacteria in a controlled environment. Let a_r denote the number of bacteria there are on the rth day. We define the rate of growth on the rth day to be $a_r - 2a_{r-1}$. If it is known that the rate of growth doubles every day, determine a_r, given that $a_0 = 1$.

7.15 Consider the operation of a factory whose average profit in every two successive months is equal to the average new order in that period. Let a_r denote the new order received and b_r denote the monthly profit in the rth month.

(a) Given that $a_r = 2^r$ for all $r \geq 0$ and $b_0 = 0$, determine b_r.

(b) Repeat part (a) for

$$a_r = \begin{cases} 2^r & 0 \leq r \leq 10 \\ 2^{10} & r \geq 11 \end{cases}$$

7.16 A particle is moving in the horizontal direction. The distance it travels in each second is equal to two times the distance it traveled in the previous second. Let a_r denote the position of the particle in the rth second. Determine a_r, given that $a_0 = 3$ and $a_3 = 10$.

7.17 How many r-digit binary sequences that have no adjacent 0s are there?

7.18 Let a_r denote the total assets of a bank at the end of the rth month which is equal to the sum of the total deposit in the rth month and 1.1 times the total assets at the end of the $(r-1)$st month. Given that the total deposit is a constant 100 (in thousands of dollars), determine a_r if $a_0 = 0$.

7.19 Let a_r denote the number of nonoverlapping regions into which the interior of a convex r-gon is divided by its diagonals. Suppose no three diagonals meet at one point.

(a) Show that

$$a_r - a_{r-1} = \frac{(r-1)(r-2)(r-3)}{6} + r - 2 \qquad r \geq 3$$

and $a_0 = a_1 = a_2 = 0$.

(b) Determine $A(z)$.

(c) Determine a_r from $A(z)$.

7.20 There are two kinds of particles inside a nuclear reactor. In every second, an α particle will split into three β particles, and a β particle will split into an α particle and two β particles. If there is a single α particle in the reactor at $t = 0$, how many particles are there altogether at $t = 100$?

7.21 Let d_r denote the number of ways of permuting r integers $\{1, 2, \ldots, r\}$ so that the integer i will not be in the ith position for $1 \leq i \leq r$.

(a) Use a combinatorial argument to show that

$$d_r = (r-1)(d_{r-1} + d_{r-2})$$

(b) Show that

$$d_r = r! \left[1 - \frac{1}{1!} + \frac{1}{2!} - \cdots + (-1)^r \frac{1}{r!} \right]$$

satisfies the recurrence relation in part (a).

7.22 A sequence of binary digits is fed to a counter at the rate of 1 digit/s. The counter is designed to register 1's in the input sequence. However, it is so slow that it is locked for exactly 7 s following each registration, during which time the input digits are ignored. Let a_r denote the number of binary sequence of length r at the end of which the counter is not locked. Find the difference equation which a_r satisfies, the boundary conditions, and the generating function $A(z)$.

7.23 For the graph in Fig. 7P.2, determine the number of directed paths of length n that start from vertex a and end at vertex d.

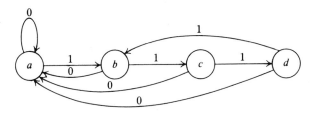

Figure 7P.2

7.24 (a) Let $A(z)$ denote the generating function of a numeric function **a**. Show that

$$a_0 = \lim_{z \to 0} A(z)$$

(This result is known as the *initial-value theorem*.)

(b) Check the initial-value theorem for the numeric function **a**, where $a_r = 2^{r+1}$ for all r.

7.25 Solve the difference equation

$$a_r^2 - 2a_{r-1}^2 = 1$$

given that $a_0 = 2$. (*Hint*: Let $b_r = a_r^2$.)

7.26 Solve the difference equation

$$ra_r + ra_{r-1} - a_{r-1} = 2^r$$

given that $a_0 = 273$. (*Hint*: let $b_r = ra_r$.)

7.27 (a) Solve the difference equation

$$a_r^2 - 2a_{r-1} = 0$$

given that $a_0 = 4$. (*Hint*: Let $b_r = \lg a_r$, where lg denotes logorithm base 2.)

(b) Solve the difference equation

$$a_r = \sqrt{a_{r-1} + \sqrt{a_{r-2} + \sqrt{a_{r-3} + \sqrt{\cdots}}}}$$

given that $a_0 = 4$. (*Hint*: Compute a_r^2.)

7.28 Solve the difference equation

$$a_r - ra_{r-1} = r! \qquad \text{for } r \geq 1$$

given that $a_0 = 2$. (*Hint*: Let $b_r = a_r/r!$.)

CHAPTER

EIGHT

GROUPS AND RINGS

8.1 INTRODUCTION

Consider the operation of a vending machine† which delivers a pack of gum when two dimes are deposited, a candy bar when a dime and a quarter are deposited, and a pack of cigarettes when two quarters are deposited. Let $A = \{$dime, quarter$\}$ and $B = \{$gum, candy, cigarettes$\}$. The operation of the vending machine can be described formally as a function from $A \times A$ to B, which is shown in Fig. 8.1a. The table in Fig. 8.1a can be set up in an alternative way as that shown in Fig. 8.1b, where the rows indicate the first coin deposited and the columns indicate the second coin deposited.

As another example, suppose that the hair color of a child is determined by that of her parents, as shown in the table of Fig. 8.2. Clearly, the relationship between the hair color of a child and that of her parents can be described by a function from $A \times A$ to A, where $A = \{$light, dark$\}$.

Let A and B be two sets. A function from $A \times A$ to B is called a *binary operation* on the set A. We shall encounter most frequently functions from $A \times A$ to A. A function from $A \times A$ to A is said to be a binary operation that is *closed*. In the example of the vending machine, a binary operation on the set $\{$dime, quarter$\}$ was defined. This binary operation is not a closed operation. In the example of the hair color of children, a binary operation on the set $\{$light, dark$\}$ was defined. Moreover, this binary operation is closed. Intuitively,

---

† To be precise, we describe here only the operation of a vending machine when two coins, which are dimes and quarters, are deposited. The operation of real vending machines is less restrictive.

Coins deposited	Merchandise delivered
(dime, dime)	gum
(dime, quarter)	candy
(quarter, dime)	candy
(quarter, quarter)	cigarettes

1st coin deposited \ 2nd coin deposited	dime	quarter
dime	gum	candy
quarter	candy	cigarettes

Merchandise delivered

(a)   (b)

Figure 8.1

Father \ Mother	light	dark
light	light	dark
dark	dark	dark

Child

Figure 8.2

a binary operation specifies a way in which two elements are "combined" to yield a third element.

A binary operation can be described using functional notation. That is, let the function from $A \times A$ to A be named f. Clearly, $f(a_1, a_2)$† will denote the image of the ordered pair (a_1, a_2) in $A \times A$. An alternative way is to write $f(a_1, a_2)$ as $a_1 f a_2$.‡ In this case, instead of the customary letter names for functions, we use "operator symbols" such as $\star, *, +, \cdot, \square, \oplus, \ldots$ as the names of binary operations on a set. Thus, we can write

$$\star(a_1, a_2) \qquad +(\text{dime}, \text{quarter}) \qquad \square(\text{light}, \text{dark})$$

or

$$a_1 \star a_2 \qquad \text{dime} + \text{quarter} \qquad \text{light} \, \square \, \text{dark}$$

The definition of a binary operation can be extended immediately. A ternary operation on a set A is a function from $(A \times A) \times A$ to B for some set B, and an m-ary operation on a set A is a function from A^m to B for some set B.§

A set together with a number of operations on the set is called an *algebraic system*. We shall use a notation such as $(A, \star, *, \square)$ for an algebraic system, where A is a set, and $\star, *,$ and $\square$ are operations on A. Our examples of the vending machine and the hair color of children are both examples of algebraic systems with one operation. Consider a self-service snack bar with two vending

---

† To be precise, we should write $f[(a_1, a_2)]$. However, there is no confusion in our notation.

‡ $f(a_1, a_2)$ is referred to as the *prefix* notation, and $a_1 f a_2$ is referred to as the *infix* notation.

§ We write A^m to denote $\underbrace{((A \times A) \times A) \times \cdots \times A.}_{m \text{ times}}$

machines. We can describe the items one can purchase by the algebraic system ({dime, quarter}, $\star$, $*$), where the binary operations $\star$ and $*$ are described in Fig. 8.3. Consider the set of natural numbers N together with the usual addition and multiplication operation of integers, $+$ and $\cdot$. Clearly, $(N, +, \cdot)$ is an algebraic system with two operations. Also, let $\square$ be a binary operation on N such that $\square(a, b)$ is equal to 0 or 1 depending on whether the sum of a and b is even or odd, and $\triangle$ be a ternary operation on N such that $\triangle(a, b, c)$ is equal to the maximum of a, b, and c. $(N, \square, \triangle)$ is also an algebraic system with two operations.

$\star$	dime	quarter		$*$	dime	quarter
dime	gum	candy		dime	potato chips	cookies
quarter	candy	cigarettes		quarter	cookies	sandwich

Figure 8.3

8.2 GROUPS

Let $\star$ be a binary operation on a set A. The operation $\star$ is said to be *associative* if

$$(a \star b) \star c = a \star (b \star c)$$

for all a, b, c in A.† Let A be a set of people and $\triangle$ be a binary operation such that $a \triangle b$ is equal to the taller one of a and b. (Assume that no two people in A are of the same height.) We note that $\triangle$ is an associative operation. On the other hand, let N be the set of all natural numbers and $\square$ be an operation such that $a \square b$ is equal to the value of $a^2 + b$. We ask the reader to check that the operation $\square$ is not associative. Intuitively, when an associative operation is to be carried out a certain number of times, the order in which the operations are carried out is not important.

Let $(A, \star)$ be an algebraic system where $\star$ is a binary operation on A. $(A, \star)$ is called a *semigroup* if the following conditions are satisfied:

1. $\star$ is a closed operation.‡
2. $\star$ is an associative operation.

Let A be the set of all positive even integers $\{2, 4, 6, \ldots\}$ and $+$ be the ordinary addition operation of integers. Since $+$ is a closed operation on A, and

† It follows that when $\star$ is an associative operation we can write $(a \star b) \star c$ as $a \star b \star c$ without any possible confusion.

‡ Note that it is possible to have an associative operation that is not closed. Consider the algebraic system $(\{1, 2, 3\}, +)$, where $+$ is the ordinary addition operation of integers.

is also an associative operation, $(A, +)$ is a semigroup. Let S be a finite alphabet. Let A denote the set of all nonempty strings of letters from S. (For example, let $S = \{\alpha, \beta, \gamma\}$. We have $A = \{\alpha, \beta, \gamma, \alpha\alpha, \alpha\beta, \alpha\gamma, \ldots, \alpha\alpha\alpha, \alpha\alpha\beta, \ldots\}$.) Let $\cdot$ be a binary operation on A such that for any two strings a and b in A, $a \cdot b$ yields a string which is the concatenation of strings a and b. (For example, $\alpha\alpha \cdot \alpha\gamma\beta = \alpha\alpha\alpha\gamma\beta$.) We note that $(A, \cdot)$ is a semigroup.

Let $(A, \star)$ be an algebraic system where $\star$ is a binary operation on A. An element in A, e, is said to be a *left identity* if for all x in A, $e \star x = x$. For example, for the algebraic system shown in Fig. 8.4a, both β and δ are left identities. An element in A, e, is said to be a *right identity* if for all x in A, $x \star e = x$. For example, for the algebraic system shown in Fig. 8.4b, α is a right identity. An element in A is said to be an *identity* if it is both a left identity and a right identity.

$\star$	α	β	γ	δ
α	δ	α	β	γ
β	α	β	γ	δ
γ	α	β	γ	γ
δ	α	β	γ	δ

(a)

$\star$	α	β	γ	δ
α	α	β	δ	γ
β	β	α	γ	δ
γ	γ	δ	α	β
δ	δ	δ	β	γ

(b)

Figure 8.4

Suppose e_1 is a left identity and e_2 is a right identity of an algebraic system $(A, \star)$. Since e_1 is a left identity, $e_1 \star e_2 = e_2$. Since e_2 is a right identity, $e_1 \star e_2 = e_1$. Thus, we have $e_1 = e_2$. We conclude that if e is a left identity, then either e is also a right identity or there is no right identity at all. Similarly, if e is a right identity, then either e is also a left identity or there is no left identity at all. *It follows that, with respect to a binary operation, there is at most one identity.*

Intuitively, an identity is a "neutral" element in that, when it is "combined" with another element, its effect on the outcome is nil. For example, let $(A, \star)$ be an algebraic system, where A is a set of colored lights and $\star$ is a binary operation such that $a \star b$ is the resultant colored light when light a is combined with light b. Clearly, white light is the identity of the algebraic system. As another example, let $(N, +)$ be an algebraic system, where N is the set of natural numbers and $+$ is the ordinary addition operation of integers. Clearly, 0 is the identity of the algebraic system.

Let $(A, \star)$ be an algebraic system, where $\star$ is a binary operation on A. $(A, \star)$ is called a *monoid* if the following conditions are satisfied:

1. $\star$ is a closed operation.
2. $\star$ is an associative operation.
3. There is an identity.

For example, let A be a set of people of different heights. Let $\triangle$ be a binary operation such that $a \triangle b$ is equal to the taller one of a and b. We note that $(A, \triangle)$ is a monoid where the identity is the shortest person in A.

Let $(A, \star)$ be an algebraic system with an identity e. Let a be an element in A. An element b is said to be a *left inverse* of a if $b \star a = e$. An element b is said to be a *right inverse* of a if $a \star b = e$. For example, for the algebraic system in Fig. 8.5, α is an identity, so β is a left inverse of γ, and δ is a right inverse of γ. An element b is said to be an *inverse* of a if it is both a left and a right inverse of a. Clearly, if b is an inverse of a, a is also an inverse of b. Intuitively, an inverse of an element "cancels" the effect of the element when they are "combined." For example, let $(A, \star)$ be an algebraic system, where A is a set of chemicals containing acids, alkalis, and water, and $\star$ is a binary operation giving the product of the combination of two chemicals. In this case, water can be considered as a neutral chemical, and the inverse of an acid is an alkali, if their combination yields water.

$\star$	α	β	γ	δ
α	α	β	γ	δ
β	β	δ	α	γ
γ	γ	β	β	α
δ	δ	α	γ	δ

Figure 8.5

Let $(A, \star)$ be an algebraic system, where $\star$ is a binary operation. $(A, \star)$ is called a *group* if the following conditions are satisfied:

1. $\star$ is a closed operation.
2. $\star$ is an associative operation.
3. There is an identity.
4. Every element in A has a left inverse.

We observe that *because of associativity, a left inverse of an element is also a right inverse of the element in a group.* Let b be a left inverse of a and c be a left inverse of b. Let e denote the identity.† Since

$$(b \star a) \star b = e \star b = b$$

we have

$$c \star ((b \star a) \star b) = c \star b = e$$

From

$$c \star ((b \star a) \star b) = ((c \star b) \star a) \star b$$
$$= (e \star a) \star b$$
$$= a \star b$$

---

† We remind the reader that the identity is unique.

we have

$$a \star b = e$$

Thus, b is also a right inverse of a. Consequently, we can refer to the inverses of the elements without making a distinction of left or right inverses.

Furthermore, we observe that *associativity also implies the uniqueness of the inverse of every element.* Suppose that both b and c are inverses of a. That is,

$$b \star a = e \qquad \text{and} \qquad c \star a = e$$

It follows that

$$(b \star a) \star b = (c \star a) \star b$$

or

$$b \star (a \star b) = c \star (a \star b)$$

or

$$b = c$$

From now on, we shall use a^{-1} to denote *the* inverse of a.

We have encountered many examples of groups before. Consider the algebraic system $(I, +)$, where I is the set of all integers and $+$ is the ordinary addition operation of integers. It is clear that $(I, +)$ is a group with 0 being the identity and the inverse of n being $-n$. Let $G = \{\text{EVEN}, \text{ODD}\}$ and a binary operation $\oplus$ be defined as in Fig. 8.6.

$\oplus$	EVEN	ODD
EVEN	EVEN	ODD
ODD	ODD	EVEN

Figure 8.6

We can check immediately that $(G, \oplus)$ is a group where EVEN is the identity and both EVEN and ODD are their own inverses. Consider the rotation of geometric figures in a plane. Let $R = \{0°, 60°, 120°, 180°, 240°, 300°\}$ denote six possible ways to rotate geometric figures drawn on a plane, namely, to rotate the figures by $0°, 60°, 120°, \ldots, 300°$. Let $\star$ be a binary operation so that for a and b in R, $a \star b$ is the overall angular rotation corresponding to the successive rotations by a and then by b. $(R, \star)$ is a group with $0°$ being the identity, the inverse of $60°$ being $300°$, the inverse of $180°$ being itself, and so on. Let $Z_n = \{0, 1, 2, \ldots, n - 1\}$. Let $\oplus$ be a binary operation on Z_n such that for a and b in Z_n

$$a \oplus b = \begin{cases} a + b & \text{if} \quad a + b < n \\ a + b - n & \text{if} \quad a + b \geq n \end{cases}$$

It can readily be checked that $(Z_n, \oplus)$ is a group for any n. $(Z_n, \oplus)$ is usually referred to as the *group of integers modulo n*.

Let $\star$ be a binary operation on A. The operation $\star$ is said to be *commutative* if

$$a \star b = b \star a$$

for all a, b in A. Let A be a set of people and $\triangle$ be a binary operation such that $a \triangle b$ is equal to the taller one of a and b and is equal to a if a and b are of the same height. Clearly, $\triangle$ is not a commutative operation. (Is $\triangle$ still an associative operation?)

A group $(A, \star)$ is called a *commutative* group or an *abelian* group,† if $\star$ is a commutative operation. For example, $(Z_n, \oplus)$ is a commutative group.

A group $(A, \star)$ is said to be *finite* if A is a finite set, and *infinite* if A is an infinite set. The size of A is often referred to as the *order* of the group.

8.3 SUBGROUPS

Let $(A, \star)$ be an algebraic system and B be a subset of A. The algebraic system $(B, \star)$ is said to be a subsystem of $(A, \star)$. The notion of a subsystem is a very natural one. Suppose $(A, \star)$ is an algebraic system describing the interaction of a set of atomic particles. If we are only interested in the interaction of some of the particles, we can consider only a subsystem of $(A, \star)$. Let $(N, +)$ be an algebraic system describing the addition of natural numbers. Clearly, $(E, +)$ is a subsystem of $(N, +)$ if E is the set of all even numbers. Also, consider the example of the rotation of geometric figures in a plane. We note that $(\{0°, 120°, 240°\}, \star)$ is a subsystem of the algebraic system $(\{0°, 60°, 120°, 180°, 240°, 300°\}, \star)$. So is $(\{0°, 180°\}, \star)$.

Let $(A, \star)$ be a group, and B be a subset of A. $(B, \star)$ is said to be a *subgroup* of A if $(B, \star)$ is also a group by itself. Suppose we want to check whether $(B, \star)$ is a subgroup for a given subset B of A. We note that:

1. We should test whether $\star$ is a closed operation on B.
2. $\star$ is known to be an associative operation.
3. Since there is only one element e in A such that $e \star x = x \star e = x$ for all x in A, we must check that e is in B. In other words, the identity of $(A, \star)$ must be in B as the identity of $(B, \star)$.
4. Since the inverse of every element in A is unique, for every element b in B, we must check that its inverse is also in B.

For example, let $(I, +)$ be an algebraic system, where I is the set of all integers and $+$ is the ordinary addition operation of integers. Clearly, $(I, +)$ is a group. Moreover, $(E, +)$ is a subgroup, where E is the set of all even integers. Also, in the example of the rotation of geometric figures in the plane, both $(\{0°, 120°, 240°\}, \star)$ and $(\{0°, 180°\}, \star)$ are subgroups of the group $(\{0°, 60°, 120°, 180°, 240°, 300°\}, \star)$.

---

† In honor of the Norwegian mathematician N. H. Abel (1802–1829).

We want to show that:

Theorem 8.1 Let $(A, \star)$ be a group and B a subset of A. If B is a finite set, then $(B, \star)$ is a subgroup of $(A, \star)$ if $\star$ is a closed operation on B.

PROOF Let a be an element in B. If $\star$ is a closed operation on B, the elements $a, a^2, a^3, \ldots$ are all in B.† Because B is a finite set, according to the pigeonhole principle, we must have $a^i = a^j$ for some i and j, $i < j$. That is, $a^i = a^i \star a^{j-i}$. It follows that a^{j-i} is the identity of $(A, \star)$, and it is included in the subset B. If $j - i > 1$, according to $a^{j-i} = a \star a^{j-i-1}$, we can conclude that a^{j-i-1} is the inverse of a, and it is included in the subset B. If $j - i = 1$, we have $a^i = a^i \star a$. Thus, a must be the identity element and is its own inverse. Consequently, that $\star$ is a closed operation on B guarantees that $(B, \star)$ is a subgroup. $\qquad\square$

8.4 GENERATORS AND EVALUATION OF POWERS

Let $(A, \star)$ be an algebraic system in which A is a set of colors and the binary operation $\star$ gives the color that the combination of two colors yields, e.g., red $\star$ yellow = orange. Suppose we are given a subset of the colors in A; we might want to know all the colors we can obtain by trying all possible combinations of the given colors. Also, consider the group $(\{0°, 60°, 120°, 180°, 240°, 300°\}, \star)$ that describes the rotation of geometric figures in the plane. Suppose we can rotate the figures only by 120° each time. Successive rotations by 120° will yield the rotations $\{0°, 120°, 240°\}$. On the other hand, suppose we can rotate the figures only by 60° each time. Successive rotations by 60° will yield all the rotations in $\{0°, 60°, 120°, 180°, 240°, 300°\}$.

Let $(A, \star)$ be an algebraic system, and $B = \{a_1, a_2, \ldots\}$ be a subset of A. Let B_1 denote the subset of A which contains B as well as all elements $a_i \star a_j$ for a_i and a_j in B. B_1 is called the set *generated directly* by B. Similarly, let B_2 denote the set generated directly by $B_1, \ldots$, and B_{i+1} denote the set generated directly by B_i. Let B^* denote the union of $B, B_1, B_2, \ldots$. The algebraic system $(B^*, \star)$ is called the *subsystem generated* by B, and an element is said to be generated by B if it is in B^*. Note that $\star$ is a closed operation on B^*. Thus, for a group $(A, \star)$, if B^* is finite, then $(B^*, \star)$ is a subgroup. If $B^* = A$, B is called a *generating set* or a *set of generators* of the algebraic system $(A, \star)$. In the example on the combination of colors, a generating set is a subset of the set of colors whose combinations will yield all the colors in the original set. In the example on the rotation of geometric figures in the plane, $\{60°\}$ is a generating set, so is $\{120°, 180°\}$.

A group that has a generating set consisting of a single element is known as

---

† We use a^m to denote $\underbrace{a \star a \star a \cdots \star a.}_{m \text{ times}}$

a *cyclic group.* Figure 8.7a shows a cyclic group for which $\{\beta\}$ is a generating set. Note that $\{\delta\}$ is also a generating set. For the example of the rotation of geometric figures in the plane, the group $(\{0°, 60°, 120°, 180°, 240°, 300°\}, \star)$ is also a cyclic group. We ask the reader to verify that the group in Fig. 8.7b is not a cyclic group.

$\star$	α	β	γ	δ
α	α	β	γ	δ
β	β	γ	δ	α
γ	γ	δ	α	β
δ	δ	α	β	γ

(a)

$\star$	α	β	γ	δ
α	α	β	γ	δ
β	β	α	δ	γ
γ	γ	δ	α	β
δ	δ	γ	β	α

(b)

Figure 8.7

Let $(A, \star)$ be a cyclic group and $\{a\}$ be a generating set of $(A, \star)$. Clearly, the elements in A can be expressed as $a, a^2, a^3, \ldots$. Since, because of associativity, $a^i \star a^j = a^j \star a^i = a^{i+j}$, it follows immediately that *any cyclic group is commutative.*

We digress for a moment to present an interesting problem in connection with the concept of generating sets of algebraic systems. Let B be a generating set of an algebraic system $(A, \star)$. For an element a in A, we would want to know the various ways of generating the element a. By generating the element a we mean to obtain a through successive combinations of the elements in the generating set. A way of generating a can be specified by a sequence of elements in a

$$a_1 \quad a_2 \quad a_3 \quad \cdots \quad a_r$$

such that $a_r = a$, and each a_i, $1 \leq i \leq r$, can be expressed as $a_j \star a_k$, where a_j and a_k are either elements from B or elements preceding a_i in the sequence. Since a sequence of r elements corresponds to generating the element a in r applications of the operation $\star$ to elements in the generating set and elements that have already been generated, it will be interesting to determine short sequences that generate a given element.

Our problem is motivated by that of finding efficient procedures to evaluate the power x^n for a given x and a positive integer n. Consider the algebraic system $(I, +)$, where I is the set of all positive integers and $+$ is the ordinary addition operation of integers. Clearly, $B = \{1\}$ is a generating set of the system. For a given integer n, we would like to know the various ways in which n can be generated. For example, the following sequences show some of the ways to generate the integer 9:

$$2 \quad 3 \quad 4 \quad 5 \quad 6 \quad 7 \quad 8 \quad 9$$
$$2 \quad 3 \quad 4 \quad 5 \quad 9$$
$$2 \quad 4 \quad 8 \quad 9$$

In the literature, a sequence of elements in I that leads to the generation of an integer n is called an *addition chain* for n. The connection between an addition chain for n and a procedure for evaluating x^n for a given value of x becomes obvious once we recall that $x^j \cdot x^k = x^{j+k}$.

The problem of determining a shortest addition chain for a given integer n is extremely interesting and has been studied rather extensively. However, a thorough discussion of the subject is beyond the scope of this book. As an illustration, we present here two simple procedures for determining addition chains for n.

If n can be factored as pq, we can determine addition chains for p and q first and then combine them to obtain an addition chain for pq. Let

$$p_1 \quad p_2 \quad \cdots \quad p_{i-1} \quad p$$

and

$$q_1 \quad q_2 \quad \cdots \quad q_{j-1} \quad q$$

be addition chains for p and q. Clearly,

$$q_1 \quad q_2 \quad \cdots \quad q_{j-1} \quad q \quad p_1 q \quad \cdots \quad p_{i-1} q \quad pq$$

is an addition chain for n.

Consider the example $n = 45$, which can be written as 5×9. Since

$$2 \quad 3 \quad 5$$

is an addition chain for 5, and

$$2 \quad 4 \quad 8 \quad 9$$

is an addition chain for 9, we obtain

$$2 \quad 4 \quad 8 \quad 9 \quad 18 \quad 27 \quad 45$$

as an addition chain for 45. Alternatively, since

$$9 = 3 \times 3$$

and since

$$2 \quad 3$$

is an addition chain for 3, we obtain

$$2 \quad 3 \quad 6 \quad 9$$

as an addition chain for 9, and

$$2 \quad 3 \quad 6 \quad 9 \quad 18 \quad 27 \quad 45$$

as an addition chain for 45.

The second procedure is a recursive one. If n is an even number, we can determine an addition chain for $n/2$, and then add $n/2$ to $n/2$ to obtain n. Thus, if

$$a_1 \quad a_2 \quad \cdots \quad n/2$$

is an addition chain for $n/2$, then

$$a_1 \quad a_2 \quad \cdots \quad n/2 \quad n$$

is an addition chain for n. If n is an odd number, we can determine an addition chain for $(n-1)/2$, add $(n-1)/2$ to $(n-1)/2$ to obtain $n-1$, and then add 1 to $n-1$ to obtain n. Thus, if

$$a_1 \quad a_2 \quad \cdots \quad (n-1)/2$$

is an addition chain for $(n-1)/2$, then

$$a_1 \quad a_2 \quad \cdots \quad (n-1)/2 \quad n-1 \quad n$$

is an addition chain for n. Now, this procedure can be applied to determine an addition chain for $n/2$ or $(n-1)/2$. For example,

$$2 \quad 4 \quad 5 \quad 10 \quad 11 \quad 22 \quad 44 \quad 45$$

is an addition chain for the integer 45 determined by this procedure.

8.5 COSETS AND LAGRANGE'S THEOREM

Let us extend the example of the vending machine in Sec. 8.1 such that two coins from the set {nickel, dime, quarter, half-dollar, dollar} are to be deposited in each purchase. If we have already deposited a quarter in the machine, we might want to know the items we will receive if the second coin to be deposited is either a dime, or a quarter, or a half-dollar. Also, consider the example on the rotation of geometric figures in the plane. Suppose an initial rotation of either 0, or 120, or 240° is followed by a subsequent rotation of 60°. We want to know all the possible total angular rotations. Let $(A, \star)$ be an algebraic system, where $\star$ is a binary operation. Let a be an element in A and H be a subset of A. The *left coset of H with respect to a*, which we shall denote $a \star H$, is the set of elements $\{a \star x | x \in H\}$. Similarly, the *right coset of H with respect to a*, which we shall denote $H \star a$, is the set of elements $\{x \star a | x \in H\}$. Clearly, in the example of the vending machine, we want to determine the left coset of the set {dime, quarter, half-dollar} with respect to a quarter, and in the example of rotation of geometric figures, we want to determine the right coset of the set $\{0°, 120°, 240°\}$ with respect to $60°$.

We can say a great deal more about cosets when we restrict ourselves to cosets in groups. Let $(A, \star)$ be a group and $(H, \star)$ be a subgroup of $(A, \star)$. We have:

Theorem 8.2 Let $a \star H$ and $b \star H$ be two cosets of H. Either $a \star H$ and $b \star H$ are disjoint or they are identical.

PROOF Suppose $a \star H$ and $b \star H$ are not disjoint, and have f as a common element. That is, there exist h_1 and h_2 in H such that $f = a \star h_1 = b \star h_2$. We can write $a = b \star h_2 \star h_1^{-1}$. For any element x in $a \star H$, since $x = a \star h_3$ for some h_3 in H, we have $x = b \star h_2 \star h_1^{-1} \star h_3$ which is an element in $b \star H$ because $h_2 \star h_1^{-1} \star h_3$ is an element in H. In a similar way we can show that any element in $b \star H$ is also an element in $a \star H$. We thus conclude that the two sets $a \star H$ and $b \star H$ are equal.† $\qquad\qquad$ □

Let $(A, \star)$ be a group, and $(H, \star)$ be a subgroup of $(A, \star)$. Because $(A, \star)$ is a group, for any a in A and any distinct h_1 and h_2 in H, $a \star h_1 \neq a \star h_2$. It follows that the size of any coset of H is the same as that of H. Furthermore, because H contains the identity of the group, if we compute all the left (right) cosets of H, we would have exhausted all the elements in A. Consequently, we can conclude that the left cosets of H form a partition of A, in which all blocks are of the same size. Thus, the size of A is equal to the number of distinct left cosets of H times the size of H. In other words we have:

Theorem 8.3 (LaGrange) The order of any subgroup of a finite group divides the order of the group.

There are some immediate consequences of LaGrange's theorem. First, we note that a group of prime order has no nontrivial subgroup.‡ It follows that a group of prime order must be cyclic, and any set containing a single element other than the identity is a generating set.

*8.6 PERMUTATION GROUPS AND BURNSIDE'S THEOREM

We study in this section an important class of groups. A one-to-one function from a set S onto itself is called a *permutation* of the set S. We use the notation $\begin{pmatrix} abcd \\ bdca \end{pmatrix}$ for the permutation of the set $\{a, b, c, d\}$ that maps a into b, b into d, c into c, and d into a; that is, in the upper row the elements in the set are written down in an arbitrary order, and in the lower row the image of an element will be written below the element itself.

For a set of n elements, S, let A denote the set of $n!$ permutations of S. We define a binary operation $\circ$ on A to be the composition of two functions. (See Prob. 3.24.) We note that *the binary operation $\circ$ is a closed operation on A*. Let π_1 and π_2 be two permutations of the set $S = \{a, b, c, \ldots, x, y, z\}$. To show that $\pi_1 \circ \pi_2$ is also a permutation of the set S, we have only to show that no two elements in S are mapped into the same element by $\pi_1 \circ \pi_2$. Suppose that π_2 maps the element a into b and π_1 maps the element b into c. $\pi_1 \circ \pi_2$ will then map the element a into c. Let x be any element distinct from a. Since π_2 is a permutation of the set S, π_2 maps x into an element that is distinct from b, say

---

† We want to remind the reader that this does not mean $a \star h = b \star h$ for all h in H.

‡ A subgroup is said to be trivial if it contains either all elements in the group or the identity only.

y. Similarly, π_1 maps y into an element that is distinct from c, say z. Thus, $\pi_1 \circ \pi_2$ maps x into z. We conclude that $\pi_1 \circ \pi_2$ always maps two distinct elements (for example, a and x) into two distinct elements (for example, c and z) and is, therefore, a permutation of the set S. For example, let

$$\pi_1 = \begin{pmatrix} abcd \\ adbc \end{pmatrix} \qquad \pi_2 = \begin{pmatrix} abcd \\ bacd \end{pmatrix}$$

we have

$$\pi_1 \circ \pi_2 = \begin{pmatrix} abcd \\ dabc \end{pmatrix}$$

We also note that the *binary operation* $\circ$ *is associative.* That is, for any permutations π_1, π_2, and π_3 of a set, we have $(\pi_1 \circ \pi_2) \circ \pi_3 = \pi_1 \circ (\pi_2 \circ \pi_3)$. This fact can be seen as follows: Suppose π_3 maps a into b, π_2 maps b into c, and π_1 maps c into d. Since $\pi_1 \circ \pi_2$ maps b into d, $(\pi_1 \circ \pi_2) \circ \pi_3$ maps a into d. Similarly, since $\pi_2 \circ \pi_3$ maps a into c, $\pi_1 \circ (\pi_2 \circ \pi_3)$ maps a into d. For example, let

$$\pi_1 = \begin{pmatrix} abcd \\ adbc \end{pmatrix} \qquad \pi_2 = \begin{pmatrix} abcd \\ bacd \end{pmatrix} \qquad \pi_3 = \begin{pmatrix} abcd \\ bdac \end{pmatrix}$$

Then

$$(\pi_1 \circ \pi_2) \circ \pi_3 = \left[\begin{pmatrix} abcd \\ adbc \end{pmatrix} \begin{pmatrix} abcd \\ bacd \end{pmatrix} \right] \begin{pmatrix} abcd \\ bdac \end{pmatrix} = \begin{pmatrix} abcd \\ dabc \end{pmatrix} \begin{pmatrix} abcd \\ bdac \end{pmatrix} = \begin{pmatrix} abcd \\ acdb \end{pmatrix}$$

and

$$\pi_1 \circ (\pi_2 \circ \pi_3) = \begin{pmatrix} abcd \\ adbc \end{pmatrix} \left[\begin{pmatrix} abcd \\ bacd \end{pmatrix} \begin{pmatrix} abcd \\ bdac \end{pmatrix} \right] = \begin{pmatrix} abcd \\ adbc \end{pmatrix} \begin{pmatrix} abcd \\ adbc \end{pmatrix} = \begin{pmatrix} abcd \\ acdb \end{pmatrix}$$

It follows that $(A, \circ)$ is a group in which the permutation mapping every element in S into itself is the identity, and the inverse of a permutation π is one that maps $\pi(a)$ into a for all a in S. For example, for $S = \{a, b, c, d\}$, the identity of $(A, \circ)$ is $\begin{pmatrix} abcd \\ abcd \end{pmatrix}$, the inverse of $\begin{pmatrix} abcd \\ bcad \end{pmatrix}$ is $\begin{pmatrix} abcd \\ cabd \end{pmatrix}$. A subgroup of $(A, \circ)$ is usually referred to as a *permutation group* of the set S.

Let $(G, \circ)$ be a permutation group of a set $S = \{a, b, \ldots\}$. A binary relation on the set S, called the *binary relation induced by* $(G, \circ)$, is defined to be such that element a is related to element b if and only if there is a permutation in G that maps a into b. For example, let

$$G = \left\{ \begin{pmatrix} abcd \\ abcd \end{pmatrix}, \begin{pmatrix} abcd \\ bacd \end{pmatrix}, \begin{pmatrix} abcd \\ abdc \end{pmatrix}, \begin{pmatrix} abcd \\ badc \end{pmatrix} \right\}$$

The binary relation induced by $(G, \circ)$ is shown in Fig. 8.8. We note that *the binary relation on S induced by a permutation group* $(G, \circ)$ *is an equivalence relation.* Since the identity permutation is in G, every element in S is related to itself in the binary relation on S induced by $(G, \circ)$. Therefore, the reflexive law is satisfied. If there is a permutation π_1 in G that maps a into b,

	a	b	c	d
a	√	√		
b	√	√		
c			√	√
d			√	√

Figure 8.8

the inverse of π_1, which is also in G, will map b into a. Therefore, the binary relation on S induced by $(G, \circ)$ satisfies the symmetric law. If there is a permutation π_1 mapping a into b and a permutation π_2 mapping b into c, the permutation $\pi_2 \circ \pi_1$, which is also in G, will map a into c. Therefore, the binary relation on S induced by $(G, \circ)$ satisfies the transitive law.

We are now ready to prove a result due to Burnside which is usually referred to as *Burnside's theorem*. Given a set S and a permutation group $(G, \circ)$ of S, we wish to find the number of equivalence classes into which S is divided by the equivalence relation on S induced by $(G, \circ)$. This problem can be solved most directly by finding the equivalence relation and then counting the number of equivalence classes. However, when the set S contains a large number of elements, such counting becomes prohibitively tedious. Burnside's theorem enables us to find the number of equivalence classes in an alternative way by counting the number of elements that are invariant under the permutations in the group. An element is said to be *invariant* under a permutation, or is called an *invariance*, if the permutation maps the element into itself.

Theorem 8.4 (Burnside) The number of equivalence classes into which a set S is divided by the equivalence relation induced by a permutation group $(G, \circ)$ of S is given by

$$\frac{1}{|G|} \sum_{\pi \in G} \psi(\pi)$$

where $\psi(\pi)$ is the number of elements that are invariant under the permutation π.

So that we can appreciate more the meaning of Burnside's theorem, let us illustrate its application before proceeding to the proof. Let $S = \{a, b, c, d\}$, and let G be the permutation group consisting of $\pi_1 = \begin{pmatrix} abcd \\ abcd \end{pmatrix}$, $\pi_2 = \begin{pmatrix} abcd \\ bacd \end{pmatrix}$, $\pi_3 = \begin{pmatrix} abcd \\ abdc \end{pmatrix}$, and $\pi_4 = \begin{pmatrix} abcd \\ badc \end{pmatrix}$. The equivalence relation on S induced by G is shown in Fig. 8.8. Clearly, S is divided into two equivalence classes, $\{a, b\}$ and

$\{c, d\}$. To compute the number of equivalence classes according to Burnside's theorem, we note that since $\psi(\pi_1) = 4$, $\psi(\pi_2) = 2$, $\psi(\pi_3) = 2$, and $\psi(\pi_4) = 0$, the number of equivalence classes is

$$\tfrac{1}{4}(4 + 2 + 2 + 0) = 2$$

PROOF For any element s in S, let $\eta(s)$ denote the number of permutations under which s is invariant. Then

$$\sum_{\pi \in G} \psi(\pi) = \sum_{s \in S} \eta(s)$$

because both $\sum_{\pi \in G} \psi(\pi)$ and $\sum_{s \in S} \eta(s)$ count the total number of invariances under all the permutations in G. [One way to count the invariances is to go through the permutations one by one and count the number of invariances under each permutation. This gives $\sum_{\pi \in G} \psi(\pi)$ as the total count. Another way to count the invariances is to go through the elements one by one and count the number of permutations under which an element is invariant. That gives $\sum_{s \in S} \eta(s)$ as the total count.]

Let a and b be two elements in S that are in the same equivalence class. We want to show that there are exactly $\eta(a)$ permutations mapping a into b. Since a and b are in the same equivalence class, there is at least one such permutation which we shall denote by π_x. Let $\{\pi_1, \pi_2, \pi_3, \ldots\}$ be the set of the $\eta(a)$ permutations under which a is invariant. Then, the $\eta(a)$ permutations in the set $\{\pi_x \circ \pi_1, \pi_x \circ \pi_2, \pi_x \circ \pi_3, \ldots\}$ are permutations that map a into b. First, we see that these permutations are all distinct because, if $\pi_x \circ \pi_1 = \pi_x \circ \pi_2$, we have

$$\pi_x^{-1} \circ (\pi_x \circ \pi_1) = \pi_x^{-1} \circ (\pi_x \circ \pi_2)$$

This gives $\pi_1 = \pi_2$, which is impossible. Secondly, we see that no other permutation in G maps a into b. Suppose that there is a permutation π_y that maps a into b. Then, $\pi_x^{-1} \circ \pi_y$ is a permutation that maps a into a, because π_x^{-1} maps b into a. Since $\pi_x^{-1} \circ \pi_y$ is a permutation in the set $\{\pi_1, \pi_2, \pi_3, \ldots\}$, $\pi_x \circ (\pi_x^{-1} \circ \pi_y) = \pi_y$ is a permutation in the set $\{\pi_x \circ \pi_1, \pi_x \circ \pi_2, \pi_x \circ \pi_3, \ldots\}$. Therefore, we conclude that there are exactly $\eta(a)$ permutations in G that map a into b.

Let $a, b, c, \ldots, h$ be the elements in S that are in one equivalence class. All the permutations in G can be categorized as those that map a into a, those that map a into b, those that map a into c, $\ldots$, and those that map a into h. Since we have shown that there are exactly $\eta(a)$ permutations in each of these categories we have

$$\eta(a) = \frac{|G|}{\text{number of elements in the equivalence class containing } a}$$

Using a similar argument, we obtain

$$\eta(b) = \eta(c) = \cdots = \eta(h)$$

$$= \frac{|G|}{\text{number of elements in the equivalence class containing } a}$$

and, therefore,

$$\eta(a) + \eta(b) + \eta(c) + \cdots + \eta(h) = |G|$$

It follows that, for any equivalence class of elements in S,

$$\sum_{\substack{\text{all } s \text{ in the equivalence class}}} \eta(s) = |G|$$

and

$$\sum_{s \in S} \eta(s) = \left(\begin{array}{c}\text{number of equivalence classes}\\ \text{into which } S \text{ is divided}\end{array}\right) \times |G|$$

Therefore, we have

Number of equivalence classes into which S is divided

$$= \frac{1}{|G|} \sum_{s \in S} \eta(s) = \frac{1}{|G|} \sum_{\pi \in G} \psi(\pi) \qquad \square$$

We now consider some illustrative examples:

Example 8.1 We want to find the number of distinct strings of length 2 that are made up of blue beads and yellow beads. The two ends of a string are not marked, and two strings are, therefore, indistinguishable if interchanging the ends of one will yield the other. Let b and y denote blue and yellow beads, respectively. Let bb, by, yb, and yy denote the four different strings of length 2 when equivalence between strings is not taken into consideration. The problem is to find the number of equivalence classes into which the set $S = \{bb, by, yb, yy\}$ is divided by the equivalence relation induced by the permutation group $(\{\pi_1, \pi_2\}, \circ)$, where

$$\pi_1 = \begin{pmatrix} bb & by & yb & yy \\ bb & by & yb & yy \end{pmatrix} \qquad \pi_2 = \begin{pmatrix} bb & by & yb & yy \\ bb & yb & by & yy \end{pmatrix}$$

The permutation π_1 merely indicates that every string is equivalent to itself, and the permutation π_2 specifies the equivalence between strings when the two ends of a string are interchanged. According to Burnside's theorem, the number of distinct strings is

$$\tfrac{1}{2}(4 + 2) = 3$$

Similarly, for the case of distinct strings of length 3 made up of blue beads and yellow beads, we have the set $S = \{bbb, bby, byb, ybb, byy, yby, yyb, yyy\}$ and the permutation group $(G, \circ)$, $G = \{\pi_1, \pi_2\}$, where π_1 is the identity permutation and π_2 is the permutation that maps a string into one that is obtained from the former by interchanging its ends; for example, bbb is mapped into bbb, bby is mapped into ybb, byb is mapped into byb, and so on. The number of elements that are invariant under π_1 is eight. The number of elements that are invariant under π_2 is four, since a string will be mapped into itself under π_2 if the beads at the two ends of a string are of the same color, and there are four such strings. Therefore, the number of distinct strings is equal to

$$\tfrac{1}{2}(8 + 4) = 6$$ □

Example 8.2 Suppose we want to find the number of distinct bracelets of five beads made up of yellow, blue, and white beads. Two bracelets are said to be indistinguishable if the rotation of one will yield another. However, to simplify the problem, we assume that the bracelets cannot be flipped over. Let S be the set of the 3^5 (= 243) distinct bracelets when rotational equivalence is not considered. Let $(\{\pi_1, \pi_2, \pi_3, \pi_4, \pi_5\}, \circ)$ be a permutation group, where π_1 is the identity permutation and π_2 is the permutation that maps a bracelet into one which is the former rotated clockwise by one bead

position (for example,). π_3, π_4, and π_5

are, similarly, permutations that map a bracelet into one which is the former rotated clockwise by two, three, and four bead positions, respectively.

The number of elements that are invariant under π_1 is 243. The number of elements that are invariant under π_2 is three because only when all five beads in a bracelet are of the same color will its rotation by one bead position yield the same bracelet. Similarly, the number of elements that are invariant under each of π_3, π_4, and π_5 is also three. Therefore, the number of distinct bracelets is

$$\tfrac{1}{5}(243 + 3 + 3 + 3 + 3) = 51$$

The result on the number of distinct bracelets leads immediately to a very interesting proof of what is known as *Fermat's little theorem* in number theory. For a prime number p, let us determine the number of distinct bracelets of p beads made up of beads of a distinct colors when rotational equivalence is considered. The number of distinct bracelets is

$$\frac{1}{p}(a^p + a + a + \cdots + a) = \frac{1}{p}[a^p + (p-1)a]$$

Since the number of distinct bracelets is a whole number, p divides $a^p + (p-1)a$, or p divides $a^p - a$. If p does not divide a, then p must divide $a^{p-1} - 1$, which is exactly Fermat's theorem. □

Example 8.3 Suppose that we are to print all the 5-digit numbers on slips of paper with one number on each slip. Clearly, there are 10^5 such slips. (For numbers smaller than 10,000, leading zeros are always filled in.) However, since the digits 0, 1, 6, 8, and 9 become 0, 1, 9, 8, and 6 when they are read upside down, there are pairs of numbers that can share the same slip if the slips will be read either right side up or upside down. For example, we can make up one slip for both the numbers 89166 and 99168. The question is then how many distinct slips will we have to make up for the 10^5 numbers. Here S is the set of the 10^5 numbers and $(\{\pi_1, \pi_2\}, \circ)$ is a permutation group of S, where π_1 is the identity permutation, and π_2 is a permutation that maps a number into itself if it is not readable as a number when turned upside down (for example, 13765 is mapped into 13765) and maps a number into the number obtained by reading the former upside down whenever it is possible (for example, 89166 is mapped into 99168). The number of invariances under π_1 is 10^5. The number of invariances under π_2 is $(10^5 - 5^5) + 3 \times 5^2$ because there are $10^5 - 5^5$ numbers that contain one or more of the digits 2, 3, 4, 5, and 7 and cannot be read upside down, and because there are 3×5^2 numbers that will read the same either right side up or upside down, for example, 16891 (the center digit of these numbers must be 0 or 1 or 8, the last digit must be the first digit turned upside down, and the fourth digit must be the second digit turned upside down). Therefore, the number of distinct slips to be made up is

$$\tfrac{1}{2}(10^5 + 10^5 - 5^5 + 3 \times 5^2) = 10^5 - \tfrac{1}{2} \times 5^5 + \tfrac{3}{2} \times 5^2 \qquad \square$$

8.7 CODES AND GROUP CODES

The coding problem is that of representing distinct messages by distinct sequences of letters from a given alphabet. For example, messages such as "emergency," "help is on the way," "all is clear," and so on, can be represented by sequences of dots and dashes. We assume in our discussion in this section that the alphabet is the binary alphabet $\{0, 1\}$. A sequence of letters from an alphabet is often referred to as a *word*. A *code* is a collection of words that are to be used to represent distinct messages. A word in a code is also called a *codeword*. A *block code* is a code consisting of words that are of the same length. One of the criteria for choosing a block code to represent a set of messages is its ability to correct errors. Suppose a codeword is transmitted from its origin to its destination. In the course of transmission, interferences such as noises might cause some of the 1s in the codeword to be received as 0s, and some of the 0s to be received as 1s. Consequently, the received word might no longer be the transmitted one, and it is our desire to recover the transmitted word if at all possible. This is what we mean by error correction.

Let A denote the set of all binary sequences of length n. Let $\oplus$ be a binary operation on A such that for $\mathbf{x}$ and $\mathbf{y}$ in A, $\mathbf{x} \oplus \mathbf{y}$ is a sequence of length n that has 1s in those positions $\mathbf{x}$ and $\mathbf{y}$ differ and has 0s in those positions $\mathbf{x}$ and $\mathbf{y}$ are the same. For example, let $\mathbf{x} = 00101$ and $\mathbf{y} = 10110$, then $\mathbf{x} \oplus \mathbf{y} = 10011$. We ask the reader to show that $(A, \oplus)$ is a group. Note that the all zero word is the identity and every word is its own inverse in $(A, \oplus)$.

Let $\mathbf{x}$ be a word in A. We define the weight of $\mathbf{x}$, denoted $w(\mathbf{x})$, to be the number of 1s in **x**. Thus, the weight of 1110000 is 3, and so is that of 1001100. For **x** and **y** in A, we define the *distance* between **x** and **y**, denoted $d(\mathbf{x}, \mathbf{y})$, to be the weight of $\mathbf{x} \oplus \mathbf{y}$, $w(\mathbf{x} \oplus \mathbf{y})$. For example, the distance between 1110000 and 1001100 is 4, and the distance between 1110000 and 0001111 is 7. Note that the distance between two words is exactly the number of positions at which they differ.

It is obvious that for any **x** and **y**, $d(\mathbf{x}, \mathbf{y}) = d(\mathbf{y}, \mathbf{x})$. We now show that for any **x**, **y**, **z** in A

$$d(\mathbf{x}, \mathbf{y}) \leqq d(\mathbf{x}, \mathbf{z}) + d(\mathbf{z}, \mathbf{y})$$

Clearly,

$$w(\mathbf{u} \oplus \mathbf{v}) \leqq w(\mathbf{u}) + w(\mathbf{v})$$

Thus, we have

$$w(\mathbf{x} \oplus \mathbf{y}) = w(\mathbf{x} \oplus \mathbf{z} \oplus \mathbf{z} \oplus \mathbf{y})$$
$$\leqq w(\mathbf{x} \oplus \mathbf{z}) + w(\mathbf{z} \oplus \mathbf{y})$$

or

$$d(\mathbf{x}, \mathbf{y}) \leqq d(\mathbf{x}, \mathbf{z}) + d(\mathbf{z}, \mathbf{y})$$

Let G be a block code. We define the *distance* of G to be the minimum distance between any pair of distinct codewords in G. The distance of a block code is closely related to its ability to correct errors, as we shall show. Suppose that corresponding to the transmission of a codeword in G, the word **y** was received. Our question is to determine from **y** the codeword that was transmitted. To motivate our discussion, let us assume the simple case that **y** turns out to be one of the codewords in G. In this case, one probably would jump to the obvious conclusion that the transmitted word was indeed **y**. We note that even such an obvious conclusion needs clarification. Since we assume that in the course of transmission, errors can occur in any of the positions, every one of the codewords in G could have been the transmitted word. When we decided that the transmitted word was **y**, we have tacitly assumed that when a word was transmitted, it was more likely that no error had occurred than that some errors had occurred. Let us not commit ourselves to such an assumption for the time being and state in a general way how we are to determine the transmitted word corresponding to a received word **y**. Let $\mathbf{x}_1, \mathbf{x}_2, \ldots, \mathbf{x}_N$ denote the codewords in G. We shall compute the conditional probability $P(\mathbf{x}_i|\mathbf{y})$ for $i = 1, 2, \ldots, N$, where $P(\mathbf{x}_i|\mathbf{y})$ is the probability that $\mathbf{x}_i$ was the transmitted word given that **y** was the received word. If $P(\mathbf{x}_k|\mathbf{y})$ is the largest of all conditional probabilities computed, we shall conclude that $\mathbf{x}_k$ was the transmitted word. Such a criterion for determining the transmitted word is known as the *maximum-likelihood decoding criterion.*

The computation of the conditional probability $P(\mathbf{x}_i|\mathbf{y})$ can be quite involved since the probability depends on many factors in the communication system. There is, however, another criterion that can be used to determine the transmitted word. We compute $d(\mathbf{x}_i, \mathbf{y})$ for $i = 1, 2, \ldots, N$, and conclude that $\mathbf{x}_k$ was the transmitted word if $d(\mathbf{x}_k, \mathbf{y})$ is the smallest among all distances computed. This is known as the *minimum-distance decoding criterion.* If we assume that the occurrences of errors in the positions are independent, and that the probability of the occurrence of an error is p, then $P(\mathbf{x}_i|\mathbf{y}) = (1-p)^{n-t}p^t$, where t is the distance between $\mathbf{x}_i$ and $\mathbf{y}$. For $p < \frac{1}{2}$, the smaller $d(\mathbf{x}_i, \mathbf{y})$ is, the larger $P(\mathbf{x}_i|\mathbf{y})$ will be.† Consequently, the minimum-distance decoding criterion is equivalent to the maximum-likelihood decoding criterion. (In this case, the conclusion that the transmitted word was $\mathbf{y}$ when the received word $\mathbf{y}$ is a codeword is justified.)

We note immediately that *a code of distance $2t + 1$ can correct t or fewer transmission errors when the minimum-distance decoding criterion is followed.* Suppose a codeword $\mathbf{x}$ was transmitted and the word $\mathbf{y}$ was received. If no more than t errors have occurred in the course of transmission, we have

$$d(\mathbf{x}, \mathbf{y}) \leq t$$

Let $\mathbf{x}_1$ be another codeword. Since

$$d(\mathbf{x}, \mathbf{x}_1) \geq 2t + 1$$

and

$$d(\mathbf{x}, \mathbf{x}_1) \leq d(\mathbf{x}, \mathbf{y}) + d(\mathbf{y}, \mathbf{x}_1)$$

we have

$$d(\mathbf{y}, \mathbf{x}_1) \geq t + 1$$

Therefore, the minimum-distance decoding criterion will indeed select $\mathbf{x}$ as the transmitted word.

We study now a class of block codes known as *group codes*. A subset G of A is called a group code if $(G, \oplus)$ is a subgroup of $(A, \oplus)$, where A is the set of binary sequences of length n.

We show now that the distance of G is equal to the minimum weight of the nonzero words in G. This result makes it much easier to compute the distance of a group code since it is no longer necessary to compute the distance between every pair of distinct words in G exhaustively. Suppose $\mathbf{x}$ is a nonzero word of minimum weight in G. Since

$$w(\mathbf{x}) = d(\mathbf{x}, \mathbf{0})‡$$

and since $\mathbf{0}$ is in G

$$w(\mathbf{x}) \geq \min_{\mathbf{y}, \mathbf{z} \in G} [d(\mathbf{y}, \mathbf{z})] \tag{8.1}$$

Consequently, $w(\mathbf{x})$ is larger than or equal to the distance of the code G. On the other hand, since

$$d(\mathbf{y}, \mathbf{z}) = w(\mathbf{y} \oplus \mathbf{z})$$

and since for any $\mathbf{y}$ and $\mathbf{z}$ in G, $\mathbf{y} \oplus \mathbf{z}$ is also in G, we have

$$d(\mathbf{y}, \mathbf{z}) \geq \min_{\substack{\mathbf{x} \in G \\ \mathbf{x} \neq \mathbf{0}}} [w(\mathbf{x})] \tag{8.2}$$

---

† See Prob. 8.42 for the case $p > \frac{1}{2}$.
‡ $\mathbf{0}$ denotes the all zero word.

From (8.1) we obtain

$$\min_{\substack{x \in G \\ x \neq 0}} [w(\mathbf{x})] \geq \min_{y, z \in G} [d(\mathbf{y}, \mathbf{z})] \tag{8.3}$$

From (8.2) we obtain

$$\min_{y, z \in G} [d(\mathbf{y}, \mathbf{z})] \geq \min_{\substack{x \in G \\ x \neq 0}} [w(\mathbf{x})] \tag{8.4}$$

Combining (8.3) and (8.4) we obtain

$$\min_{\substack{x \in G \\ x \neq 0}} [w(\mathbf{x})] = \min_{y, z \in G} [d(\mathbf{y}, \mathbf{z})]$$

For group codes, there is an efficient way to determine the transmitted word corresponding to a received word according to the minimum-distance decoding criterion. Let $(G, \oplus)$ be a group code. Let $\mathbf{y}$ be a received word. Since $d(\mathbf{x}_i, \mathbf{y}) = w(\mathbf{x}_i \oplus \mathbf{y})$, the weights of the words in the coset $G \oplus \mathbf{y}$ are the distances between the codewords in G and $\mathbf{y}$. Let $\mathbf{e}$ denote the word† of smallest weight in $G \oplus \mathbf{y}$. According to the minimum-distance decoding criterion, $\mathbf{e} \oplus \mathbf{y}^{-1} = \mathbf{e} \oplus \mathbf{y}$ is the transmitted codeword. Since this argument is valid for all $\mathbf{y}$ in the coset $G \oplus \mathbf{y}$, our decoding procedure can be stated as:

1. Determine all cosets of G.
2. For each coset, pick the word of the smallest weight,‡ which we shall refer to as the leader of the coset.
3. For a received word $\mathbf{y}$, $\mathbf{e} \oplus \mathbf{y}$ is the transmitted word, where $\mathbf{e}$ is the leader of the coset containing $\mathbf{y}$.

As an example, let $G = \{0000, 0011, 1101, 1110\}$. It can easily be checked that $(G, \oplus)$ is a group. The rows in Fig. 8.9 are the distinct cosets of G. We ask the reader to check that according to the minimum distance decoding criterion the received word 1011 will be decoded as 0011, the received word 1010 will be decoded as 1110, and the received word 1111 will be decoded either as 1101 or 1110 depending on whether 0010 or 0001 was chosen as the leader of the coset containing the word 1111.

$$
\begin{array}{cccc}
0000 & 0011 & 1101 & 1110 \\
1000 & 1011 & 0101 & 0110 \\
0100 & 0111 & 1001 & 1010 \\
0010 & 0001 & 1111 & 1100
\end{array}
$$

Figure 8.9

† Or one of the words of smallest weight.
‡ Or one of the words of smallest weight.

8.8 ISOMORPHISMS AND AUTOMORPHISMS

Let $(A, \star)$ be the algebraic system shown in Fig. 8.10a. Since the names of the elements of A, a, b, c, and d as well as the name of the operation on A, $\star$ are abstract names, there is no reason we cannot change them to other abstract names. For example, by changing a, b, c, d to α, β, γ, δ and $\star$ to $*$, we obtain the algebraic system $(B, *)$ in Fig. 8.10b. Clearly, one would agree that the two systems $(A, \star)$ and $(B, *)$ are "essentially the same." We say that an algebraic system $(B, *)$ is *isomorphic* to the algebraic system $(A, \star)$ if we can obtain $(B, *)$ from $(A, \star)$ by renaming the elements and/or the operation in $(A, \star)$. In a more formal but equivalent way, we say that $(B, *)$ is isomorphic to $(A, \star)$ if there exists a one-to-one onto function f from A to B such that for all a_1 and a_2 in A

$$f(a_1 \star a_2) = f(a_1) * f(a_2)$$

The function f is called an *isomorphism* from $(A, \star)$ to $(B, *)$, and $(B, *)$ is called an *isomorphic image* of A. For example, the function f such that

$$f(a) = \alpha$$

$$f(b) = \beta$$

$$f(c) = \gamma$$

$$f(d) = \delta$$

is an isomorphism from the algebraic system $(A, \star)$ in Fig. 8.10a to the algebraic system $(B, *)$ in Fig. 8.10b. Note that the function g such that

$$g(a) = \delta$$

$$g(b) = \gamma$$

$$g(c) = \beta$$

$$g(d) = \alpha$$

is also an isomorphism from $(A, \star)$ to $(B, *)$.

$\star$	a	b	c	d
a	a	b	c	d
b	b	a	a	c
c	b	d	d	c
d	a	b	c	d

(a)

$*$	α	β	γ	δ
α	α	β	γ	δ
β	β	α	α	γ
γ	β	δ	δ	γ
δ	α	β	γ	δ

(b)

Figure 8.10

As another example, we note that in Fig. 8.11 the algebraic systems $(B, \oplus)$, $(C, *)$, and $(D, +)$ are all isomorphic to the algebraic system $(A, \star)$. As a matter of fact, the system $(B, \oplus)$ corresponds to the addition of even and odd numbers, the system $(C, *)$ corresponds to the rotation of geometric figures in the plane by 0 and 180°, and the system $(D, +)$ corresponds to the situation of purchasing two items in a five-and-ten store, where the element 0¢ in D stands for amounts that are multiples of 10¢, the element 5¢ stands for amounts that are multiples of 10¢ plus 5¢, and the binary operation + determines whether the total purchase price is a multiple of 10¢ or is a multiple of 10¢ plus 5¢. Indeed, the notion of isomorphism between two algebraic systems links abstract algebraic systems to physical situations we encounter in practice. Consequently, properties of abstract algebraic systems will have direct interpretation in terms of physical situations. From an applications point of view, this indeed is *the* reason to study abstract algebraic systems.

An isomorphism from an algebraic system $(A, \star)$ to $(A, \star)$ is called an *automorphism* on $(A, \star)$. For example, the function f such that

$$f(a) = d$$
$$f(b) = c$$
$$f(c) = b$$
$$f(d) = a$$

is an automorphism on the algebraic system $(A, \star)$ in Fig. 8.10a. A physical interpretation of an automorphism on an algebraic system is a way in which the elements in the system interchange their roles.

$\star$	a	b
a	a	b
b	b	a

$(A, \star)$

$\oplus$	EVEN	ODD
EVEN	EVEN	ODD
ODD	ODD	EVEN

$(B, \oplus)$

$*$	0°	180°
0°	0°	180°
180°	180°	0°

$(C, *)$

$+$	0¢	5¢
0¢	0¢	5¢
5¢	5¢	0¢

$(D, +)$

Figure 8.11

Example 8.4 As an illustrative example on the notion of isomorphism between groups, we shall show that there is a unique group, up to isomorphism, of order p for any prime p. We recall that for any integer n, $(Z_n, \oplus)$ is a group. Let $(G, \star)$ be a group of order p. Since any group of prime order is cyclic, the elements in G can be represented as

$a^0, a, a^2, \ldots, a^{p-1}$ for any element a in G that is not the identity.† The function $f(a^i) = i$ is clearly an isomorphism from $(G, \bigstar)$ to $(Z_p, \oplus)$. Consequently, we conclude that any group of order p is isomorphic to $(Z_p, \oplus)$. □

8.9 HOMOMORPHISMS AND NORMAL SUBGROUPS

The notion of isomorphic algebraic systems can be generalized immediately. Let $(A, \bigstar)$ and $(B, *)$ be two algebraic systems. Let f be a function from A onto B such that for any a_1 and a_2 in A

$$f(a_1 \bigstar a_2) = f(a_1) * f(a_2)$$

f is called a *homomorphism* from $(A, \bigstar)$ to $(B, *)$, and $(B, *)$ is called a *homomorphic image* of $(A, \bigstar)$.

For example, for the two algebraic systems in Fig. 8.12a and b the function f such that

$$f(\alpha) = 1 \qquad f(\beta) = 1 \qquad f(\gamma) = 1$$

$$f(\delta) = 0 \qquad f(\epsilon) = 0$$

$$f(\zeta) = -1$$

is a homomorphism from the algebraic system $(\{\alpha, \beta, \gamma, \delta, \epsilon, \zeta\}, \bigstar)$ to the algebraic system $(\{1, 0, -1\}, *)$.

$\bigstar$	α	β	γ	δ	ϵ	ζ
α	α	β	α	α	γ	δ
β	β	α	γ	β	γ	ϵ
γ	α	γ	α	β	γ	ϵ
δ	α	β	β	δ	ϵ	ζ
ϵ	γ	γ	γ	ϵ	ϵ	ζ
ζ	δ	ϵ	ϵ	ζ	ζ	ζ

(a)

$*$	1	0	-1
1	1	1	0
0	1	0	-1
-1	0	-1	-1

(b)

Figure 8.12

There is an alternative way to look at the notion of a homomorphism from one algebraic system to another. Let $(A, \bigstar)$ be an algebraic system and R be an equivalence relation on A. R is called a *congruence relation* on A (with respect to $\bigstar$) if (a_1, a_2) and (b_1, b_2) in R implies that $(a_1 \bigstar b_1, a_2 \bigstar b_2)$ is also in R. For example, for the algebraic system in Fig. 8.13a, the equivalence relation R in Figure 8.13b is a congruence relation. On the other hand, for the algebraic system in Fig. 8.14a, the equivalence relation R in Fig. 8.14b is not a

† a^0 stands for the identity of $(G, \bigstar)$.

congruence relation. [For instance, although (a, b) and (c, d) are in R, $(a \star c, b \star d)$, which is equal to (d, a), is not in R.] The equivalence classes into which A is divided are called *congruence classes*. We can define a new algebraic system

$\star$	a	b	c	d
a	a	a	d	c
b	b	a	c	d
c	c	d	a	b
d	d	d	b	a

(a)

Figure 8.13

$\star$	a	b	c	d
a	a	a	d	c
b	b	a	d	a
c	c	b	a	b
d	c	d	b	a

Figure 8.14

$(B, *)$ as follows: Let $B = \{A_1, A_2, \ldots, A_r\}$ be the partition of A induced by R.[†] Let the binary operation $*$ be such that for any A_i and A_j in B, $A_i * A_j$ is equal to the congruence class A_k containing the element $a_1 \star a_2$, where a_1 is any element in A_i and a_2 is any element in A_j. Note that because R is a congruence relation, the operation $*$ is well defined. (That is, the congruence class $A_i * A_j$ is uniquely determined no matter what elements a_1 in A_i and a_2 in A_j we happen to pick.) We note immediately that $(B, *)$ is a homomorphic image of $(A, \star)$, since the function

$$f(a) = A_i \quad \text{if} \quad a \in A_i$$

is a homomorphism from $(A, \star)$ to $(B, *)$.[‡] Intuitively, a homomorphic image of an algebraic system can be viewed as a gross description of the behavior of the system when some of the characters distinguishing some of the elements in the system are ignored. Consequently, elements that become indistinguishable can be placed in one congruence class, and the system's behavior can be described by how these congruence classes interact.

As an example, consider the algebraic system in Fig. 8.12a as a description of the interaction of six different kinds of particles $\{\alpha, \beta, \gamma, \delta, \epsilon, \zeta\}$. Suppose α, β and γ are positively charged particles; δ and ϵ are neutral particles; and ζ is

† Let us remind the reader that a partition is a set of subsets and that it is perfectly all right that the elements of the set B in the algebraic system $(B, *)$ happen to be sets themselves.

‡ We ask the reader to show the converse, namely, a homomorphism from $(A, \star)$ to $(B, *)$ induces a congruence relation on A.

a negatively charged particle. If we let $\{1, 0, -1\}$ denote the three kinds of particles, Fig. 8.12b shows how the three kinds of particles interact. Indeed, as was pointed out above, the function f such that

$$f(\alpha) = 1 \qquad f(\beta) = 1 \qquad f(\gamma) = 1$$
$$f(\delta) = 0 \qquad f(\epsilon) = 0$$
$$f(\zeta) = -1$$

is a homomorphism from the algebraic system $(\{\alpha, \beta, \gamma, \delta, \epsilon, \zeta\}, \star)$ to the algebraic system $(\{1, 0, -1\}, *)$.

As another example, consider the algebraic system $(I, \cdot)$, where I is the set of all integers and $\cdot$ is the ordinary multiplication operation of integers. Suppose we are not interested in distinguishing all the integers in I but rather in making a distinction among positive integers, negative integers, and zero. We note immediately that the algebraic system $(B, \odot)$ in Fig. 8.15 is a homomorphic image of $(I, \cdot)$ and that the function f such that

$$f(n) = \begin{cases} \text{POSITIVE} & \text{if } n \text{ is a positive integer} \\ \text{NEGATIVE} & \text{if } n \text{ is a negative integer} \\ \text{ZERO} & \text{if } n = 0 \end{cases}$$

is indeed a homomorphism from $(I, \cdot)$ to $(B, \odot)$.

$\odot$	POSITIVE	NEGATIVE	ZERO
POSITIVE	POSITIVE	NEGATIVE	ZERO
NEGATIVE	NEGATIVE	POSITIVE	ZERO
ZERO	ZERO	ZERO	ZERO

Figure 8.15

Given a group $(G, \star)$, one would like to know how to determine a homomorphic image of $(G, \star)$, and furthermore, all homomorphic images of $(G, \star)$.[†] We recall our discussion in Sec. 8.5 that a subgroup of $(G, \star)$ induces a partition of G into cosets of the subgroup. It is therefore a reasonable guess to ask whether this partition divides G into congruence classes. The answer to this question is negative, although an additional condition on the subgroup will do the trick.

Let H be a subgroup of G. H is said to be a *normal subgroup* if, for any element a in G, the left coset $a \star H$ is equal to the right coset $H \star a$. (Note that, if G is a commutative group, any subgroup of G is normal.) For example, for the group $(A, \star)$ shown in Fig. 8.16a, $(H, \star)$ is a normal subgroup, where

[†] It is not difficult to see that a homomorphic image of a group is always a group. See Prob. 8.36.

$H = \{\alpha, \beta, \gamma\}$. For instance,

$$\delta \star H = \{\delta \star \alpha, \delta \star \beta, \delta \star \gamma\} = \{\delta, \zeta, \epsilon\}$$

$$H \star \delta = \{\alpha \star \delta, \beta \star \delta, \gamma \star \delta\} = \{\delta, \epsilon, \zeta\}$$

$\star$	α	β	γ	δ	ϵ	ζ
α	α	β	γ	δ	ϵ	ζ
β	β	γ	α	ϵ	ζ	δ
γ	γ	α	β	ζ	δ	ϵ
δ	δ	ζ	ϵ	α	γ	β
ϵ	ϵ	δ	ζ	β	α	γ
ζ	ζ	ϵ	δ	γ	β	α

$(A, \star)$

(a)

$*$	$\{\alpha, \beta, \gamma\}$	$\{\delta, \epsilon, \zeta\}$
$\{\alpha, \beta, \gamma\}$	$\{\alpha, \beta, \gamma\}$	$\{\delta, \epsilon, \zeta\}$
$\{\delta, \epsilon, \zeta\}$	$\{\delta, \epsilon, \zeta\}$	$\{\alpha, \beta, \gamma\}$

$(B, *)$

(b)

Figure 8.16

We want to show now that the distinct left (right) cosets of a normal subgroup H are congruence classes of G. Let $a \star H$ and $b \star H$ be two cosets. We want to show that for all elements a_1 in $a \star H$ and all elements b_1 in $b \star H$, the elements $a_1 \star b_1$ are all in one coset of H. Let

$$a_1 = a \star h_1$$

$$b_1 = b \star h_2$$

for some h_1 and h_2 in H. We have

$$
\begin{aligned}
a_1 \star b_1 &= (a \star h_1) \star (b \star h_2) \\
&= (a \star h_1) \star (h_3 \star b) \quad \text{for some } h_3 \in H \\
&= a \star h_4 \star b \quad \text{for some } h_4 \in H \\
&= a \star b \star h_5 \quad \text{for some } h_5 \in H
\end{aligned}
$$

Thus, $a_1 \star b_1$ is in the coset $(a \star b) \star H$.

For example, for the group $(A, \star)$ shown in Fig. 8.16a and the normal subgroup $(H, \star)$, where $H = \{\alpha, \beta, \gamma\}$, the congruence classes are $\{\alpha, \beta, \gamma\}$ and $\{\delta, \epsilon, \zeta\}$. Consequently, we have the homomorphic image $(B, *)$ shown in Fig. 8.16b, where $B = \{\{\alpha, \beta, \gamma\}, \{\delta, \epsilon, \zeta\}\}$. As another example, for the group $(I, +)$, where I is the set of all integers and $+$ is the ordinary addition operation of integers, $(E, \oplus)$ is a normal subgroup where E is the set of all even integers. Corresponding to the subgroup $(E, \oplus)$, we have the homomorphic image shown in Fig. 8.17, where EVEN is the congruence class of all even integers and ODD is the congruence class of all odd integers.

Our next question is whether exhausting all normal subgroups of $(A, \star)$ will have exhausted all homomorphic images of $(A, \star)$. It turns out that this

$\oplus$	EVEN	ODD
EVEN	EVEN	ODD
ODD	ODD	EVEN

Figure 8.17

indeed is the case as we shall show. Let f be a homomorphism from $(A, \star)$ to $(B, *)$. We want to show that f corresponds to a partition of A into congruence classes induced by a normal subgroup. Let H be all the elements in A whose images under f are the identity of B, which we shall denote ϵ.

We show first that $(H, \star)$ is a subgroup of $(A, \star)$:

1. We want to show that $\star$ is closed on H. For any a and b in H, since

$$f(a \star b) = f(a) * f(b) = \epsilon * \epsilon = \epsilon$$

$a \star b$ is also in H.

2. We want to show that e, the identity of A, is in H. For an arbitrary element a in A, we have

$$f(a \star e) = f(a) * f(e)$$

or

$$f(a) = f(a) * f(e)$$

Since $(B, *)$ is a group, $f(e)$ must be the identity of $(B, *)$. Consequently, e is in H.

3. We want to show that a^{-1} is in H for any element a in H. Since

$$f(a \star a^{-1}) = f(a) * f(a^{-1})$$

or

$$f(e) = f(a) * f(a^{-1})$$

or

$$\epsilon = \epsilon * f(a^{-1})$$

or

$$\epsilon = f(a^{-1})$$

we conclude that a^{-1} is in H.

We show now that $(H, \star)$ is a normal subgroup: For any a in A and any h in H

$$\begin{aligned}
f(a \star h \star a^{-1}) &= f(a) * f(h) * f(a^{-1}) \\
&= f(a) * \epsilon * f(a^{-1}) \\
&= f(a) * f(a^{-1}) \\
&= f(a \star a^{-1}) \\
&= f(e) \\
&= \epsilon
\end{aligned}$$

Thus, $a \star h \star a^{-1}$ is in H. That is, $a \star h \star a^{-1} = h_1$ or $a \star h = h_1 \star a$ for some h_1 in H. We conclude that H is normal.

Finally, we show that if a and b are in the same coset of H, then $f(a) = f(b)$. Since b can be written as

$$b = a \star h$$

for some h in H, we have

$$f(b) = f(a) * f(h) = f(a)$$

Conversely, we want to show that, if $f(a) = f(b)$, then a and b are in the same coset of H. Since

$$f(a^{-1} \star b) = f(a^{-1}) * f(b)$$
$$= f(a^{-1}) * f(a)$$
$$= \epsilon$$

$a^{-1} \star b$ is in H, that is, $a^{-1} \star b = h$ or $b = a \star h$ for some h in H.

8.10 RINGS, INTEGRAL DOMAINS, AND FIELDS

We have so far studied algebraic systems with one binary operation. We turn now to study briefly several classes of algebraic systems with two binary operations. It is clear that given two algebraic systems $(A, \star)$ and $(A, *)$, we can always "combine" them to yield an algebraic system with two binary operations $(A, \star, *)$. On the other hand, an algebraic system with two operations can be more than merely a "conglomerate" of two algebraic systems with one operation if the two operations are related in some way. For example, let $(A, \star, *)$ be an algebraic system such that for any a and b in A, $a \star b = b * a$. We observe that the two operations $\star$ and $*$ are related in that either both of them are commutative operations or both of them are not. (Why?)

A most natural and important way in which two binary operations can be related is the property of *distributivity*. Let $(A, \star, *)$ be an algebraic system with two binary operations. The operation $*$ is said to *distribute* over the operation $\star$ if for any a, b, and c in A,

$$a * (b \star c) = (a * b) \star (a * c)$$

and

$$(b \star c) * a = (b * a) \star (c * a)$$

For example, for the algebraic system $(\{\alpha, \beta\}, \star, *)$ where $\star$ and $*$ are as defined in Fig. 8.18, the operation $*$ is distributive over the operation $\star$, while the operation $\star$ is *not* distributive over the operation $*$.†

---

† For example, $\beta \star (\alpha * \beta) = \beta$ and $(\beta \star \alpha) * (\beta \star \beta) = \alpha$.

★	α	β
α	α	β
β	β	α

*	α	β
α	α	α
β	α	β

Figure 8.18

We are particularly interested in the classes of algebraic systems with two operations which are known as *rings, integral domains,* and *fields*. An algebraic system $(A, +, \cdot)$ is called a *ring* if the following conditions are satisfied:

1. $(A, +)$ is an abelian group.
2. $(A, \cdot)$ is a semigroup.
3. The operation $\cdot$ is distributive over the operation $+$.

Let Z_n be the set of integers $\{0, 1, 2, \ldots, n-1\}$. Let $\oplus$ be a binary operation on Z_n such that

$$a \oplus b = \begin{cases} a + b & \text{if} & a + b < n \\ a + b - n & \text{if} & a + b \geq n \end{cases}$$

Let $\odot$ be a binary operation on Z_n such that

$$a \odot b = \text{the remainder of } ab \text{ divided by } n$$

It was pointed out in Sec. 8.2 that $(Z_n, \oplus)$ is an abelian group. Furthermore, it is not difficult to see that $\odot$ is a closed and an associative operation.[†] Consequently, $(Z_n, \odot)$ is a semigroup. We leave it to the reader to show that $\odot$ distributes over $\oplus$ and thus to conclude that $(Z_n, \oplus, \odot)$ is a ring.

We shall consistently refer to the two operations in a ring $(A, +, \cdot)$ as the addition and the multiplication operation. We shall also refer to $a + b$ as the *sum* of a and b, and $a \cdot b$ as the *product* of a and b. The identity of the abelian group $(A, +)$ is referred to as the *additive identity* and will be denoted 0. The inverse of an element a of the group $(A, +)$ is referred to as the *additive inverse* of a and will be denoted $-a$. We sometimes write $a - b$ to mean the sum of a and the additive inverse of b.

We now show that for any element a in a ring $(A, +, \cdot)$, $0 \cdot a = a \cdot 0 = 0$. Since

$$0 \cdot a = (0 + 0) \cdot a = 0 \cdot a + 0 \cdot a$$

we have

$$0 \cdot a = 0$$

That $a \cdot 0 = 0$ can be proved in a similar way.[‡]

[†] See Problem 8.4.

[‡] Note, indeed, how "useful" the property of distributivity is. See also Theorem 9.9.

Let $(A, +, \cdot)$ be an algebraic system with two binary operations. $(A, +, \cdot)$ is called an *integral domain* if:

1. $(A, +)$ is an abelian group.
2. The operation $\cdot$ is commutative. Furthermore, if $c \neq 0$ and $c \cdot a = c \cdot b$, then $a = b$, where 0 denotes the additive identity.
3. The operation $\cdot$ is distributive over the operation $+$.

For example, let A be the set of all integers, and $+$ and $\cdot$ be the ordinary addition and multiplication operations of integers. We note that $(A, +)$ is an abelian group with 0 being the identity and $-n$ being the inverse of n for any integer n. We also note that the operation $\cdot$ is commutative. Furthermore, for any nonzero integer c, $c \cdot a = c \cdot b$ implies $a = b$. Since $\cdot$ is distributive over $+$, it follows that $(A, +, \cdot)$ is an integral domain.

Let $(A, +, \cdot)$ be an algebraic system with two binary operations. $(A, +, \cdot)$ is called a *field* if:

1. $(A, +)$ is an abelian group.
2. $(A - \{0\}, \cdot)$ is an abelian group.
3. The operation $\cdot$ is distributive over the operation $+$.

We ask the reader to check that $(F, +, \cdot)$ is a field, where F is the set of all rational numbers, and $+$ and $\cdot$ are the ordinary addition and multiplication operations of rational numbers. Other examples of fields are $(R, +, \cdot)$, where R is the set of all real numbers and $+$ and $\cdot$ are the ordinary addition and multiplication operations of real numbers, and $(C, +, \cdot)$, where C is the set of all complex numbers, and $+$ and $\cdot$ are the ordinary addition and multiplication operations of complex numbers. (One probably realizes now, if not earlier, that when we studied addition, subtraction, multiplication, and division of rational, real, and complex numbers in grade school, we studied the operations in the fields of rational, real, and complex numbers. Furthermore, subtraction is not really an "independent" operation in that it is equivalent to addition of the additive inverse of an element. Similarly, division is equivalent to multiplication by the multiplicative inverse of an element.)

As another example, we show that the algebraic system $(Z_n, \oplus, \odot)$ is a field if and only if n is a prime. It is clear that both $\oplus$ and $\odot$ are commutative operations. We recall that $(Z_n, \oplus)$ is a group for any n. Let us show that $(Z_n - \{0\}, \odot)$ is a group if and only if n is a prime. If n is not a prime, $n = ab$ for some a and b in $Z_n - \{0\}$. That is, $a \odot b = 0$. Since the operation $\odot$ is not closed on the set $Z_n - \{0\}$, $(Z_n - \{0\}, \odot)$ cannot possibly be a group. Let us examine the case when n is a prime. First of all, for any a and b in $Z_n - \{0\}$, $a \odot b$ is not equal to 0. Therefore, the operation $\odot$ is closed on the set $Z_n - \{0\}$. Furthermore, it is clear that the operation $\odot$ is an associative operation and that 1 is the identity of $(Z_n - \{0\}, \odot)$. Finally, let us show that for any a and any two distinct elements b and c in $Z_n - \{0\}$, $a \odot b$ is not equal to $a \odot c$. Assume that

$a \odot b$ is equal to $a \odot c$. That is,

$$ab = kn + r$$

$$ac = ln + r$$

for some k, l and r. Without loss of generality, suppose that $b > c$ and $k > l$. We have

$$ab - ac = kn - ln$$

or

$$a(b - c) = (k - l)n \tag{8.5}$$

Since both a and $b - c$ are less than n, (8.5) is an impossibility when n is a prime. It now follows from the pigeonhole principle that, for every a in $Z_n - \{0\}$, there exists a b in $Z_n - \{0\}$ such that

$$a \odot b = 1$$

That is, b is the inverse of a in $(Z_n - \{0\}, \odot)$. We thus conclude that $(Z_n - \{0\}, \odot)$ is an abelian group. We leave it to the reader to show that the operation $\odot$ is distributive over the operation $\oplus$. For a prime n, $(Z_n, \oplus, \odot)$ is usually referred to as the *field of integers modulo n.*

The identity of the group $(A - \{0\}, \cdot)$ is referred to as the *multiplicative identity*, which is commonly denoted 1. The *inverse* of an element a in the group $(A - \{0\}, \cdot)$ is referred to as the *multiplicative inverse* of a and will be denoted $1/a$. We sometimes write a/b to mean the product of a and the multiplicative inverse of b.

*8.11 RING HOMOMORPHISMS

We introduced the notion of homomorphisms on algebraic systems with one binary operation in Sec. 8.9. We study now homomorphisms on algebraic systems with two operations and, in particular, on rings, in further detail. Let $(A, +, \cdot)$ and $(B, \oplus, \odot)$ be two algebraic systems. An onto function f from A to B is said to be a *homomorphism from $(A, +, \cdot)$ to $(B, \oplus, \odot)$* if for any a and b in A

$$f(a + b) = f(a) \oplus f(b)$$

$$f(a \cdot b) = f(a) \odot f(b)$$

Furthermore, $(B, \oplus, \odot)$ is called a *homomorphic image* of $(A, +, \cdot)$.

Alternatively, let R be a congruence relation on A with respect to both the operations $+$ and $\cdot$. That is, R is an equivalence relation on A and if (a_1, a_2) and (b_1, b_2) are in R, then $(a_1 + b_1, a_2 + b_2)$ and $(a_1 \cdot b_1, a_2 \cdot b_2)$ are also in R. Let $B = \{A_1, A_2, \ldots, A_r\}$ be the set of congruence classes into which A is divided. We define two binary operations $\oplus$ and $\odot$ on B such that $A_i \oplus A_j$ is equal to the congruence class $a_1 + a_2$ is in, and $A_i \odot A_j$ is equal to the congruence class

$a_1 \cdot a_2$ is in, for any a_1 in A_i and a_2 in A_j. Following our discussion in Sec. 8.9, $(B, \oplus, \odot)$ is a homomorphic image of $(A, +, \cdot)$.

For example, let N be the set of natural numbers and $+$ and $\cdot$ be the ordinary addition and multiplication operations of natural numbers. Consider the algebraic system $(\{\text{EVEN}, \text{ODD}\}, \oplus, \odot\})$, where the operations $\oplus$ and $\odot$ are as defined in Fig. 8.19. We note that $(\{\text{EVEN}, \text{ODD}\}, \oplus, \odot)$ is a homomorphic image of $(N, +, \cdot)$. Furthermore, the function f such that

$$f(n) = \begin{cases} \text{EVEN} & \text{if } n \text{ is an even number} \\ \text{ODD} & \text{if } n \text{ is an odd number} \end{cases}$$

is the corresponding homomorphism from $(N, +, \cdot)$ to $(\{\text{EVEN}, \text{ODD}\}, \oplus, \odot)$.

$\oplus$	EVEN	ODD
EVEN	EVEN	ODD
ODD	ODD	EVEN

$\odot$	EVEN	ODD
EVEN	EVEN	EVEN
ODD	EVEN	ODD

Figure 8.19

Let us investigate now the problem of determining the homomorphic images of a ring.[†] Let $(A, +, \cdot)$ be a ring and $(H, +)$ be a subgroup of $(A, +)$. Let R denote the equivalence relation that H induces on A. Because $(A, +)$ is an abelian group, R is a congruence relation with respect to $+$. We want to know under what condition R is also a congruence relation with respect to $\cdot$. Suppose that (a, b) and (c, d) are in R. That is,

$$a = b + h_1$$
$$c = d + h_2$$

for some h_1 and h_2 in H. It follows that

$$a \cdot c = (b + h_1) \cdot (d + h_2)$$
$$= (b \cdot d) + (b \cdot h_2) + (h_1 \cdot d) + (h_1 \cdot h_2)$$

We note that, *if* both $b \cdot h_2$ and $h_1 \cdot d$ are in H, then

$$(b \cdot h_2) + (h_1 \cdot d) + (h_1 \cdot h_2)$$

is also in H, and we shall have $(a \cdot c, b \cdot d)$ in R.

Such an observation motivates the definition of an ideal in a ring. Let $(A, +, \cdot)$ be a ring and H be a subset of A. H is said to be an *ideal* if the following conditions are satisfied:

1. $(H, +)$ is a subgroup of $(A, +)$.
2. For any a in A and h in H, $a \cdot h$ and $h \cdot a$ are both in H.[‡]

[†] It is not hard to show that a homomorphic image of a ring is always a ring. (See Prob. 8.36.)

[‡] Intuitively, any element in A is "swallowed" into the ideal when multiplied by an element in the ideal.

It follows immediately from our discussion that an ideal of a ring $(A, +, \cdot)$ induces a congruence relation on A with respect to both the operations $+$ and $\cdot$, and consequently defines a homomorphic image of $(A, +, \cdot)$.

On the other hand, we shall show that every homomorphism of a ring corresponds to an ideal of the ring. Let f be a homomorphism from a ring $(A, +, \cdot)$ to $(B, \oplus, \odot)$. Let H be the set of elements of A that are mapped into the additive identity of $(B, \oplus, \odot)$ under f. [We remind the reader that, since it is a homomorphic image of $(A, +, \cdot)$, $(B, \oplus, \odot)$ is a ring.] It has been shown in Sec. 8.9 that H is a normal subgroup of $(A, +)$. To show that H is an ideal, we note that for any a in A and h in H,

$$f(a \cdot h) = f(a) \odot f(h) = f(a) \odot 0 = 0$$

where 0 denotes the additive identity of $(B, \oplus, \odot)$. Thus, $a \cdot h$ is in H. In a similar way, we can show that $h \cdot a$ is in H. Thus, we can conclude that H is an ideal.

For example, consider the ring $(Z_6, \oplus, \odot)$ shown in Fig. 8.20a. We can check immediately that $\{0, 2, 4\}$ is an ideal. The two congruence classes are $\{0, 2, 4\}$ and $\{1, 3, 5\}$, and the homomorphic image is shown in Fig. 8.20b.

$\oplus$	0	1	2	3	4	5
0	0	1	2	3	4	5
1	1	2	3	4	5	0
2	2	3	4	5	0	1
3	3	4	5	0	1	2
4	4	5	0	1	2	3
5	5	0	1	2	3	4

$\odot$	0	1	2	3	4	5
0	0	0	0	0	0	0
1	0	1	2	3	4	5
2	0	2	4	0	2	4
3	0	3	0	3	0	3
4	0	4	2	0	4	2
5	0	5	4	3	2	1

(a)

$+$	$\{0, 2, 4\}$	$\{1, 3, 5\}$
$\{0, 2, 4\}$	$\{0, 2, 4\}$	$\{1, 3, 5\}$
$\{1, 3, 5\}$	$\{1, 3, 5\}$	$\{0, 2, 4\}$

$\cdot$	$\{0, 2, 4\}$	$\{1, 3, 5\}$
$\{0, 2, 4\}$	$\{0, 2, 4\}$	$\{0, 2, 4\}$
$\{1, 3, 5\}$	$\{0, 2, 4\}$	$\{1, 3, 5\}$

(b)

Figure 8.20

As another example, consider the ring $(I, +, \cdot)$, where I is the set of all integers, and $+$ and $\cdot$ are the ordinary addition and multiplication operations of integers. Let H be the set of all multiples of n $\{\ldots, -2n, -n, 0, n, 2n, 3n, \ldots\}$ for any integer n. Clearly, H is an ideal. Moreover, a corresponding homomorphic image is $(Z_n, \oplus, \odot)$.

*8.12 POLYNOMIAL RINGS AND CYCLIC CODES

Let $(A, \star, *)$ be an algebraic system where $\star$ is an m-ary operation and $*$ is an n-ary operation. We sometimes refer to the elements in A as *constants*. A *variable* (or an *indeterminant*) is a symbolic name that is not the name of any element in A. We define *algebraic expressions* in $(A, \star, *)$ according to the following rules:

1. Any constant is an algebraic expression.
2. Any variable is an algebraic expression.
3. If $e_1, e_2, \ldots, e_m$ are algebraic expressions, then $\star(e_1, e_2, \ldots, e_m)$ is an algebraic expression.
4. If $e_1, e_2, \ldots, e_n$ are algebraic expressions, then $*(e_1, e_2, \ldots, e_n)$ is an algebraic expression.

For example, if x and y are variables, the following are algebraic expressions in $(Z_4, \oplus, \odot)$:

$$2$$
$$2 \oplus x$$
$$(3 \odot (x \oplus x)) \odot (x \odot y)$$

Let $(A, +, \cdot)$ be a ring. An algebraic expression in $(A, +, \cdot)$ of the form

$$a_0 + (a_1 \cdot x) + (a_2 \cdot x^2) + \cdots + (a_{n-1} \cdot x^{n-1}) + (a_n \cdot x^n)\dagger \qquad n \geqq 0 \qquad (8.6)$$

is called a *polynomial*, where $a_0, a_1, \ldots, a_{n-1}, a_n$ are constants and x is a variable. Note that because $+$ and $\cdot$ are associative operations, the expression in (8.6) is unambiguous. We shall use notation such as $A(x), B(x), \ldots$ for polynomials. The *degree* of a polynomial is the largest n for which a_n is not equal to 0. Following the convention of omitting the multiplication operation symbol $\cdot$ and the parentheses in (8.6), we rewrite (8.6) as

$$a_0 + a_1 x + a_2 x^2 + \cdots + a_{n-1} x^{n-1} + a_n x^n$$

We discuss briefly the construction of polynomial rings. Let $(F, +, \cdot)$ be a field. Let $F[x]$ denote the set of all polynomials in $(F, +, \cdot)$ with the variable x. We define an algebraic system $(F[x], \boxplus, \boxdot)$, where $\boxplus$ and $\boxdot$ are two binary operations defined as follows: For

$$A(x) = a_0 + a_1 x + a_2 x^2 + \cdots + a_n x^n$$
$$B(x) = b_0 + b_1 x + b_2 x^2 + \cdots + b_m x^m$$

we have

$$A(x) \boxplus B(x) = C(x) = c_0 + c_1 x + c_2 x^2 + \cdots + c_r x^r$$

where

$$c_i = a_i + b_i \ddagger$$

$\dagger$ We use x^i to denote $\underbrace{x \cdot x \cdot \cdots \cdot x}_{i \text{ times}}$.

$\ddagger$ We assume that $a_i = 0$ for $i > n$ and $b_i = 0$ for $i > m$.

and

$$A(x) \boxed{\cdot} B(x) = D(x) = d_0 + d_1 x + d_2 x^2 + \cdots + d_r x^r$$

where

$$d_i = (a_0 \cdot b_i) + (a_1 \cdot b_{i-1}) + \cdots + (a_i \cdot b_0)$$

It is routine to check that $(F[x], \boxed{+}, \boxed{\cdot})$ is a ring, which is commonly called a *polynomial ring.*

Let $(F, +, \cdot)$ be a field. Let $F_n[x]$ denote the set of polynomials of degree less than n in $F[x]$. Let $G(x)$ be a polynomial of degree n in $F[x]$. Let $(F_n[x], \triangle, \triangle)$ be an algebraic system, where $\triangle$ and $\triangle$ are two binary operations defined as follows: For

$$A(x) = a_0 + a_1 x + a_2 x^2 + \cdots + a_{n-1} x^{n-1}$$

$$B(x) = b_0 + b_1 x + b_2 x^2 + \cdots + b_{n-1} x^{n-1}$$

we have

$$A(x) \triangle B(x) = A(x) \boxed{+} B(x)$$

and

$$A(x) \triangle B(x) = \text{the remainder of } A(x) \boxed{\cdot} B(x)$$
$$\text{divided by the polynomial } G(x)\dagger$$

Again, it is routine to check that $(F_n[x], \triangle, \triangle)$ is a ring, which is commonly referred to as the *ring of polynomials modulo $G(x)$.*

For example, let $(F, +, \cdot)$ be the field of integers modulo 2. There are four polynomials in $F_2[x]$, namely, 0, 1, x, $1 + x$. Let $G(x)$ be $1 + x^2$. The operations $\triangle$ and $\triangle$ are defined in Fig. 8.21.

Construction of polynomial rings leads immediately to the construction of Galois fields. A detailed discussion of the subject will be beyond the scope of this book.

$\triangle$	0	1	x	$1+x$
0	0	1	x	$1+x$
1	1	0	$1+x$	x
x	x	$1+x$	0	1
$1+x$	$1+x$	x	1	0

$\triangle$	0	1	x	$1+x$
0	0	0	0	0
1	0	1	x	$1+x$
x	0	x	1	$1+x$
$1+x$	0	$1+x$	$1+x$	0

Figure 8.21

As an illustrative example, let us consider a class of codes, known as cyclic codes. We recall our discussion in Sec. 8.7 that a block code is a collection of words of the same length. A block code is called a *cyclic code* if for every codeword $a_0 a_1 a_2 \cdots a_{n-2} a_{n-1}$, the sequence $a_{n-1} a_0 a_1 a_2 \cdots a_{n-2}$ is also a

---

† Division of polynomials is carried out in the same way as was done in high school algebra with, of course, addition and multiplication being the field operations in $(F, +, \cdot)$.

codeword. In other words, for a cyclic code, a *cyclic shift* of a codeword is also a codeword. We show now a very compact way to describe cyclic codes.

Let $(F, +, \cdot)$ be the field of integers modulo 2. First of all, we note that a binary sequence $a_0 a_1 a_2 \cdots a_{n-1}$ can be conveniently represented as a polynomial

$$a_0 + a_1 x + a_2 x^2 + \cdots + a_{n-1} x^{n-1}$$

in the polynomial ring $(F[x], \boxplus, \boxdot)$. Let $(F_n[x], \triangle, \triangle)$ be the ring of polynomials modulo $1 + x^n$. Let I be an ideal of $(F_n[x], \triangle, \triangle)$. We want to show that the polynomials in I constitute a cyclic code. Because $(I, \triangle)$ is a group, the polynomials in I indeed constitute a group code. Let $A(x)$ be a polynomial in I. According to the definition of an ideal, $x \triangle A(x)$ is also a polynomial in I. Let

$$A(x) = a_0 + a_1 x + a_2 x^2 + \cdots + a_{n-1} x^{n-1}$$

$$x \triangle A(x) = \text{remainder of } \frac{a_0 x + a_1 x^2 + \cdots + a_{n-1} x^n}{1 + x^n}$$

$$= \text{remainder of } \frac{a_0 x + a_1 x^2 + \cdots + a_{n-1} + a_{n-1}(1 + x^n)}{1 + x^n}$$

$$= a_{n-1} + a_0 x + a_1 x^2 + \cdots + a_{n-2} x^{n-1}$$

Therefore, the polynomials in I constitute a cyclic code.

On the other hand, let I denote a set of polynomials corresponding to the codewords in a cyclic code. Clearly, $(I, \triangle)$ is an abelian group. If $A(x)$ is in I, $x \triangle A(x)$ is in I, $x^i \triangle A(x)$ is in I, and $b_j x^j \triangle A(x)$ is in I, because the polynomials in I form a cyclic code. It follows that $B(x) \triangle A(x)$ is in I for any $B(x)$ in $F_n[x]$.

A great deal more can be said about describing cyclic codes as ideals in polynomial rings. See, for example, the references cited in Sec. 8.13.

8.13 REMARKS AND REFERENCES

Some useful general references on algebraic systems are Birkhoff and MacLane[2], Cohn[3], Herstein[4], Paley and Weichsel[8]. See chap. 4 of Knuth[5] for further information on addition chains. See chap. 5 of Liu[7] on Pólya's theory of counting which generalizes Burnside's theorem in Sec. 8.6. For further discussion on algebraic coding theory, see Berlekamp[1], Lin[6], and Peterson and Weldon[9].

[1] Berlekamp, E. R.: "Algebraic Coding Theory," McGraw-Hill Book Company, New York, 1968.
[2] Birkhoff, G. and S. MacLane: "A Survey of Modern Algebra," 3rd ed., The Macmillan Company, New York, 1965.
[3] Cohn, P. M.: "Universal Algebra," Harper and Row, New York, 1965.
[4] Herstein, I. N.: "Topics in Algebra," Blaisdell Publishing Company, Waltham, Mass., 1964.
[5] Knuth, D. E.: "The Art of Computer Programming, Vol. 2, Seminumerical Algorithms," Addison-Wesley Publishing Company, Reading, Mass., 1969.

[6] Lin, S.: "An Introduction to Error-correcting Codes," Prentice-Hall, Inc., Englewood Cliffs, N.J., 1970.

[7] Liu, C. L.: "Introduction to Combinatorial Mathematics," McGraw-Hill Book Company, New York, 1968.

[8] Paley, H., and P. M. Weichsel: "A First Course in Abstract Algebra," Holt, Rinehart and Winston, New York, 1966.

[9] Peterson, W. W., and E. J. Weldon, Jr.: "Error-correcting Codes," 2d ed., MIT Press, Cambridge, Mass., 1972.

PROBLEMS

8.1 Let N be the set of all natural numbers. For each of the following determine whether $*$ is an associative operation:

    (a) $a * b = \max(a, b)$

    (b) $a * b = \min(a, 2)$

    (c) $a * b = a + b + 3$

    (d) $a * b = a + 2b$

    (e) $a * b = \begin{cases} \min(a, b) & \text{if } \min(a, b) < 10 \\ \max(a, b) & \text{if } \min(a, b) \geq 10 \end{cases}$

8.2 Let $(A, \star)$ be an algebraic system where $\star$ is a binary operation such that, for any a and b in A, $a \star b = a$.

    (a) Show that $\star$ is an associative operation.

    (b) Can $\star$ ever be a commutative operation?

8.3 Let $(A, \star)$ be an algebraic system such that for all a, b, c, d in A

$$a \star a = a$$

$$(a \star b) \star (c \star d) = (a \star c) \star (b \star d)$$

Show that

$$a \star (b \star c) = (a \star b) \star (a \star c)$$

8.4 Let Z_n denote the set of integers $\{0, 1, 2, \ldots, n - 1\}$. Let $\odot$ be a binary operation on Z_n such that

$$a \odot b = \text{the remainder of } ab \text{ divided by } n$$

    (a) Construct the table for the operation $\odot$ for $n = 4$.

    (b) Show that $(Z_n, \odot)$ is a semigroup for any n.

8.5 Let $(A, *)$ be a semigroup. Let a be an element in A. Consider a binary operation $\square$ on A such that, for every x and y in A,

$$x \square y = x * a * y$$

Show that $\square$ is an associative operation.

8.6 Let $(A, *)$ be a semigroup. Furthermore, for every a and b in A, if $a \neq b$, then $a * b \neq b * a$.

    (a) Show that for every a in A,

$$a * a = a$$

    (b) Show that for every a, b in A,

$$a * b * a = a$$

    (c) Show that for every a, b, c in A,

$$a * b * c = a * c$$

(*Hint:* Note that $a * b = b * a$ implies $a = b$.)

8.7 Let $(A, *)$ be a semigroup. Show that, for a, b, c in A, if $a * c = c * a$ and $b * c = c * b$, then $(a * b) * c = c * (a * b)$.

8.8 Let $(\{a, b\}, *)$ be a semigroup where $a * a = b$. Show that:
  (a) $a * b = b * a$
  (b) $b * b = b$

8.9 Let $(A, *)$ be a commutative semigroup. Show that if $a * a = a$ and $b * b = b$, then $(a * b) * (a * b) = a * b$.

8.10 Let $(A, *)$ be a semigroup. Show that, if A is a finite set, there exists a in A such that $a * a = a$.

8.11 Let $(A, *)$ be a semigroup. Furthermore, let there be an element a in A such that for every x in A there exist u and v in A satisfying the relation

$$a * u = v * a = x$$

Show that there is an identity element in A.

8.12 Let $(A, *)$ be a semigroup and e be a left identity. Furthermore, for every x in A there exists $\hat{x}$ in A such that $\hat{x} * x = e$.
  (a) Show that for any a, b, c in A, if $a * b = a * c$, then $b = c$.
  (b) Show that $(A, *)$ is a group by showing that e is an identity element. (*Hint:* Note that $\hat{x} * x * \hat{x} * x = e$.)

8.13 Let $(A, \star)$ be an algebraic system such that for all a, b in A

$$(a \star b) \star a = a$$

$$(a \star b) \star b = (b \star a) \star a$$

  (a) Show that $a \star (a \star b) = a \star b$ for all a and b.
  (b) Show that $a \star a = (a \star b) \star (a \star b)$ for all a and b.
  (c) Show that $a \star a = b \star b$ for all a and b.
  (d) Let e denote the element $a \star a$. Show that $e \star a = a$ and $a \star e = e$.
  (e) Show that $a \star b = b \star a$ if and only if $a = b$.
  (f) Let $(A, \star)$ satisfy the additional condition

$$a \star b = (a \star b) \star b$$

Show that $\star$ is idempotent and commutative.

8.14 A *central groupoid* is an algebraic system $(A, \star)$, where $\star$ is a binary operation such that

$$(a \star b) \star (b \star c) = b$$

for all a, b, c in A.
  (a) Show that

$$a \star ((a \star b) \star c) = a \star b$$

$$(a \star (b \star c)) \star c = b \star c$$

in a central groupoid.
  (b) Let $(A, \star)$ be an algebraic system where $\star$ is a binary operation such that

$$(a \star ((b \star c) \star d)) \star (c \star d) = c$$

for all a, b, c, d in A. Show that $(A, \star)$ is a central groupoid. [*Hint:* Use the result $(a \star ((d \star (b \star c)) \star d)) \star ((b \star c) \star d) = b \star c$ to show $(b \star c) \star (c \star d) = c$.]
  (c) Construct a directed graph with four vertices in which there is exactly one path of length 2 between any two vertices.
  (d) Let $G = (V, E)$ be a directed graph in which there is exactly one path of length 2 between any two vertices. For any two vertices a and b in V, let (a, c) and (c, b) be the two edges in the path from a to b. We define an algebraic system $(V, \star)$ such that $a \star b = c$. Show that $(V, \star)$ is a central groupoid.

(*e*) Let $(A, \star)$ be a central groupoid. We construct a directed graph $G = (A, E)$ such that there is an edge (a, c) in E if and only if there is b in A such that $a \star b = c$. Show that there is exactly one path of length 2 between any two vertices in A.

(*f*) Let $(A, \star)$ be a central groupoid and $G = (A, E)$ be the corresponding directed graph as defined in part (*e*). For $a \in A$, let

$$R(a) = \{b \,|\, (a, b) \in E\}$$

$$L(a) = \{b \,|\, (b, a) \in E\}$$

Show that for any a and b in A, $R(a)$ and $L(b)$ have the same number of elements. Show that for any a and b in A, $R(a)$ and $R(b)$ have the same number of elements.

(*g*) Let $(A, \star)$ be a central groupoid. Show that A has m^2 elements for some integer m. [*Hint:* Suppose there are m elements in $R(a)$. What can be said about the sets $R(b_1)$, $R(b_2)$, ..., $R(b_m)$ for $b_1, b_2, ..., b_m$ in $R(a)$?].

8.15 Let $(A, \star)$ be a group. Show that for any $a, b, c, d, a_1, b_1, c_1, d_1$ in A, if

$$a \star c = a_1 \star c_1$$

$$a \star d = a_1 \star d_1$$

$$b \star c = b_1 \star c_1$$

then

$$b \star d = b_1 \star d_1$$

[*Hint:* Note that $b \star d = b \star (c \star c^{-1}) \star (a^{-1} \star a) \star d$.]

8.16 (*a*) Show that every group containing exactly two elements is isomorphic to $(Z_2, \oplus)$.

(*b*) Show that every group containing exactly three elements is isomorphic to $(Z_3, \oplus)$.

(*c*) How many nonisomorphic groups that contain exactly four elements are there?

8.17 Let $(A, \star)$ and $(B, *)$ be two algebraic systems. The cartesian product of $(A, \star)$ and $(B, *)$ is an algebraic system $(A \times B, \square)$, where $\square$ is a binary operation such that for any (a_1, b_1) and (a_2, b_2) in $A \times B$

$$(a_1, b_1) \,\square\, (a_2, b_2) = (a_1 \star a_2, b_1 * b_2)$$

Show that the cartesian product of two groups is a group.

8.18 Let $(A, \cdot)$ be a group.

(*a*) Show that $(ab)^{-1} = b^{-1}a^{-1}$.

(*b*) Show that $(a_1 a_2 \cdots a_{r-1} a_r)^{-1} = a_r^{-1} a_{r-1}^{-1} \cdots a_2^{-1} a_1^{-1}$.

(*c*) Show that $(a^i b^j)^{-1} = b^{-j} a^{-i}$. [$b^{-j}$ denotes $(b^{-1})^j$ and a^{-i} denotes $(a^{-1})^i$.]

8.19 Let $(A, *)$ be a monoid such that for every x in A, $x * x = e$, where e is the identity element. Show that $(A, *)$ is an abelian group.

8.20 Let $(G, \star)$ be a group and H be a nonempty subset of G. Show that $(H, \star)$ is a subgroup if for any a and b in H, $a \star b^{-1}$ is also in H.

8.21 Show that any subgroup of a cyclic group is cyclic.

8.22 The *order* of an element a in a group is defined to be the least positive integer m such that $a^m = e$. (If no positive power of a equals e, the order of a is defined to be infinite.) Show that, in a finite group, the order of an element divides the order of the group.

8.23 Let H_1 and H_2 be subgroups of a group G, neither of which contains the other. Show that there exists an element of G belonging neither to H_1 nor H_2.

8.24 Let $(H, \cdot)$ be a subgroup of a group $(G, \cdot)$. Let $N = \{x \,|\, x \in G, xHx^{-1} = H\}$. Show that $(N, \cdot)$ is a subgroup of $(G, \cdot)$.

8.25 Let $(A, \star)$ be a group where A has an even number of elements. Show that there is an element a in A such that $a \star a = e$, where e is the identity element and $a \neq e$.

8.26 Let $(A, \star)$ be a group. Let B be a subset of A such that $2|B| > |A|$. Show that, for any a in A, $a = b_1 \star b_2$ for some b_1 and b_2 in B. [*Hint:* Consider the set $C = \{a \star b^{-1} \,|\, b \in B\}$.]

8.27 Let $(H, \cdot)$ and $(K, \cdot)$ be subgroups of a group $(G, \cdot)$. Let

$$HK = \{h \cdot k \mid h \in H, k \in K\}$$

Show that $(HK, \cdot)$ is a subgroup of $(G, \cdot)$ if and only if $HK = KH$.

8.28 Let $(A, \star)$ be a group. Show that $(A, \star)$ is an abelian group if and only if $a^2 \star b^2 = (a \star b)^2$ for all a and b in A.

8.29 Let $(A, \star)$ be a group. Show that $(A, \star)$ is an abelian group if $a^3 \star b^3 = (a \star b)^3$, $a^4 \star b^4 = (a \star b)^4$, and $a^5 \star b^5 = (a \star b)^5$ for all a and b in A.

8.30 Let $(G, \star)$ be a group of even order. Let $(H, \star)$ be a subgroup of $(G, \star)$ where $|H| = |G|/2$. Show that $(H, \star)$ is a normal subgroup.

8.31 Let $(H, \star)$ be a subgroup of a group $(G, \star)$. Show that $(H, \star)$ is a normal subgroup if and only if $a \star H \star a^{-1} \subseteq H$ for every $a \in G$.

8.32 Let $(G, \star)$ be a group. Let $H = \{a \mid a \in G$ and $a \star b = b \star a$ for all $b \in G\}$. Show that H is a normal subgroup.

8.33 (*a*) Let $(H, \star)$ and $(K, \star)$ be subgroups of a group $(G, \star)$. Show that $(H \cap K, \star)$ is also a subgroup.

(*b*) Show that if $(H, \star)$ and $(K, \star)$ are normal subgroups then $(H \cap K, \star)$ is also a normal subgroup.

8.34 Let $(H, \star)$ and $(K, \star)$ be subgroups of a group $(G, \star)$. Show that the function f from $H \times K$ to G such that $f[(h, k)] = h \star k$ for all h in H and k in K is an isomorphism from the cartesian product of $(H, \star)$ and $(K, \star)$ to $(G, \star)$ if and only if:

1. H and K are normal subgroups;
2. $\{h \star k \mid h \in H, k \in K\} = G$;
3. $H \cap K = \{e\}$, where e is the identity element.

8.35 Let $(H, \star)$ be a normal subgroup of a group $(G, \star)$. Show that the homomorphic image of $(G, \star)$ induced by the subgroup $(H, \star)$ is an abelian group if and only if $a \star b \star a^{-1} \star b^{-1}$ is in H for all a and b in G.

8.36 Let $(A, \star)$ and $(B, *)$ be two algebraic systems and f be a homomorphism from $(A, \star)$ to $(B, *)$.

(*a*) Show that, if $\star$ is an associative operation, so is $*$.

(*b*) Show that, if e is an identity element in $(A, \star)$, then $f(e)$ is an identity element in $(B, *)$.

(*c*) Show that, if b is an inverse of a in $(A, \star)$, the $f(b)$ is an inverse of $f(a)$ in $(B, *)$.

(*d*) Show that a homomorphic image of a ring is a ring.

8.37 Let f_1 and f_2 be homomorphisms from an algebraic system $(A, \star)$ to another algebraic system $(B, *)$. Let g be a function from A to B such that

$$g(a) = f_1(a) * f_2(a)$$

for all a in A. Show that g is a homomorphism from $(A, \star)$ to $(B, *)$ if $(B, *)$ is a commutative semigroup.

8.38 Let f and g be homomorphisms from a group $(G, \star)$ to a group $(H, *)$. Show that $(C, \star)$ is a subgroup of $(G, \star)$, where

$$C = \{x \in G \mid f(x) = g(x)\}$$

8.39 A rod divided into six segments is to be colored with one or more of four different colors. In how many ways can this be done?

8.40 In how many distinct ways can the sectors of the circles in Fig. 8P.1 be painted with three colors?

8.41 (*a*) Determine the number of distinct 2×2 chessboards whose cells are painted white and black. Two chessboards are considered distinct if one cannot be obtained from another through rotation.

(*b*) Repeat part (*a*) for 4×4 chessboards.

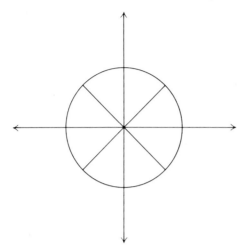

Figure 8P.1

8.42 (*a*) Show that for $p < \frac{1}{2}$, $(1-p)^{n-t_1}p^{t_1} > (1-p)^{n-t_2}p^{t_2}$ for $t_1 < t_2$.

(*b*) Show that for $p > \frac{1}{2}$, $(1-p)^{n-t_1}p^{t_1} < (1-p)^{n-t_2}p^{t_2}$ for $t_1 < t_2$.

8.43 (*a*) Consider the code $G = \{00000, 11111\}$. Set up the coset table to show that G can indeed correct all single- and double-transmission errors.

(*b*) A code G contains 16 codewords: 0000000, 1111111, 1101000 and all its cyclic shifts, 0010111 and all its cyclic shifts. Show that $(G, \oplus)$ is a group code. Set up the coset table to show that G can correct all single-transmission errors.

8.44 Let $(A, \star, *)$ be an algebraic system with e_1 and e_2 being the identity elements with respect to the operations $\star$ and $*$, respectively. Given that the operations $\star$ and $*$ distribute over each other show that $x \star x = x$ and $x * x = x$ for all x in A. (*Hint:* Show that $e_1 = e_1 * e_1$ and $e_2 = e_2 \star e_2$.)

8.45 Let $(A, \star, *)$ be an algebraic system, where

$$a \star b = a$$

for all a, b in A, and $*$ is an arbitrary binary operation. Show that $*$ distributes over $\star$.

8.46 Let $(A, +, \cdot)$ be a ring such that $a \cdot a = a$ for all a in A.

(*a*) Show that $a + a = 0$ for all a, where 0 is the additive identity.

(*b*) Show that the operation $\cdot$ is commutative.

8.47 Show that an integral domain that has a finite number of elements is a field.

8.48 An $n \times n$ *Latin square* is an $n \times n$ arrangement of n distinct symbols so that no symbol appears more than once in any row or any column. For a Latin square A, we shall use a_{ij} to denote the symbol in the ith row and the jth column of A. Two $n \times n$ Latin squares A and B are said to be *orthogonal* if for no i, j, k, l such that $a_{ij} = a_{kl}$ and $b_{ij} = b_{kl}$. A set of Latin squares is said to be orthogonal if every two of them are orthogonal.

(*a*) Construct a 4×4 Latin square.

(*b*) Construct a pair of 3×3 orthogonal Latin squares.

(*c*) Let $(\{b_0, b_1, \ldots, b_{n-1}\}, +, \cdot)$ be a field where b_0 is the additive identity. Let us construct $(n-1)$ $n \times n$ squares $A_1, A_2, \ldots, A_{n-1}$ such that

$$a_{ij}^{(r)} = b_r \cdot b_i + b_j \qquad i, j = 0, 1, 2, \ldots, n-1$$

$$r = 1, 2, \ldots, n-1$$

where $a_{ij}^{(r)}$ is the ijth entry of the square A_r. Show that each square is a Latin square. Show that the set is an orthogonal set.

8.49 A ring $(A, +, \cdot)$ is said to be a *commutative ring with unity* if $(A, \cdot)$ is a commutative monoid.

(a) An ideal H of a ring is said to be a *prime ideal* if, for any two elements a and b that are not in H, $a \cdot b$ is not in H either. Show that the homomorphic image of a commutative ring with unity induced by an ideal is an integral domain if and only if the ideal is prime.

(b) An ideal H is said to be a *maximal ideal* if the only ideals of the ring that contain H are H and the ring itself. Show that the homomorphic image of a commutative ring with unity induced by an ideal is a field if and only if the ideal is maximal.

8.50 (a) Let $(F, +, \cdot)$ be the field of integers modulo 2 and $(F[x], \boxplus, \boxdot)$ the corresponding ring of polynomials. Construct the ring of polynomials modulo $1 + x + x^2$.

(b) Let $(F, +, \cdot)$ be the field of integers modulo 3 and $(F[x], \boxplus, \boxdot)$ the corresponding ring of polynomials. Construct the ring of polynomials modulo $2 + 2x + x^2$.

8.51 Let $(R, +, \cdot)$ be the field of real numbers. Let $(R[x], \boxplus, \boxdot)$ be the corresponding ring of polynomials, and $(R_2[x], \triangle, \triangle)$ be the ring of polynomials modulo $1 + x^2$.

(a) For $a + bx$ and $c + dx$ in $R_2[x]$ determine $(a + bx) \triangle (c + dx)$ and $(a + bx) \triangle (c + dx)$.

(b) Do you see any similarities between $(R_2, \triangle, \triangle)$ and the field of complex numbers?

BOOLEAN ALGEBRAS

9.1 LATTICES AND ALGEBRAIC SYSTEMS

We recall that a lattice is a partially ordered set in which every two elements have a unique least upper bound and a unique greatest lower bound. For example, Fig. 9.1 shows a lattice. There is a natural way to define an algebraic system with two operations corresponding to a given lattice. Let $(A, \leq)$ be a lattice. We define an algebraic system $(A, \vee, \wedge)$, where $\vee$ and $\wedge$ are two binary operations on A such that for a and b in A, $a \vee b$ is equal to the least upper bound of a and b, and $a \wedge b$ is equal to the greatest lower bound of a and b. We shall refer to $(A, \vee, \wedge)$ as the algebraic system defined by the lattice $(A, \leq)$. For example, the algebraic system defined by the lattice in Fig. 9.1 is shown in Fig. 9.2. The binary operation $\vee$ is often referred to as the *join* operation and the binary operation $\wedge$ is often referred to as the *meet* operation. Consequently, the least upper bound of a and b is also referred to as the *join* of a and b, and the greatest lower bound of a and b is also referred to as the *meet* of a and b.

For example, let $\mathcal{P}(S)$ be the power set of a given set S. We recall that $(\mathcal{P}(S), \subseteq)$ is a lattice. Consequently, it defines a corresponding algebraic system $(\mathcal{P}(S), \vee, \wedge)$. For $S = \{a, b, c\}$, the lattice $(\mathcal{P}(S), \subseteq)$ is shown in Fig. 9.3a and the tables for the operations $\vee$ and $\wedge$ are shown in Fig. 9.3b. We recognize that the join of two subsets of S is the union of the two subsets and the meet of two subsets of S is the intersection of the two subsets. [When we introduced the notions of union and intersection of sets in Chap. 1, we intentionally did not bring up the notion of binary operations. Indeed, all the results in Sec. 1.2 on the combination of sets to yield new sets can be phrased conveniently in terms of binary operations on $\mathcal{P}(S)$.]

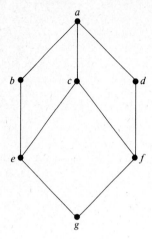

Figure 9.1

As another example, let $(N, \leq)$ be a partially ordered set, where N is the set of natural numbers and $\leq$ is the "less than or equal to" relationship among natural numbers. One can readily check that $(N, \leq)$ is a lattice, and it defines an algebraic system $(N, \vee, \wedge)$ such that $a \vee b = \max(a, b)$ and $a \wedge b = \min(a, b)$.

Also, as another example, let N^+ be the set of all positive integers, let $|$ be a binary relation on N^+ such that $a|b$ if and only if a divides b. We ask the reader to check that $(N^+, |)$ is a lattice. Furthermore, the join of two elements a and b is the least common multiple of a and b, and the meet of two elements a and b is the greatest common divisor of a and b.

We have the following results:

Theorem 9.1 For any a and b in a lattice $(A, \leq)$

$$a \leq a \vee b \tag{9.1}$$

$$a \wedge b \leq a \tag{9.2}$$

PROOF Because the join of a and b is an upper bound of a, $a \leq a \vee b$. Because the meet of a and b is a lower bound of a, $a \wedge b \leq a$. □

$\vee$	a	b	c	d	e	f	g		$\wedge$	a	b	c	d	e	f	g
a	a	a	a	a	a	a	a		a	a	b	c	d	e	f	g
b	a	b	a	a	b	a	b		b	b	b	e	g	e	g	g
c	a	a	c	a	c	c	c		c	c	e	c	f	e	f	g
d	a	a	a	d	a	d	d		d	d	g	f	d	g	f	g
e	a	b	c	a	e	c	e		e	e	e	e	g	e	g	g
f	a	a	c	d	c	f	f		f	f	g	f	f	g	f	g
g	a	b	c	d	e	f	g		g	g	g	g	g	g	g	g

Figure 9.2

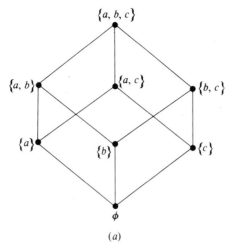

(*a*)

∨	{*a, b, c*}	{*a, b*}	{*a, c*}	{*b, c*}	{*a*}	{*b*}	{*c*}	∅
{*a, b, c*}	{*a, b, c*}	{*a, b, c*}	{*a, b, c*}	{*a, b, c*}	{*a, b, c*}	{*a, b, c*}	{*a, b, c*}	{*a, b, c*}
{*a, b*}	{*a, b, c*}	{*a, b*}	{*a, b, c*}	{*a, b, c*}	{*a, b*}	{*a, b*}	{*a, b, c*}	{*a, b*}
{*a, c*}	{*a, b, c*}	{*a, b, c*}	{*a, c*}	{*a, b, c*}	{*a, c*}	{*a, b, c*}	{*a, c*}	{*a, c*}
{*b, c*}	{*a, b, c*}	{*a, b, c*}	{*a, b, c*}	{*b, c*}	{*a, b, c*}	{*b, c*}	{*b, c*}	{*b, c*}
{*a*}	{*a, b, c*}	{*a, b*}	{*a, c*}	{*a, b, c*}	{*a*}	{*a, b*}	{*a, c*}	{*a*}
{*b*}	{*a, b, c*}	{*a, b*}	{*a, b, c*}	{*b, c*}	{*a, b*}	{*b*}	{*b, c*}	{*b*}
{*c*}	{*a, b, c*}	{*a, b, c*}	{*a, c*}	{*b, c*}	{*a, c*}	{*b, c*}	{*c*}	{*c*}
∅	{*a, b, c*}	{*a, b*}	{*a, c*}	{*b, c*}	{*a*}	{*b*}	{*c*}	∅

∧	{*a, b, c*}	{*a, b*}	{*a, c*}	{*b, c*}	{*a*}	{*b*}	{*c*}	∅
{*a, b, c*}	{*a, b, c*}	{*a, b*}	{*a, c*}	{*b, c*}	{*a*}	{*b*}	{*c*}	∅
{*a, b*}	{*a, b*}	{*a, b*}	{*a*}	{*b*}	{*a*}	{*b*}	∅	∅
{*a, c*}	{*a, c*}	{*a*}	{*a, c*}	{*c*}	{*a*}	∅	{*c*}	∅
{*b, c*}	{*b, c*}	{*b*}	{*c*}	{*b, c*}	∅	{*b*}	{*c*}	∅
{*a*}	{*a*}	{*a*}	{*a*}	∅	{*a*}	∅	∅	∅
{*b*}	{*b*}	{*b*}	∅	{*b*}	∅	{*b*}	∅	∅
{*c*}	{*c*}	∅	{*c*}	{*c*}	∅	∅	{*c*}	∅
∅	∅	∅	∅	∅	∅	∅	∅	∅

Figure 9.3

(*b*)

Theorem 9.2 For any a, b, c, d in a lattice $(A, \le)$, if

$$a \le b \quad \text{and} \quad c \le d$$

then

$$a \vee c \le b \vee d \tag{9.3}$$

$$a \wedge c \le b \wedge d \tag{9.4}$$

PROOF Since

$$b \leq b \vee d \qquad d \leq b \vee d$$

by transitivity

$$a \leq b \vee d \qquad c \leq b \vee d$$

In other words, $b \vee d$ is an upperbound of a and c. Since $a \vee c$ is the least upper bound of a and c, we have

$$a \vee c \leq b \vee d$$

Since

$$a \wedge c \leq a \qquad a \wedge c \leq c$$

by transitivity

$$a \wedge c \leq b \qquad a \wedge c \leq d$$

In other words, $a \wedge c$ is a lower bound of b and d. Since $b \wedge d$ is the greatest lower bound of b and d, we have

$$a \wedge c \leq b \wedge d \qquad\qquad \square$$

9.2 PRINCIPLE OF DUALITY

The principle of duality is an important concept that appears in many different places and contexts. We begin with some simple illustrative examples. In the United States, all passenger cars have the driver's seat at the left front. Consequently, we observe traffic regulations such as the following:

Cars always travel on the right side of a street.
On a multilane divided highway, left lanes are passing lanes.
Cars may make right turns at a red light.

However, since all passenger cars in England have the driver's seat at the right front, they observe traffic regulations such as these:

Cars always travel on the left side of a street.
On a multilane divided highway, right lanes are passing lanes.
Cars may make left turns at a red light.

We note how the concepts of left and right can be interchanged in these two situations so that meaningful traffic rules in the United States become meaningful traffic rules in England, and conversely. We say that the concepts of left and right are dual concepts.

As another example, let us tell a story which we freely adapt from the Chinese fiction *Flowers in the Mirror*.† There was a country called The Land of

---

† The original title is *Ching Hua Yuan*. It was written by Li Ju-Chen (1763–1830). An English translation by Lin Tai-Yi was published by the University of California Press, 1965.

Gentlemen. The following conversation is a typical example of the behavior of its people:

Customer:	I want to buy this vase. It is most beautiful.
Store-keeper:	The vase does not really look that good. Please note that there are some defects in the paint job.
Customer:	You would not call these defects, would you? The vase is a perfect piece of art work. How much do you want for it?
Store-keeper:	Usually I will sell a vase like this for $10. Since you have been a good customer of ours for many years, I will sell it for $15.
Customer:	$15 is too low. I cannot allow myself to take advantage of you. How about $20?
Store-keeper:	I would be making a good profit at $15. Since you insist, let us agree on $16.
Customer:	My wife will complain that I do not pay enough for things I buy. However, I am willing to buy the vase at $18.

It is interesting to observe that the many concepts in business bargaining in The Land of Gentlemen are opposite to those we are more accustomed to. As a matter of fact, it is quite obvious how we can rephrase a conversation between a customer and a storekeeper in The Land of Gentlemen to make it sound more familiar to us, and, conversely, how we can rephrase a conversation between a customer and a storekeeper that we frequently overhear to make it sound more familiar to the people in The Land of Gentlemen by suitable substitution of the concepts of high and low prices, profitable and nonprofitable business transactions, good and bad workmanship, and so on. Again, these pairs of concepts are pairs of dual concepts.

We hope the reader is now ready for a more serious discussion. Let $(A, \leq)$ be a partially ordered set. Let $\leq_R$ be a binary relation on A such that for a and b in A, $a \leq_R b$ if and only if $b \leq a$. It is not hard to see that $(A, \leq_R)$ is also a partially ordered set. Furthermore, if $(A, \leq)$ is a lattice so is $(A, \leq_R)$.[†] We note that the lattices $(A, \leq)$ and $(A, \leq_R)$ are closely related, and so are the algebraic systems defined by them. To be specific, the join operation of the algebraic system defined by $(A, \leq)$ is the meet operation of the algebraic system defined by $(A, \leq_R)$, and the meet operation of the algebraic system defined by $(A, \leq)$ is the join operation of the algebraic system defined by $(A, \leq_R)$. Consequently, given any valid statement concerning the general properties of lattices we can obtain another valid statement by replacing the relation $\leq$ with $\geq$,[‡] the term "join operation" with the term "meet operation," and the term "meet operation" with the term "join operation." This is known as the *principle of duality for lattices.*

---

† See Prob. 3.20.

‡ Clearly, $a \leq_R b$ if and only if $a \geq b$.

As an example, we note that the relation in (9.1) can be stated as, "The join of any two elements in a lattice is larger than or equal to each of the two elements." Correspondingly, we have, "The meet of any two elements in a lattice is less than or equal to each of the two elements," which is the relation in (9.2). Similarly, the relation in (9.4) can be obtained from the relation in (9.3), and conversely, according to the principle of duality. We shall see more examples on the application of the principle of duality as we proceed to show some of the general properties of algebraic systems defined by lattices.†

9.3 BASIC PROPERTIES OF ALGEBRAIC SYSTEMS DEFINED BY LATTICES

We show now some of the basic properties possessed by algebraic systems defined by lattices. Let $(A, \vee, \wedge)$ be an algebraic system defined by a lattice $(A, \leq)$. We have:

Theorem 9.3 Both the join and meet operations are commutative.

PROOF It follows directly from the definition of the least upper bound and the greatest lower bound of two elements in a lattice. □

Theorem 9.4 Both the join and meet operations are associative.

PROOF We first show that the join operation $\vee$ is associative. That is, for a, b, c in A,

$$a \vee (b \vee c) = (a \vee b) \vee c$$

Let

$$a \vee (b \vee c) = g$$

and

$$(a \vee b) \vee c = h$$

Since g is the join of a and $(b \vee c)$,

$$a \leq g \qquad b \vee c \leq g$$

Furthermore, $b \vee c \leq g$ means that

$$b \leq g \qquad c \leq g$$

Since the join of a and b is the least upper bound of a and b, from $a \leq g$

---

† There are other important examples illustrating the principle of duality. In electrical engineering, the concepts of voltage and current, resistance and conductance, inductance and capacitance are dual concepts. (For more details, see, for example, [6].) In graph theory, the concepts of circuit and cut-set, tree and cotree are dual concepts. (For more details, see, for example, [8].)

and $b \leq g$ we obtain

$$a \vee b \leq g$$

which, together with $c \leq g$, leads to

$$(a \vee b) \vee c \leq g$$

We have thus shown that

$$h \leq g$$

In a similar manner, we can show that

$$g \leq h$$

Consequently, because of the antisymmetry property of a partial ordering relation, we conclude that

$$g = h$$

According to the principle of duality, the meet operation $\wedge$ is also associative. $\square$

Theorem 9.5 For every a in A, $a \vee a = a$ and $a \wedge a = a$.

PROOF According to (9.1),

$$a \leq a \vee a$$

Since $a \leq a$, we obtain†

$$a \vee a \leq a$$

It follows that

$$a \vee a = a$$

According to the principle of duality, we also have

$$a \wedge a = a \qquad \square$$

The results in Theorem 9.5 are known as the *idempotent property* of the join and meet operations.

Theorem 9.6 For any a and b in A,

$$a \vee (a \wedge b) = a$$

$$a \wedge (a \vee b) = a$$

PROOF Since $a \vee (a \wedge b)$ is the join of a and $a \wedge b$, we have

$$a \leq a \vee (a \wedge b) \qquad (9.5)$$

Since

$$a \leq a \qquad a \wedge b \leq a$$

† If $g \leq f$ and $h \leq f$, then $g \vee h \leq f$ because $g \vee h$ is the least upper bound of g and h.

according to (9.3), we have

$$a \vee (a \wedge b) \leq a \vee a \qquad (9.6)$$

Since

$$a \vee a = a$$

(9.6) becomes

$$a \vee (a \wedge b) \leq a \qquad (9.7)$$

Combining (9.5) and (9.7), we obtain

$$a \vee (a \wedge b) = a$$

That $a \wedge (a \vee b) = a$ follows from the principle of duality. □

The results in Theorem 9.6 are known as the *absorption property* of the join and meet operations.

For example, consider the lattice of natural numbers ordered by the "less than or equal to" relationship, $(N, \leq)$. Because the maximum of two numbers a and b is the same as the maximum of the two numbers b and a, the join operation in $(N, \vee, \wedge)$ is commutative. Similarly, the meet operation is also commutative. Because both $\max(\max(a, b), c)$ and $\max(a, \max(b, c))$ are equal to the maximum of the three numbers a, b, and c, the join operation in $(N, \vee, \wedge)$ is associative. Similarly, the meet operation is associative. The idempotent property for the join and meet operations follows from the facts that the maximum of a and a is a, and that the minimum of a and a is a. The equations

$$\max[a, \min(a, b)] = a$$

$$\min[a, \max(a, b)] = a$$

yield the absorption property.

As another example, consider the lattice of positive integers ordered by the "divides by" relationship, $(N^+, |)$. Because the least common multiple of a and b is the same as the least common multiple of b and a, the join operation in $(N^+, \vee, \wedge)$ is commutative. Similarly, the meet operation is commutative. Because both $(a \vee b) \vee c$ and $a \vee (b \vee c)$ are the least common multiple of the three numbers a, b, and c, the join operation is associative. Similarly, the meet operation is associative. The least common divisor of a and a is a, as is the greatest common divisor of a and a. Therefore, the idempotent property of the join and meet operations follows. The equations

$$\text{lcm}[a, \gcd(a, b)] = a$$

$$\gcd[a, \text{lcm}(a, b)] = a$$

yield the absorption property.

9.4 DISTRIBUTIVE AND COMPLEMENTED LATTICES

We study now classes of lattices that possess additional properties. Naturally, one can anticipate that these lattices will define algebraic systems that are even more "structured." A lattice is said to be a *distributive* lattice if the meet operation distributes over the join operation and the join operation distributes over the meet operation. That is, for any a, b, and c,

$$a \wedge (b \vee c) = (a \wedge b) \vee (a \wedge c)$$

$$a \vee (b \wedge c) = (a \vee b) \wedge (a \vee c)\dagger$$

The lattice in Fig. 9.3a is a distributive lattice. However, the lattice in Fig. 9.4 is not a distributive lattice. Note that in the lattice in Fig. 9.4,

$$b \wedge (c \vee d) = b \wedge a = b$$

$$(b \wedge c) \vee (b \wedge d) = e \vee e = e$$

Our definition of the distributivity of a lattice is slightly redundant as the following theorem shows:

Theorem 9.7 If the meet operation is distributive over the join operation in a lattice, then the join operation is also distributive over the meet operation. If the join operation is distributive over the meet operation, then the meet operation is also distributive over the join operation.

PROOF Given that

$$a \wedge (b \vee c) = (a \wedge b) \vee (a \wedge c)$$

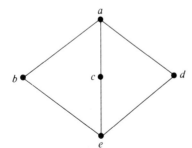

Figure 9.4

† Because of commutativity, the equations

$$(a \vee b) \wedge c = (a \wedge c) \vee (b \wedge c)$$

$$(a \wedge b) \vee c = (a \vee c) \wedge (b \vee c)$$

are implied.

we obtain

$$(a \vee b) \wedge (a \vee c) = [(a \vee b) \wedge a] \vee [(a \vee b) \wedge c]$$
$$= a \vee [(a \vee b) \wedge c]$$
$$= a \vee [(a \wedge c) \vee (b \wedge c)]$$
$$= [a \vee (a \wedge c)] \vee (b \wedge c)$$
$$= a \vee (b \wedge c)$$

By duality, we obtain the result that if the join operation $\vee$ is distributive over the meet operation $\wedge$ then the meet operation $\wedge$ is also distributive over the join operation $\vee$. □

An element a in a lattice $(A, \leq)$ is called a *universal lower bound* if for every element $b \in A$, $a \leq b$. For example, for the lattice shown in Fig. 9.5, h is a universal lower bound. It follows from the definition that if a lattice has a universal lower bound, it is unique. For if we assume that there are two universal lower bounds a and b, then

$$a \leq b \qquad \text{and} \qquad b \leq a$$

imply

$$a = b$$

An element a in a lattice $(A, \leq)$ is called a *universal upper bound* if for every element $b \in A$, $b \leq a$. Again, if a lattice has a universal upper bound, it is unique.

We shall use 0 to denote the universal lower bound and 1 to denote the universal upper bound of a lattice (if such bounds exist). Note the 0 is indeed the identity of the join operation $\vee$, and 1 the identity of the meet operation $\wedge$. For example, in the lattice $(\mathcal{P}(S), \subseteq)$ where $\mathcal{P}(S)$ is the power set of a set S and

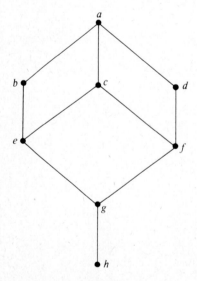

Figure 9.5

$\subseteq$ is the set inclusion relation, the empty set $\emptyset$ is the universal lower bound and the set S is the universal upper bound. We have

Theorem 9.8 Let $(A, \leq)$ be a lattice with universal upper and lower bounds 0 and 1. For any element a in A,

$$a \vee 1 = 1 \qquad a \wedge 1 = a$$
$$a \vee 0 = a \qquad a \wedge 0 = 0$$

PROOF According to (9.1), $1 \leq a \vee 1$. Since 1 is the universal upper bound, $a \vee 1 \leq 1$. Thus, $a \vee 1 = 1$.

According to (9.2), $a \wedge 1 \leq a$. Since $a \leq a$ and $a \leq 1$, according to (9.4), $a \wedge a \leq a \wedge 1$, or $a \leq a \wedge 1$. Thus, $a \wedge 1 = a$.

The other two relations can be proved similarly. $\qquad\square$

Let $(A, \leq)$ be a lattice with universal lower and upper bounds 0 and 1. For an element a in A, an element b is said to be a *complement* of a if

$$a \vee b = 1 \qquad \text{and} \qquad a \wedge b = 0$$

Note that because of commutativity, if a is a complement of b, then b is also a complement of a. For example, for the lattice in Fig. 9.1, d is a complement of e and e is a complement of d. Note that an element might have more than one complement. For example, in the lattice in Fig. 9.1, both b and e are complements of d. Yet, on the other hand, in the lattice in Fig. 9.1, c has no complement. We note, however, that 0 is the unique complement of 1 and 1 is the unique complement of 0.†

A lattice is said to be a *complemented lattice* if every element in the lattice has a complement. (Clearly, a complemented lattice must have universal lower and upper bounds.) For example, the lattice in Fig. 9.6 is a complemented lattice, where the complement of a is c, the complement of b is also c, and it follows that the complements of c are a and b.

The following theorem shows again that distributivity is a very "useful" property:

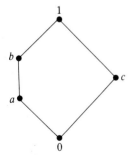

0 **Figure 9.6**

† It is clear that 0 is a complement of 1 and 1 is a complement of 0. We ask the reader to show uniqueness. (See Prob. 9.14.)

Theorem 9.9 In a distributive lattice, if an element has a complement then this complement is unique.

PROOF Suppose that an element a has two complements b and c. That is,

$$a \vee b = 1 \quad a \wedge b = 0$$
$$a \vee c = 1 \quad a \wedge c = 0$$

We have

$$
\begin{aligned}
b &= b \wedge 1 \\
&= b \wedge (a \vee c) \\
&= (b \wedge a) \vee (b \wedge c) \\
&= 0 \vee (b \wedge c) \\
&= (a \wedge c) \vee (b \wedge c) \\
&= (a \vee b) \wedge c \\
&= 1 \wedge c \\
&= c \qquad\qquad\qquad \square
\end{aligned}
$$

9.5 BOOLEAN LATTICES AND BOOLEAN ALGEBRAS

A complemented and distributive lattice is also called a *boolean lattice*. Let $(A, \leq)$ be a boolean lattice. Since every element in a boolean lattice has a unique complement, we can define a unary operation on A, denoted $^-$, so that for every a in A $\bar{a}$ is the complement of a.† The unary operation $^-$ is referred to as the *complementation* operation. Consequently, we say that a boolean lattice $(A, \leq)$ defines an algebraic system $(A, \vee, \wedge, ^-)$, where $\vee$, $\wedge$, and $^-$ are the join, meet, and complementation operations, respectively. An algebraic system defined by a boolean lattice is known as a *boolean algebra*.

We can easily construct examples of boolean algebras. Let S be a finite set. We recall that $(\mathscr{P}(S), \subseteq)$ is a lattice. Furthermore, $(\mathscr{P}(S), \subseteq)$ is a boolean lattice in which the universal upper bound is S, the universal lower bound is $\emptyset$, and the complement of any set T in $\mathscr{P}(S)$ is the set $S - T$. We now show:

Theorem 9.10 For any a and b in a boolean algebra

$$\overline{a \vee b} = \bar{a} \wedge \bar{b}$$

$$\overline{a \wedge b} = \bar{a} \vee \bar{b}$$

---

† Instead of writing $-a$, the notation $\bar{a}$ is more convenient. Thus, for example, the complement of the join of a and b will be written as $\overline{a \vee b}$.

PROOF We have,

$$(a \vee b) \vee (\bar{a} \wedge \bar{b}) = [(a \vee b) \vee \bar{a}] \wedge [(a \vee b) \vee \bar{b}]$$
$$= [(a \vee \bar{a}) \vee b] \wedge [a \vee (b \vee \bar{b})]$$
$$= 1 \wedge 1$$
$$= 1$$

and

$$(a \vee b) \wedge (\bar{a} \wedge \bar{b}) = [a \wedge (\bar{a} \wedge \bar{b})] \vee [b \wedge (\bar{a} \wedge \bar{b})]$$
$$= [(a \wedge \bar{a}) \wedge \bar{b}] \vee [(b \wedge \bar{b}) \wedge \bar{a}]$$
$$= 0 \vee 0$$
$$= 0$$

Thus, $\bar{a} \wedge \bar{b}$ is the complement of $a \vee b$. That is,

$$\overline{a \vee b} = \bar{a} \wedge \bar{b}$$

That $\overline{a \wedge b} = \bar{a} \vee \bar{b}$ follows from the principle of duality.   □

The results in Theorem 9.10 are known as *DeMorgan's laws*.

9.6 UNIQUENESS OF FINITE BOOLEAN ALGEBRAS

One might imagine that there is a great deal of freedom for constructing many different boolean algebras. As it turns out, this is not quite the case. We shall show that a finite boolean algebra† has exactly 2^n elements for some $n > 0$. Moreover, there is a unique boolean algebra of 2^n elements for every $n > 0$.

Let a and b be two elements in a lattice. We recall that a is said to *cover* b if $b < a$ ‡ and there is no element c such that $b < c$ and $c < a$. Let $(A, \leq)$ be a lattice with a universal lower bound 0. An element is called an *atom* if it covers 0. For example, for the lattice in Fig. 9.1, e and f are atoms.

Let $(A, \leq)$ be a finite lattice with a universal lower bound. We claim that, for any nonzero element§ b there exists at least one atom a such that $a \leq b$. Clearly, if b is an atom, this is immediate. If b is not an atom, because $(A, \leq)$ is a finite lattice, there must be a chain in $(A, \leq)$ such that $0 < b_i < \cdots < b_2 < b_1 < b$, where b_i is an atom. It follows that $b_i \leq b$.

We prove now a sequence of lemmas:

Lemma 9.1 In a distributive lattice, if $b \wedge \bar{c} = 0$, then $b \leq c$.

---

† A finite boolean algebra is one with a finite number of elements.
‡ We use the notation $b < a$ to mean $b \leq a$ and $b \neq a$.
§ By a nonzero element we mean an element that is not equal to the universal lower bound 0.

PROOF Since $b \wedge \bar{c} = 0$, we have

$$(b \wedge \bar{c}) \vee c = c$$

According to the distributive law, we have

$$(b \vee c) \wedge (\bar{c} \vee c) = c$$

or

$$b \vee c = c \qquad\qquad (9.8)$$

Since (9.8) means that c is the least upper bound of b and c, $b \leq c$. $\qquad \square$

Lemma 9.2 Let $(A, \vee, \wedge, ^{-})$ be a finite boolean algebra. Let b be any nonzero element in A, and $a_1, a_2, \ldots, a_k$ be all the atoms of A such that $a_i \leq b$. Then $b = a_1 \vee a_2 \vee \cdots \vee a_k$.

PROOF Since

$$a_1 \leq b \quad a_2 \leq b \quad \cdots \quad a_k \leq b$$

it follows immediately

$$a_1 \vee a_2 \vee \cdots \vee a_k \leq b$$

For notational convenience, let c denote $a_1 \vee a_2 \vee \cdots \vee a_k$.

Let us suppose that $b \wedge \bar{c} \neq 0$. In that case, there exists an atom a such that $a \leq (b \wedge \bar{c})$. Since

$$b \wedge \bar{c} \leq b$$

and

$$b \wedge \bar{c} \leq \bar{c}$$

according to the transitive law, we have

$$a \leq b \qquad\qquad (9.9)$$

and

$$a \leq \bar{c}$$

According to (9.9) a is equal to one of the atoms $a_1, a_2, \ldots, a_k$. It follows that

$$a \leq c$$

Combining $a \leq \bar{c}$ and $a \leq c$, we obtain

$$a \leq c \wedge \bar{c}$$

or

$$a \leq 0$$

which is impossible.

It follows that we must have $b \wedge \bar{c} = 0$. In that case, according to Lemma 9.1, $b \leq c$. By the antisymmetry law, we have

$$b = a_1 \vee a_2 \vee \cdots \vee a_k \qquad\qquad \square$$

Lemma 9.3 Let $(A, \vee, \wedge, \bar{\ })$ be a finite boolean algebra. Let b be any nonzero element in A, and $a_1, a_2, \ldots, a_k$ be all the atoms of A such that $a_i \leq b$. Then $b = a_1 \vee a_2 \vee \cdots \vee a_k$ is the *unique* way to represent b as a join of atoms.

PROOF Suppose that we have an alternative representation

$$b = a_{j_1} \vee a_{j_2} \vee \cdots \vee a_{j_t}$$

Clearly, since b is the least upper bound of $a_{j_1}, a_{j_2}, \ldots, a_{j_t}$

$$a_{j_1} \leq b \quad a_{j_2} \leq b \quad \cdots \quad a_{j_t} \leq b$$

In other words, if b is expressed as a join of atoms, they must be atoms that are less than or equal to b.

Now consider an atom a_{j_u}, $1 \leq u \leq t$. Since $a_{j_u} \leq b$, we have

$$a_{j_u} \wedge b = a_{j_u}$$

That is,

$$a_{j_u} \wedge (a_1 \vee a_2 \vee \cdots \vee a_k) = a_{j_u}$$

or

$$(a_{j_u} \wedge a_1) \vee (a_{j_u} \wedge a_2) \vee \cdots \vee (a_{j_u} \wedge a_k) = a_{j_u}$$

Then, for some a_i, $1 \leq i \leq k$,

$$a_{j_u} \wedge a_i \neq 0$$

Since both a_{j_u} and a_i are atoms, we must have $a_{j_u} = a_i$.

Thus each atom in the alternative representation is an atom in the original one, and the lemma follows. $\square$

It follows from Lemmas 9.1, 9.2, and 9.3 that there is a one-to-one correspondence between the elements of a boolean lattice and the subsets of the atoms. As a matter of fact, this one-to-one correspondence is an isomorphism from $(A, \leq)$ to $(\mathscr{P}(S), \subseteq)$, where S is the set of atoms. We thus have:

Theorem 9.11 Let $(A, \vee, \wedge, \bar{\ })$ be a finite boolean algebra. Let S be the set of atoms. Then $(A, \vee, \wedge, \bar{\ })$ is isomorphic to the algebraic system defined by the lattice $(\mathscr{P}(S), \subseteq)$.

It follows immediately from Theorem 9.11 that there exists a unique finite boolean algebra of 2^n elements for any $n > 0$. Furthermore, there are no other finite boolean algebras.

9.7 BOOLEAN FUNCTIONS AND BOOLEAN EXPRESSIONS

Let $(A, \vee, \wedge, \bar{\ })$ be a boolean algebra. Let us consider the functions from A^n to A. For example, the table in Fig. 9.7 shows a function f from $\{0, 1\}^3$ to $\{0, 1\}$, and

	f
(0, 0)	1
(0, 1)	0
(0, 2)	0
(0, 3)	3
(1, 0)	1
(1, 1)	1
(1, 2)	0
(1, 3)	3
(2, 0)	2
(2, 1)	0
(2, 2)	1
(2, 3)	1
(3, 0)	3
(3, 1)	0
(3, 2)	0
(3, 3)	2

	f
(0, 0, 0)	0
(0, 0, 1)	0
(0, 1, 0)	1
(0, 1, 1)	0
(1, 0, 0)	1
(1, 0, 1)	1
(1, 1, 0)	0
(1, 1, 1)	1

Figure 9.7   **Figure 9.8**

the table in Fig. 9.8 shows a function f from $\{0, 1, 2, 3\}^2$ to $\{0, 1, 2, 3\}$. Although a function can always be described exhaustively in tabular form, we are interested in alternative ways of describing functions. We recall that there is the possibility of specifying a function by a "closed form expression." Let us indeed pursue such an approach.

Let $(A, \vee, \wedge, \bar{\ })$ be a boolean algebra. A *boolean expression* over $(A, \vee, \wedge, \bar{\ })$ is defined as follows:†

1. Any element of A is a boolean expression.
2. Any variable name is a boolean expression.
3. If e_1 and e_2 are boolean expressions, then $\bar{e_1}$, $e_1 \vee e_2$, $e_1 \wedge e_2$ are boolean expressions.

For example,

$$0 \vee x$$

$$((\overline{2 \wedge 3}) \vee (x_1 \vee \bar{x_2})) \wedge \overline{(x_1 \wedge x_3)}$$

are boolean expressions over the boolean algebra $(\{0, 1, 2, 3\}, \vee, \wedge, \bar{\ })$. A boolean expression that contains n *distinct* variables is usually referred to as a boolean expression of n variables.

Let $E(x_1, x_2, \ldots, x_n)$ be a boolean expression of n variables over a boolean algebra $(A, \vee, \wedge, \bar{\ })$. By an assignment of values to the variables $x_1, x_2, \ldots, x_n$, we mean an assignment of elements of A to be the values of the variables. For an

† See Sec. 8.12 for the definition of an algebraic expression over an algebraic system.

assignment of values to the variables we can evaluate the expression $E(x_1, x_2, \ldots, x_n)$ by substituting the variables in the expression by their values. For example, for the boolean expression

$$E(x_1, x_2, x_3) = (x_1 \vee x_2) \wedge (\bar{x_1} \vee \bar{x_2}) \wedge \overline{(x_2 \vee x_3)}$$

over the boolean algebra $(\{0, 1\}, \vee, \wedge, ^-)$ the assignment of values $x_1 = 0$, $x_2 = 1$, $x_3 = 0$ yields

$$E(0, 1, 0) = (0 \vee 1) \wedge (\bar{0} \vee \bar{1}) \wedge \overline{(1 \vee 0)}$$
$$= 1 \wedge 1 \wedge 0$$
$$= 0$$

Two boolean expressions of n variables are said to be *equivalent* if they assume the same value for every assignment of values to the n variables. For example, the reader can readily check that the two boolean expressions

$$(x_1 \wedge x_2) \vee (x_1 \wedge \bar{x_3})$$

$$x_1 \wedge (x_2 \vee \bar{x_3})$$

are equivalent. We write

$$E_1(x_1, x_2, \ldots, x_n) = E_2(x_1, x_2, \ldots, x_n)$$

to mean the two expressions $E_1(x_1, x_2, \ldots, x_n)$ and $E_2(x_1, x_2, \ldots, x_n)$ are equivalent. Thus, when we say to manipulate or to simplify a boolean expression we always mean to manipulate or to simplify it into an equivalent form. Because elements of A will be assigned as values of the variables in a boolean expression, all equalities we derived in previous sections involving elements of a boolean algebra can be applied to manipulate and simplify boolean expressions. For example, we note that

$$(x_1 \wedge x_2) \vee (x_1 \wedge \bar{x_3}) = x_1 \wedge (x_2 \vee \bar{x_3})$$
$$(x_1 \wedge x_2) \vee (x_1 \wedge \bar{x_2}) = x_1$$
$$x_1 \wedge x_2 = (x_1 \wedge x_2) \wedge 1$$
$$= (x_1 \wedge x_2) \wedge (x_3 \vee \bar{x_3})$$
$$= (x_1 \wedge x_2 \wedge x_3) \vee (x_1 \wedge x_2 \wedge \bar{x_3})$$

It is not difficult to see how we can specify a function from A^n to A by a boolean expression $E(x_1, x_2, \ldots, x_n)$. Namely, let each assignment of values to the variables $x_1, x_2, \ldots, x_n$ be an ordered n-tuple in the domain A^n, and let the corresponding value of $E(x_1, x_2, \ldots, x_n)$ be the corresponding image in the range A. For example, the reader can check immediately that the boolean expression

$$(\bar{x_1} \wedge x_2 \wedge \bar{x_3}) \vee (x_1 \wedge \bar{x_2}) \vee (x_1 \wedge x_3)$$

over the boolean algebra $(\{0, 1\}, \vee, \wedge, ^-)$ defines the function f in Fig. 9.7.

On the other hand, we ask whether every function from A^n to A can be specified by a boolean expression over $(A, \vee, \wedge, \bar{})$. The answer is negative. For example, there is no boolean expression over the boolean algebra of four elements that defines the function in Fig. 9.8. (See Problem 9.24.) A function from A^n to A is called a *boolean function* if it can be specified by a boolean expression (of n variables).

We note, however, that for the case of the two-valued boolean algebra any function from $\{0, 1\}^n$ to $\{0, 1\}$ is a boolean function. As a matter of fact, we shall show two ways to obtain a boolean expression that specifies a given function from $\{0, 1\}^n$ to $\{0, 1\}$. A boolean expression of n variables $x_1, x_2, \ldots, x_n$ is said to be a *minterm* if it is of the form

$$\tilde{x}_1 \wedge \tilde{x}_2 \wedge \cdots \wedge \tilde{x}_n \tag{9.10}$$

where we use $\tilde{x}_i$ to denote either x_i or $\bar{x}_i$. A boolean expression over $(\{0, 1\}, \vee, \wedge, \bar{})$ is said to be in *disjunctive normal form* if it is a join of minterms. For example,

$$(\bar{x}_1 \wedge \bar{x}_2 \wedge \bar{x}_3) \vee (\bar{x}_1 \wedge x_2 \wedge \bar{x}_3) \vee (x_1 \wedge x_2 \wedge x_3) \tag{9.11}$$

is a boolean expression in disjunctive normal form. Furthermore, there are three minterms in the expression, namely, $\bar{x}_1 \wedge \bar{x}_2 \wedge \bar{x}_3, \bar{x}_1 \wedge x_2 \wedge \bar{x}_3$, and $x_1 \wedge x_2 \wedge x_3$.

A boolean expression of n variables $x_1, x_2, \ldots, x_n$ is said to be a *maxterm* if it is of the form

$$\tilde{x}_1 \vee \tilde{x}_2 \vee \cdots \vee \tilde{x}_n \tag{9.12}$$

where, again we use $\tilde{x}_i$ to denote either x_i or $\bar{x}_i$. A boolean expression over $(\{0, 1\}, \vee, \wedge, \bar{})$ is said to be in *conjunctive normal form* if it is a meet of maxterms. For example,

$$(x_1 \vee x_2 \vee \bar{x}_3) \wedge (x_1 \vee \bar{x}_2 \vee \bar{x}_3) \wedge (\bar{x}_1 \vee x_2 \vee x_3) \wedge (\bar{x}_1 \vee x_2 \vee \bar{x}_3) \wedge (\bar{x}_1 \wedge \bar{x}_2 \vee x_3) \tag{9.13}$$

is a boolean expression in conjunctive normal form consisting of five maxterms.

Given a function from $\{0, 1\}^n$ to $\{0, 1\}$, we can obtain a boolean expression in disjunctive normal form corresponding to this function by having a minterm corresponding to each ordered n-tuple of 0s and 1s for which the value of the function is 1. Specifically, for each such n-tuple, we have a minterm

$$\tilde{x}_1 \wedge \tilde{x}_2 \wedge \cdots \wedge \tilde{x}_n$$

in which $\tilde{x}_i$ is x_i if the ith component of the n-tuple is 1 and is $\bar{x}_i$ if the ith component of the n-tuple is 0.† For example, the boolean expression in (9.11) corresponds to the function f in Fig. 9.9.

Similarly, given a function from $\{0, 1\}^n$ to $\{0, 1\}$, we can obtain a boolean expression in conjunctive normal from corresponding to this function by having a maxterm corresponding to each ordered n-tuple of 0s and 1s at which

---

† We ask the reader to convince herself that the boolean expression obtained indeed specifies the given function.

the value of the function is 0. Specifically, for each such n-tuple we have a maxterm

$$\tilde{x}_1 \vee \tilde{x}_2 \vee \cdots \vee \tilde{x}_n$$

in which $\tilde{x}_i$ is x_i if the ith component of the n-tuple is 0 and is $\overline{x}_i$ if the ith component of the n-tuple is 1. For example, the boolean expression in (9.13) corresponds to the function f in Figure 9.9.

			f
0	0	0	1
0	0	1	0
0	1	0	1
0	1	1	0
1	0	0	0
1	0	1	0
1	1	0	0
1	1	1	1

Figure 9.9

9.8 PROPOSITIONAL CALCULUS

We recall our discussion in Sec. 1.7 on propositions: A proposition is a statement that may either be true (T) or be false (F). Furthermore, propositions can be combined to yield new propositions. In particular, we introduced the notion of disjunction, conjunction, and negation of propositions. We show now that our discussion can be formulated in the framework of an algebraic system.

Consider an algebraic system $(\{F, T\}, \vee, \wedge, \overline{})$ where the definitions of the operations $\vee$, $\wedge$, and $\overline{}$ are shown in Fig. 9.10. The reader can check immediately that $(\{F, T\}, \vee, \wedge, \overline{})$ is a boolean algebra of two elements. Within such an algebraic framework, an atomic proposition is a variable

$\vee$	F	T
F	F	T
T	T	T

$\wedge$	F	T
F	F	F
T	F	T

$\overline{}$	
F	T
T	F

Figure 9.10

that can assume either the value F (false) or the value T (true). A tautology corresponds to the constant T and a contradiction corresponds to the constant F. The disjunction of two propositions p and q can be represented by the algebraic expression $p \vee q$. Note that the definition of the binary operation $\vee$ in Fig. 9.10 is indeed consistent with the definition of the disjunction of two propositions as shown in the truth table of Fig. 1.3. Similarly, the conjunction of two propositions p and q can be represented by the algebraic expression $p \wedge q$, and the negation of a proposition p can be represented by the algebraic expression $\bar{p}$. Again, the definitions of the operation $\wedge$ and $^-$ are consistent with that of the conjunction and negation of propositions as shown in Fig. 1.3. It follows that a compound proposition can be represented by a boolean expression. Furthermore, the truth table of a compound proposition is exactly a tabular description of the value of the corresponding boolean expression for all possible combinations of the values of the atomic propositions. We conclude that all our results on manipulation and simplification of boolean expressions can be applied to manipulate and simplify compound propositions as the following examples illustrate.

> **Example 9.1** Consider the statement, "I will go to the ball game either if there is no examination tomorrow or if there is an examination tomorrow and the game is a championship match." Let p denote the proposition, "There is an examination tomorrow" and q denote the proposition, "The game is a championship match." Clearly, I will go to the ball game if the proposition
>
> $$\bar{p} \vee (p \wedge q)$$
>
> is true. However, the equality
>
> $$\bar{p} \vee (p \wedge q) = \bar{p} \vee q$$
>
> enables us to simplify our statement to, "I will go to the ball game either if there is no examination tomorrow or if the game is a championship match."
> □

> **Example 9.2** Consider the following instructions given to a maintenance technician:
>
> 1. The electric power should be turned on if it is not the case that nobody is in the office and the automatic monitoring system is not in operation.
> 2. The automatic monitoring system will be in operation if and only if nobody is in the office or a large payroll is left in the office.
>
> Let p denote the proposition, "nobody is in the office," q denote the proposition "the automatic monitoring system is in operation," and r denote the proposition "a large payroll is left in the office." Clearly, the electric power should be turned on if the proposition
>
> $$\overline{p \wedge \bar{q}}$$
>
> is true. However, since
>
> $$q = p \vee r$$

we have

$$\overline{(p \wedge \overline{q})} = \overline{p \wedge \overline{(p \vee r)}}$$

$$= \overline{p \wedge \overline{(\overline{p} \wedge \overline{r})}}$$

$$= T$$

Consequently, we can conclude that the instructions are completely superfluous and can be replaced by the simple instruction of always leaving the electric power on. □

Example 9.3 Consider the directed graph in Fig. 9.11. We want to determine a minimal subset of the edges which, if removed, would destroy all the directed circuits in the graph. (By a *minimal* set of edges, we mean a set such that the removal of any proper subset of it will not destroy all the directed circuits.) We note that in the graph in Fig. 9.11 there are five directed circuits, namely, (a, d, g, f, b), (a, e, f, b), (c, d, g, f), (c, e, f), and (g, i, h). In order to destroy the directed circuit (a, d, g, f, b), we should remove at least one of the edges a, d, g, f, b. In order to destroy the directed circuit (a, e, f, b), we should remove at least one of the edges a, e, f, b. Consequently, in order to destroy all five directed circuits, we should remove at least one edge from each of them. From the algebraic expression,

$$(a \vee d \vee g \vee f \vee b) \wedge (a \vee e \vee f \vee b) \wedge (c \vee d \vee g \vee f) \wedge (c \vee e \vee f) \wedge (g \vee i \vee h)$$

$$= (a \wedge d \wedge e \wedge h) \vee (e \wedge g) \vee \cdots$$

we note that the removal of the subset $\{a, d, e, h\}$, or $\{e, g\}$, ... will destroy all directed circuits in the graph. □

Finally, we recall that we defined in Sec. 1.7 two additional ways of combining propositions, namely, $p \rightarrow q$ and $p \leftrightarrow q$ for given propositions p and q. It is clear that we can augment our algebraic system by defining two operations $\rightarrow$ and $\leftrightarrow$ as shown in Fig. 9.12. However, such an augmentation is unnecessary in that we can always replace algebraic expressions involving the operations $\rightarrow$ and $\leftrightarrow$ by equivalent algebraic expressions involving only the operations $\wedge$, $\vee$, and $\overline{}$. Specifically, from the truth tables for $p \rightarrow q$ and $p \leftrightarrow q$, we obtain

$$p \rightarrow q = \overline{p} \vee q$$

$$p \leftrightarrow q = (p \wedge q) \vee (\overline{p} \wedge \overline{q})$$

for the boolean expressions corresponding to the compound propositions $p \rightarrow q$ and $p \leftrightarrow q$.†

---

† Conceptually, the situation is similar to that of not defining subtraction as another operation in a ring since subtraction of an element is equivalent to the addition of the additive inverse of the element.

Figure 9.11

$\rightarrow$	F	T
F	T	T
T	F	T

$\leftrightarrow$	F	T
F	T	F
T	F	T

Figure 9.12

As an example, we note that

$$p \rightarrow q = \bar{p} \vee q$$
$$= \bar{\bar{q}} \vee \bar{p}$$
$$= \bar{q} \rightarrow \bar{p}$$

Therefore, the propositions, "If the temperature is above 30°C, we shall go to the beach," "Either the temperature is not above 30°C, or we shall go to the beach," and, "If we do not go to the beach, the temperature is not above 30°C" are equivalent. Similarly, the propositions, "If a group is cyclic, it is abelian," "Either a group is not cyclic or it is abelian," and, "If a group is not abelian, it is not cyclic" are equivalent.

As another example, we note that

$$p \leftrightarrow q = (\bar{p} \vee q) \wedge (p \vee \bar{q})†$$
$$= (p \rightarrow q) \wedge (q \rightarrow p)$$

Therefore, the statement, "There will be a $2 service charge if and only if the monthly balance of the checking account is less than $100" is equivalent to the statement, "If there is a $2 service charge then it must be the case that the monthly balance of the checking account is less than $100, and if the monthly balance of the checking account is less than $100 then there will be a $2 service charge."

At this point, we recall an observation in Sec. 1.7 that there is a strong similarity between composition of propositions to yield new propositions and composition of sets to yield new sets. Indeed, since many of the properties and results on sets and set operations and on propositions and proposition operations are that of boolean algebras, such a similarity is not a coincidence.

---

† This is the conjunctive normal form for the boolean expression corresponding to the compound statement $p \leftrightarrow q$.

9.9 DESIGN AND IMPLEMENTATION OF DIGITAL NETWORKS

Suppose we want to design an electronic circuit that will sound a buzzer in an automobile if the temperature of the engine exceeds 200°F or if the automobile is in gear and the driver did not have the seat-belt buckled. Clearly, we have the relationship

$$b = p \vee (q \wedge \bar{r})$$

where b is the proposition "sound the buzzer," p is the proposition "the temperature of the engine exceeds 200°F," q is the proposition "the automobile is in gear," and r is the proposition "the driver's seat-belt is buckled." To build an electronic circuit that will behave as described, we must first decide upon a convention for representing propositions by electric signals. For example, we might represent a proposition by an electric voltage. If the proposition is true, it will be represented by a high voltage (say 6 V), and if the proposition is false, it will be represented by a low voltage (say 0 V). Thus, in the foregoing example, corresponding to the proposition p, a high-voltage signal will appear if the temperature of the engine exceeds 200°F, and a low-voltage signal will appear if the temperature of the engine does not exceed 200°F. Similarly, corresponding to the proposition b, a high-voltage signal, which can be used to trigger the buzzer, will appear if the conditions to sound the buzzer are met, and a low-voltage signal, which will not be sufficient to trigger the buzzer, will appear if the conditions to sound the buzzer are not met. Once we have decided upon a convention for representing propositions by electric signals, we can design electronic circuits corresponding to the operations $\vee$, $\wedge$, and $^-$. Throughout our discussion in the remainder of this section, we shall assume the convention of representing propositions by voltages as described in the foregoing.

An *OR-gate* is a circuit that has two inputs and one output as shown schematically in Fig. 9.13a. The output voltage of an OR-gate is high if either one or both of the input voltages is high, and the output voltage is low if both of the input voltages are low. Clearly, the output signal of an OR-gate corresponds to a proposition which is the disjunction of the propositions corresponding to the input signals.

An *AND-gate* is a circuit that has two inputs and one output as shown schematically in Fig. 9.13b. The output voltage of an AND-gate is high if both of the input voltages are high, and the output voltage is low if either one or both of the input voltages is low. Clearly, the output signal of an AND-gate corresponds to a proposition which is the conjunction of the propositions corresponding to the input signals.

A *NOT-gate*, or an *inverter*, is a circuit that has one input and one output, as shown schematically in Fig. 9.13c. Its output voltage is high if the input voltage is low, and its output voltage is low if the input voltage is high. The output signal of a NOT-gate corresponds to a proposition which is the negation of the proposition corresponding to the input signal.

x	y	Output
L	L	L
L	H	H
H	L	H
H	H	H

OR gate

(a)

x	y	Output
L	L	L
L	H	L
H	L	L
H	H	H

AND gate

(b)

x	Output
L	H
H	L

NOT gate

(c)

Figure 9.13

It is quite obvious that we can interconnect these devices to form an electronic circuit that realizes any given boolean expression. For example, the circuit in Fig. 9.14 realizes the boolean expression

$$p \vee (q \wedge \bar{r})$$

corresponding to the problem of the buzzer in an automobile presented earlier.

Figure 9.14

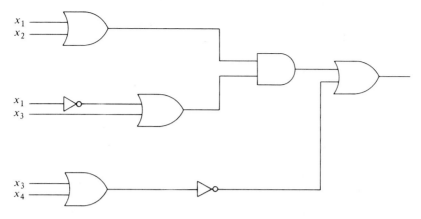

Figure 9.15

As another example, the circuit in Fig. 9.15 realizes the boolean expression

$$[(x_1 \vee x_2) \wedge (\overline{x_1} \vee x_3)] \vee \overline{(x_3 \vee x_4)}$$

We hope that our very brief discussion has illustrated the essential ideas in connection with the design and implementation of digital networks. It is not difficult to visualize how the behavior of complex computational and control devices can be specified as compound propositions and then realized with electronic components. The reader is encouraged to find out many further details on the design and implementation of digital networks from a book on switching theory and logical design.

9.10 SWITCHING CIRCUITS

We show now another example of the application of boolean algebras. Consider a simple on-off electric switch that can be placed either in the *off* position or in the *on* position, as shown in Fig. 9.16a. When the switch is in the off position we say that there is an *open circuit* between the two terminals 1 and 2. When the switch is in the on position we say that there is a *closed circuit* between the two terminals 1 and 2. As is well known to the reader, we can use an on-off switch to turn on or off any electric appliance, as illustrated in Fig. 9.16b.

Switches can be interconnected in various ways to perform complex control functions. For example, an air-conditioning system might be controlled by the thermostats in two rooms. If the temperature in either room exceeds 25°C, the thermostat in that room will close a switch to turn on the air-conditioning system. As another example, the door to the vault of a bank might be controlled electronically so that it cannot be opened unless two separate keys are used to close two electric switches simultaneously.

The two most fundamental ways of interconnecting switches are parallel

Figure 9.16

connection and series connection. Two switches can be interconnected in *parallel*, as shown in Fig. 9.17a, and in *series*, as shown in Fig. 9.17b. It is immediately clear that in a parallel connection of two switches, there is an open circuit between terminals 1 and 2 if both switches are in the off position, and there is a closed circuit between terminals 1 and 2 if either one or both of the switches are in the on position. Similarly, in a series connection of two switches, there is an open circuit between terminals 1 and 2 if either one or both of the switches are in the off position, and there is a closed circuit between terminals 1 and 2 if both switches are in the on position. Thus, in the example on the air-conditioning system we will connect the two switches controlled by the thermostats in parallel. On the other hand, in the example on the door to the vault of a bank, we will connect the two switches controlled by the two keys in series.

Since we can view a parallel connection or a series connection of two switches as a "super" switch that is either open or closed, we can interconnect them with other switches to form complex switching networks, such as the one shown in Fig. 9.18, where symbolic names are used to denote different switches.

Figure 9.17

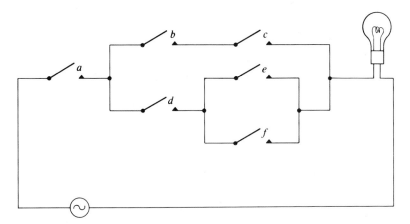

Figure 9.18

One immediately realizes that the light in the network in Fig. 9.18 can be turned on by various settings of the switches in the network. For example, the light will be turned on if switches *a*, *b*, *c* are closed, or if switches *a*, *d*, *f* are closed, and so on. On the other hand, the light will be turned off if switch *a* is open or if switches *b*, *e*, *f*, are open, and so on. To study the behavior of complex switching networks, we define now an algebraic system. Let ({open, closed}, **PAR**, **SER**) be an algebraic system where **PAR** and **SER** are two binary operations whose definitions are shown in Fig. 9.19. Clearly, in this algebraic system, a switch is a variable that can either assume the value *open* or the value *closed*. The binary operation **PAR** corresponds to the parallel connection of two switches, and the binary operation **SER** corresponds to the series connection of two switches. Consequently, a switching network consisting of parallel and series connection of switches can be described by an algebraic expression in our algebraic system. For example, the switching network in Fig. 9.18 can be described by the algebraic expression:

$$a \text{ SER } ((b \text{ SER } c) \text{ PAR } (d \text{ SER } (e \text{ PAR } f)))$$

PAR	open	closed
open	open	closed
closed	closed	closed

SER	open	closed
open	open	open
closed	open	closed

Figure 9.19

We recognize immediately that our algebraic system is isomorphic to the algebraic system ({0, 1}, ∨, ∧) which is obtained from the two-valued boolean algebra by dropping the negation operation. First of all, the results in boolean algebras can be applied directly to the study of the behavior of switching networks. Furthermore, we see the possibility of using switches to make up circuits to realize propositions that do not include the negation operation. For example, Fig. 9.20 shows a circuit which is a "majority vote taker," corresponding to the compound proposition

$$(a \wedge b) \vee (a \wedge c) \vee (b \wedge c)$$

That is, the light in Fig. 9.20 will be turned on if and only if two or more of the three switches a, b, c are closed.

In practice, an on-off switch is usually replaced by a *relay*. A relay is an electromechanical device shown schematically in Fig. 9.21a. When the on-off switch is closed, the electromagnetic coil is energized which, in turn, closes the contacts. Consequently, there will be a closed circuit between terminals 1 and 2. When the on-off switch is open, the electromagnetic coil is not energized and the contacts will be left open. Consequently, there is an open circuit between terminals 1 and 2. We shall use the same symbolic name to refer to both the on-off switch controlling the electromagnetic coil and the contacts which the coil controls. At this point, one might say that a relay functions in exactly the same way as does a single on-off switch, and might wonder why we go through an "indirect" step of using an on-off switch to control the contacts. One reason is that the electromagnetic coil can control more than one pair of contacts. Consequently, closing the on-off switch will close more than one pair of contacts, as Fig. 9.21b shows. A simple application is to use a switch to turn on several electric devices simultaneously such as a stereo system and several lamps. Secondly, a relay can control two kinds of contacts. What we have just described are known as *normally open* contacts, meaning that they are open when the controlling on-off switch is open and closed when the on-off switch is closed. Alternatively, we could have what are known as *normally closed* contacts. They are closed when the controlling on-off switch is open and open when the controlling on-off switch is closed. A pair of normally closed contacts is shown schematically in Fig. 9.22, where we use a to denote the on-off switch and $\bar{a}$ to denote the pair of normally closed contacts controlled by a. For

Figure 9.20

Coil

Contacts

Switch
a

Battery

(a)

Contacts

Coil

Switch
a

Battery

(b)

Figure 9.21

Figure 9.22

example, one can use a single switch to turn on a stereo system and, at the same time, to turn off several lamps, or conversely, as shown in Fig. 9.23.

We can connect relay contacts in parallel and in series in exactly the same manner as on-off switches are interconnected. For example, Fig. 9.24 shows a circuit which is a two-out-of-three vote taker. Note that the light will be turned on if and only if exactly two out of three of the on-off switches *a*, *b*, and *c* are closed.

One notes immediately that circuits made up of relay contacts connected in parallel and in series can be described by algebraic expressions in the algebraic system ({open, closed}, **SER, PAR,** ¯), where ¯ is a unary operation corresponding to the usage of a pair of normally closed contacts. The algebraic system ({open, closed}, **SER, PAR,** ¯) is usually referred to as the *switching algebra.*

Figure 9.23

Figure 9.24

Needless to say, the switching algebra is a two-valued boolean algebra, as the reader has anticipated. Consequently, all the results on boolean algebras can be applied to the analysis and synthesis of relay networks. We conclude our discussion with an illustrative example:

Example 9.4 A light is controlled by the switching circuit shown in Fig. 9.25a. The circuit can be described by the algebraic expression:

$$(a \ \textbf{PAR} \ \bar{b} \ \textbf{PAR} \ d) \ \textbf{SER} \ (a \ \textbf{PAR} \ b \ \textbf{PAR} \ \bar{c}) \ \textbf{SER} \ (a \ \textbf{PAR} \ c \ \textbf{PAR} \ d)$$

in the switching algebra. This expression corresponds to the expression

$$(a \vee \bar{b} \vee d) \wedge (a \vee b \vee \bar{c}) \wedge (a \vee c \vee d)$$

in the two-valued boolean algebra $(\{0, 1\}, \vee, \wedge, \bar{\ })$. Furthermore, we note

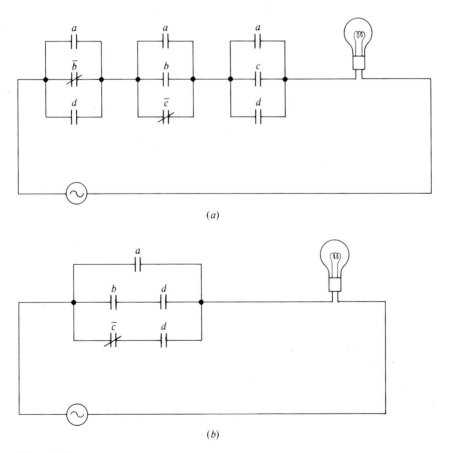

(a)

(b)

Figure 9.25

that

$$(a \vee \bar{b} \vee d) \wedge (a \vee b \vee \bar{c}) \wedge (a \vee c \vee d)$$
$$= [a \vee ((\bar{b} \vee d) \wedge (b \vee \bar{c}))] \wedge (a \vee c \vee d)$$
$$= a \vee [(\bar{b} \vee d) \wedge (b \vee \bar{c}) \wedge (c \vee d)]$$
$$= a \vee [[(\bar{b} \wedge \bar{c}) \vee (b \wedge d) \vee (\bar{c} \wedge d)] \wedge (c \vee d)]$$
$$= a \vee (b \wedge c \wedge d) \vee (\bar{b} \wedge \bar{c} \wedge d) \vee (b \wedge d) \vee (\bar{c} \wedge d)$$
$$= a \vee (b \wedge d) \vee (\bar{c} \wedge d)$$

Since the expression

$$a \vee (b \wedge d) \vee (\bar{c} \wedge d)$$

corresponds to the expression

$$a \ \textbf{PAR} \ (b \ \textbf{SER} \ d) \ \textbf{PAR} \ (\bar{c} \ \textbf{SER} \ d)$$

in the switching algebra, the circuit in Fig. 9.25b will perform the same control function as that in Fig. 9.25a. $\qquad\square$

9.11 REMARKS AND REFERENCES

Some useful general references on lattice theory and boolean algebras are Abbott[1], Birkhoff[2], Hohn[5], Rutherford[9], and Szász[10]. For further discussion on the subject of switching theory and logical design, see, for example, Caldwell[3], Hill and Peterson[4], and Kohavi[7].

[1] Abbott, J. C.: "Sets, Lattices, and Boolean Algebras," Allyn and Bacon, Inc., Boston, 1969.
[2] Birkhoff, G.: "Lattice Theory," 3d ed., Am. Math. Soc. Coll. Publ., Providence, R.I., 1967.
[3] Caldwell, S. H.: "Switching Circuits and Logical Design," John Wiley & Sons, Inc., New York, 1958.
[4] Hill, F. J., and G. R. Peterson: "Switching Theory and Logical Design," John Wiley & Sons, Inc., New York, 1968.
[5] Hohn, F. E.: "Applied Boolean Algebra," The Macmillan Company, New York, 1960.
[6] Guillemin, E. A.: "Introductory Circuit Theory," John Wiley & Sons, Inc., New York, 1953.
[7] Kohavi, Z.: "Switching and Finite Automata Theory," McGraw-Hill Book Company, New York, 1970.
[8] Liu, C. L.: "Introduction to Combinatorial Mathematics," McGraw-Hill Book Company, New York, 1968.
[9] Rutherford, D. E., "Introduction to Lattice Theory," Oliver and Boyd, London, 1965.
[10] Szász, G.: "Introduction to Lattice Theory," Academic Press, New York, 1963.

PROBLEMS

9.1 Let a and b be two elements in a lattice $(A, \leq)$. Show that $a \wedge b = b$ if and only if $a \vee b = a$.

9.2 Let a, b, c be elements in a lattice $(A, \leq)$. Show that, if $a \leq b$, then $a \vee (b \wedge c) \leq b \wedge (a \vee c)$.

9.3 Let a, b, c be elements in a lattice $(A, \leq)$. Show that

$$a \vee (b \wedge c) \leq (a \vee b) \wedge (a \vee c)$$
$$(a \wedge b) \vee (a \wedge c) \leq a \wedge (b \vee c)$$

9.4 Let a and b be two elements in a lattice $(A, \leq)$. Show that $a \wedge b < a$ and $a \wedge b < b$ if and only if a and b are incomparable.

9.5 Let $(A, \vee, \wedge)$ be an algebraic system, where $\vee$ and $\wedge$ are binary operations satisfying the absorption law. Show that $\vee$ and $\wedge$ also satisfy the idempotent law.

9.6 We study in this problem the possibility of defining a lattice by an algebraic system with two binary operations. Let $(A, \vee, \wedge)$ be an algebraic system, where $\vee$ and $\wedge$ are binary operations satisfying the commutative, associative, and absorption laws.

(a) Define a binary relation $\leq$ on A such that for any a and b in A, $a \leq b$ if and only if $a \wedge b = a$. Show that $\leq$ is a partial ordering relation.

(b) Show that $a \vee b$ is the least upper bound of a and b in $(A, \leq)$. Show that $a \wedge b$ is the greatest lower bound of a and b in $(A, \leq)$.

9.7 Show that a lattice is distributive if and only if for any elements a, b, c in the lattice

$$(a \vee b) \wedge c \leq a \vee (b \wedge c)$$

9.8 Let $(A, \leq)$ be a distributive lattice. Show that, if

$$a \wedge x = a \wedge y \quad \text{and} \quad a \vee x = a \vee y$$

for some a, then $x = y$.

9.9 Show that a lattice $(A, \leq)$ is distributive if and only if for any elements a, b, c in A,

$$(a \wedge b) \vee (b \wedge c) \vee (c \wedge a) = (a \vee b) \wedge (b \vee c) \wedge (c \vee a)$$

[*Hint*: To show that $(A, \leq)$ is distributive, consider the elements a, $b \vee c$, and $(a \vee b) \wedge (a \vee c)$.]

9.10 A lattice $(A, \leq)$ is called a *modular lattice* if for any a, b, c in A, where $a \leq c$,

$$a \vee (b \wedge c) = (a \vee b) \wedge c$$

Show that a lattice is modular if and only if the following condition holds:

$$a \vee (b \wedge (a \vee c)) = (a \vee b) \wedge (a \vee c)$$

9.11 Show that for any elements a, b, c in a modular lattice

$$(a \vee b) \wedge c = b \wedge c$$

implies

$$(c \vee b) \wedge a = b \vee a$$

9.12 Let $(A, \leq)$ be a lattice. A subset I of A is called an *ideal* if the following conditions are satisfied:

1. For any a and b in I, $a \vee b$ is in I.
2. For any a in I and any x in A, $a \wedge x$ is in I.

(a) Show that condition 2 can be replaced by:

2'. If a is in I and $x \leq a$, then x is in I.

(b) Let I be a subset of A. Show that I is an ideal if and only if for any a and b in A:

1''. If a and b are in I, then $a \vee b$ is in I.
2''. If $a \vee b$ is in I, then both a and b are in I.

9.13 Let $(A, \leq)$ be a distributive lattice. Let a and b be two arbitrary elements in A. Let $I(a, b)$ denote the set

$$\{x \,|\, x \in A, a \wedge x = b \wedge x\}$$

Show that $I(a, b)$ is an ideal. (See Prob. 9.12 for the definition of an ideal.)

9.14 Show that in a lattice with universal lower bound 0 and universal upper bound 1, 0 is the *unique* complement of 1, and 1 is the *unique* complement of 0.

9.15 Show that

$$a \vee (\bar{a} \wedge b) = a \vee b$$

$$a \wedge (\bar{a} \vee b) = a \wedge b$$

in a boolean algebra.

9.16 Let $(A, \vee, \wedge, \bar{\ })$ be a boolean algebra. Show that $(A, \oplus)$ is a commutative group, where $\oplus$ is defined as

$$a \oplus b = (a \wedge \bar{b}) \vee (\bar{a} \wedge b)$$

9.17 Let $(A, \star)$ be the algebraic system defined in Prob. 8.13 [including the condition in part (f)] with the additional property that for a, b, c in A,

$$a \star (b \star c) = b \star (a \star c)$$

$(A, \star)$ is called an *implication algebra*. We define a binary relation $\leq$ on A such that $a \leq b$ if and only if $a \star b = e$.

(a) Show that $\leq$ is a partial ordering relation.

(b) Show that $a \leq b$ if and only if $b = x \star a$ for some x in A.

(c) Show that for a and b in A, $(a \star b) \star b$ is a least upper bound of a and b.

9.18 Let $(A, \vee, \wedge, \bar{\ })$ be a boolean algebra. Let $\star$ be a binary operation defined on A such that

$$a \star b = \bar{a} \vee b$$

Show that $(A, \star)$ is an implication algebra as defined in Prob. 9.17.

9.19 Let

$$E(x_1, x_2, x_3) = (x_1 \wedge x_2) \vee (x_1 \wedge x_3) \vee (\bar{x}_2 \wedge x_3)$$

be a boolean expression over the two-valued boolean algebra. Write $E(x_1, x_2, x_3)$ in both disjunctive and conjunctive normal forms.

9.20 Let

$$E(x_1, x_2, x_3) = \overline{\overline{(x_1 \vee x_2)} \vee \overline{(x_1 \wedge x_3)}}$$

be a boolean expression over the two-valued boolean algebra. Write $E(x_1, x_2, x_3)$ in both disjunctive and conjunctive normal forms.

9.21 Let

$$E(x_1, x_2, x_3, x_4) = (x_1 \wedge x_2 \wedge \bar{x}_3) \vee (x_1 \wedge \bar{x}_2 \wedge x_4) \vee (x_2 \wedge \bar{x}_3 \wedge \bar{x}_4)$$

be a boolean expression over the two-valued boolean algebra. Write $E(x_1, x_2, x_3, x_4)$ in both disjunctive and conjunctive normal forms.

9.22 (a) Express the function in Fig. 9P.1 in disjunctive normal form.

(b) Express the function in Fig. 9P.1 in conjunctive normal form.

	f
(0, 0, 0)	1
(0, 0, 1)	0
(0, 1, 0)	1
(0, 1, 1)	0
(1, 0, 0)	0
(1, 0, 1)	1
(1, 1, 0)	0
(1, 1, 1)	1

Figure 9P.1

9.23 Simplify the following boolean expressions:

(a) $(a \wedge b) \vee (a \wedge b \wedge c) \vee (b \wedge c)$

(b) $(a \wedge b) \vee (a \wedge \bar{b} \wedge c) \vee (b \wedge c)$

(c) $(a \wedge b) \vee (\bar{a} \wedge b \wedge \bar{c}) \vee (b \wedge c)$

(d) $((a \wedge \bar{b}) \vee c) \wedge (a \vee \bar{b}) \wedge c$

9.24 We defined in Sec. 9.7 the notions of disjunctive normal form and conjunctive normal form for boolean expressions over the two-valued boolean algebra. We shall see in this problem how these notions can be extended to boolean algebras in general. Let $E(x_1, x_2, \ldots, x_n)$ be a boolean expression over a boolean algebra $(A, \vee, \wedge, \bar{\ })$. We shall use the notation $E(x_i = a)$ to denote the boolean expression obtained from $E(x_1, x_2, \ldots, x_n)$ by substituting x_i by a, where a is an element of A.

(a) For any x_i, show that

$$E(x_1, x_2, \ldots, x_n) = (\bar{x}_i \wedge E(x_i = 0)) \vee (x_i \wedge E(x_i = 1))$$

[*Hint:* The equality can be proved by induction on the length of the expression $E(x_1, x_2, \ldots, x_n)$, where the length of an expression is defined to be the total number of appearances of elements of A, variable names, the operations $\vee$, $\wedge$, $\bar{\ }$ in the expression with replications counted. Thus, expressions of length one are elements of A and the variables $x_1, x_2, \ldots, x_n$.]

(b) Use the result in part (a) to show that any boolean expression can be written as a join of expressions of the form

$$c_{\delta_1 \delta_2 \cdots \delta_n} \wedge \tilde{x}_1 \wedge \tilde{x}_2 \wedge \cdots \wedge \tilde{x}_n$$

where $c_{\delta_1 \delta_2 \cdots \delta_n}$ is an element of A, and $\tilde{x}_i$ denotes either x_i or $\bar{x}_i$. Such a way of writing a boolean expression is called a disjunctive normal form.

(c) Let

$$E(x_1, x_2) = (2 \wedge x_1) \wedge (x_1 \vee \bar{x}_2)$$

be a boolean expression over the boolean algebra $(\{0, 1, 2, 3\}, \vee, \wedge, \bar{\ })$. Rewrite $E(x_1, x_2)$ in disjunctive normal form.

(d) Let f be a function from A^n to A for some boolean algebra $(A, \vee, \wedge, \bar{\ })$. If f is a boolean function, how can we determine a boolean expression that specifies the function f? (*Hint:* Consider the disjunctive normal form of the expression.)

(e) Show that the function in Fig. 9.8 is not a boolean function.

(f) For any x_i, show that

$$E(x_1, x_2, \ldots, x_n) = (x_i \vee E(x_i = 0)) \wedge (\bar{x}_i \vee E(x_i = 1))$$

(g) Use the result in part (f) to show that any boolean expression can be written as a meet of expressions of the form

$$d_{\delta_1 \delta_2 \cdots \delta_n} \vee \tilde{x}_1 \vee \tilde{x}_2 \vee \cdots \vee \tilde{x}_n$$

where $d_{\delta_1 \delta_2 \cdots \delta_n}$ is an element of A and $\tilde{x}_i$ denotes either x_i or $\bar{x}_i$. Such a way of writing a boolean expression is called a conjunctive normal form.

9.25 A student will receive a passing grade in a course if he meets one or more of the following conditions:

1. He got a B or better in the mid-term examination and an A in the final examination.
2. He got a B or better in both the mid-term and final examinations and did not miss any homework assignment.
3. He got a B or better in at least one of the mid-term and final examinations and he has an athletic scholarship.
4. His father is the instructor of the course.

Use the following propositions as variables to write a boolean expression describing the set of conditions stated above.

    a: Getting a B in mid-term examination
    b: Getting an A in mid-term examination
    c: Getting a B in final examination
    d: Getting an A in final examination
    e: Missing some homework assignment
    f: Having an athletic scholarship
    g: Being the son of the instructor

Simplify the expression and thus obtain a simpler set of conditions.

9.26 The following table shows the contents of five tool boxes labeled a, b, c, d, e:

	Screw driver	Wrench	Pliers	Hammer	Saw
a	×	×			
b		×	×	×	
c	×		×		
d		×			×
e			×	×	

Write a boolean expression to show how selections of the tool boxes can be made so that each selection contains at least one tool of each kind.

9.27 A committee is to be selected from five candidates a, b, c, d, e. The selection must satisfy all the following conditions:

1. Either *a* or *b* must be included, but not both.
2. Either *c* or *e* or both must be included.
3. If *d* is included, then *b* must be included.
4. Either both *a* and *c* are included or neither is included.
5. If *e* is included, then *c* and *d* must be included.

How should the selection be made?

Figure 9P.2

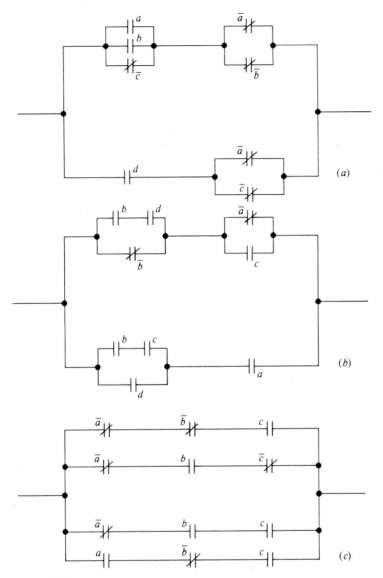

Figure 9P.3

9.28 Show how the method of boolean algebras can be applied to determine the minimal dominating sets of a graph. Illustrate the procedure by determining the minimal dominating sets of the graph in Fig. 4P.2. (See Prob. 4.10 for the definition of a dominating set of a graph.) (*Hint:* For each vertex, either the vertex itself or one of the vertices adjacent to it must be included in a dominating set.)

9.29 Simplify the circuit in Fig. 9P.2. (There is one circuit that uses only three OR-gates and AND-gates.)

9.30 (*a*) Show how an OR-gate can be replaced by a suitable interconnection of AND-gates and NOT-gates.

   (*b*) Show how an AND-gate can be replaced by a suitable interconnection of OR-gates and NOT-gates.

(c) A NOR-gate is a circuit that has two inputs and one output such that the output is equal to $\overline{x \vee y}$ for the inputs x and y. Show that we can realize any boolean expression using NOR-gates alone by showing how OR-gates, AND-gates, and NOT-gates can be replaced by suitable interconnections of NOR-gates.

(d) A NAND-gate is a circuit that has two inputs and one output such that the output is $\overline{x \wedge y}$ for the inputs x and y. Show that we can realize any boolean expression using NAND-gates alone by showing how OR-gates, AND-gates, and NOT-gates can be replaced by suitable interconnections of NAND-gates.

9.31 (a) Simplify the relay networks in Fig. 9P.3.

(b) Construct digital networks using OR-gates, AND-gates, and NOT-gates to realize the boolean expressions realized by the relay networks in Fig. 9P.3.

INDEX